Design for X :
 concurrent engineering
      1996.

2005 09 22

D1243168

Design for X

Design for X

Concurrent engineering imperatives

Edited by

G.Q. Huang

School of Engineering
University of Abertay
Dundee, UK

CHAPMAN & HALL

London · Weinheim · New York · Tokyo · Melbourne · Madras

Published by Chapman & Hall, 2–6 Boundary Row, London SE1 8HN, UK

Chapman & Hall, 2–6 Boundary Row, London SE1 8HN, UK

Chapman & Hall GmbH, Pappelallee 3, 69469 Weinheim, Germany

Chapman & Hall USA, 115 Fifth Avenue, New York, NY 10003, USA

Chapman & Hall Japan, ITP-Japan, Kyowa Building, 3F, 2-2-1 Hirakawacho, Chiyoda-ku, Tokyo 102, Japan

Chapman & Hall Australia, 102 Dodds Street, South Melbourne, Victoria 3205, Australia

Chapman & Hall India, R. Seshadri, 32 Second Main Road, CIT East, Madras 600 035, India

First edition 1996

Printed in Great Britain by T.J. Press, Padstow, Cornwall

ISBN 0 412 78750 4

A catalogue record for this book is available from the British Library

Library of Congress Catalog Card Number: 96-85633

CONTENTS

List of Contributors   vii

Preface   xi

Introduction   1

PART ONE   DFX EXPERIENCE

1. **Design for Manufacture and Assembly: The B & D Experience**   19
   Geoffrey Boothroyd

2. **Case Experience with Design for Assembly Methods**   41
   Paul G. Leaney

3. **Applying DFX Experience in Design for Environment**   72
   Carolien G. Van Hemel; Koels Keldmann

4. **Design for Competition: The Swedish DFX Experience**   96
   Margareta Norell; Sören Andersson

5. **Developing DFX Tools**   107
   George Q. Huang

6. **Implementing DFX Tools**   130
   George Q. Huang

PART TWO   DESIGN FOR LIFE CYCLE

7. **GRAI Approach to Product Development**   153
   Guy Doumeingts; Philippe Girard; Benoit Eynard

8. **Design for Dimensional Control**   173
   Paul G. Leaney

9. **Design for Assembly Cost for Printed Circuit Boards**   196
   Donald S. Remer; Frederick S. Ziegler; Mickael Bak; Patrick M. Doneen

10. **Design for Inspectability**   216
    Colin G. Drury

11. **Design for Effective Material Storage and Distribution**   230
    B. Gopalakrishnan; S. Chintala; S. Adhikari; G. Bhaskaran

12. **Design for Reliability**   245
    John. A. Stephenson; Ken M. Wallace

13. **Design for Electromagnetic Compatibility**   268
    J.F. Dawson; M.D. Ganley; M.P. Robinson; A.C. Marvin; S.J. Porter

14. **Design for Serviceability**   298
    Peter Dewhurst; Nicholas Abbatiello

15. **Ease-of-Disassembley Evaluation in Design for Recycling**   318
    Thomas A. Hanft; Ehud Kroll

PART THREE   DESIGN FOR COMPETITIVENESS

16. **Design for Quality**   335
    Edoardo Rovida; Gian Francesco Biggiogero

17. **Design for Modularity**   356
    Gunnar Erixon

18. **Design for Optimal Environmental Impact**   380
    Leigh Holloway; Ian Tranter; David W. Clegg

19. **Design for the Life Cycle: Activity Based Costing and Uncertainty**   398
    Bert Bras; Jan Emblemsvåg

PART FOUR   TRADEOFF AND INTEGRATION

20. **Design Optimisation for Product Life Cycle**   424
    Masataka Yoshimura

21. **A Meta-Methodology for the Application of DFX Guidelines**   441
    Brett Watson; David Radcliffe; Paul Dale

22. **Design for Technical Merit**   463
    Tim N.S. Murdoch; Ken M. Wallace

Index   485

CONTRIBUTORS

Nicholas Abbatiello, G.E. Plastics, Pittsfield, Massachusetts, USA

S. Adhikari, Department of Industrial Engineering, West Virginia University, Morgantown, WV 26506, USA

Sören Andersson, Department of Machine Design, Royal Institute of Technology, KTH S-10044, Stockholm, Sweden

Michael Bak, R. B. Weber & Co Inc, 1717 Embarcadero Road, STE 2000, Palo Alto, CA 94303, USA

G. Bhaskaran, Department of Mechanical Engineering, West Virginia University, Morgantown, WV 26506, USA

Gian Francesco Biggiogero, Politecnico Milano, p.ZA L. Da Vinci 32, 20133 - Milano, Italy

Geoffrey Boothroyd, Department of Industrial and Manufacturing Engineering, 103 Gillbreth Hall, University of Rhode Island, Kingston, RI 02881-0805, USA

Bert Bras, Systems Realization Laboratory, School of Mechanical Engineering, Georgia Institute of Technology, Atlanta, Georgia 30332-0405 USA

S. Chintala, Department of Industrial Engineering, West Virginia University, Morgantown, WV 26506, USA

David W. Clegg, Division of Materials and Environmental Engineering, School of Engineering, Sheffield Hallam University, Pond Street, Sheffield, S1 1WB, UK

Paul Dale, 86 Ironside Street, St Lucia, Australia 4072

John F. Dawson, Applied Electromagnetics Group, Department of Electronics, University of York, Heslington, York YO1 5DD, UK

Peter Dewhurst, Department of Industrial and Manufacturing Engineering, 103 Gillbreth Hall, University of Rhode Island, Kingston, RI 02881-0805, USA

Patrick M. Doneen, Western Digital Corporation, 19401 40th Avenue W, Suite 302, Lynnwood, WA 98036, USA

Guy Doumeingts, Laboratoire d'Automatique et de Productique, Groupe GRAI, Universite Bordeaux, 351 Curs de la Liberation, 33405 TALENCE Cedex, France

Colin G. Drury, Department of Industrial Engineering, School of Engineering and Applied Sciences, State University of New York at Buffalo, 342 Lawrence D Bell Hall, Box 602050, USA

Jan Emblemsvåg, Systems Realization Laboratory, School of Mechanical Engineering, Georgia Institute of Technology, Atlanta, Georgia 30332-0405 USA

Benoit Eynard, Laboratoire d'Automatique et de Productique, Groupe GRAI, Universite Bordeaux, 351 Curs de la Liberation, 33405 TALENCE Cedex, France

Gunnar Erixon, Department of Manufacturing Systems, Falun Borlange University, Centre of Industrial Engineering and Royal Institute of Technology, Stockholm S-100 44, Sweden

Troels Feldmann, Institute of Engineering Design, Technical University of Denmark, Lynby, Denmark

M. D. Ganley, Applied Electromagnetics Group, Department of Electronics, University of York, Heslington, York YO1 5DD, UK

Philippe Girard, Laboratoire d'Automatique et de Productique, Groupe GRAI, Universite Bordeaux, 351 Curs de la Liberation, 33405 TALENCE Cedex, France

B. Gopalakrishnan, Department of Industrial Engineering, West Virginia University, Morgantown, WV 26506, USA

Thomas A. Hanft, Department of Mechanical Engineering, Texas A & M University, College Station, TX 77843-3123, USA

Leigh Holloway, School of Engineering, Sheffield Hallam University, Pond Street, Sheffield, S1 1WB, UK

George Q. Huang, School of Engineering, University of Abertay Dundee, Bell Street, Dundee, DD1 1HG, Scotland, UK

Ehud Kroll, Department of Mechanical Engineering, Texas A & M University, College Station, TX 77843-3123, USA

Paul G. Leaney, Department of Manufacturing Engineering, Loughborough University of Technology, Leicestershire LE11 3TU, England,

A. C. Marvin, Applied Electromagnetics Group, Department of Electronics, University of York, Heslington, York YO1 5DD, UK

Tim N. S. Murdoch, Engineering Design Centre, Department of Engineering, Cambridge University, Trumpington Street, Cambridge CB2 1PZ, UK

Margareta Norell, Department of Machine Design, Royal Institute of Technology, KTH S-10044, Stockholm, Sweden; and Department of Machine Design and Materials Technology, Norwegian Institute of Technology, Trondheim, Norway

S. J. Porter, Applied Electromagnetics Group, Department of Electronics, University of York, Heslington, York YO1 5DD, UK

David Radcliffe, Department of Mechanical Engineering, University of Queensland, St Lucia, Australia 4072

Donald S. Remer, Department of Engineering, Harvey Mudd College, Claremont, CA 91711-2834, USA

Martin P. Robinson, Applied Electromagnetics Group, Department of Electronics, University of York, Heslington, York YO1 5DD, UK

Edoardo Rovida, Politecnico Milano, p.ZA L. Da Vinci 32, 20133 - Milano, Italy

John A. Stephenson, Engineering Design Centre, Department of Engineering, Cambridge University, Trumpington Street, Cambridge CB2 1PZ, UK

Ian Tranter, School of Engineering, Sheffield Hallam University, Pond Stree, Sheffield, S1 1WB, UK

Carolien van Hemel, Section of Environmental Product Development, Faculty of Industrial Design Engineering, Delft University of Technology, Jaffalaan 9, 2628 BX Delft, The Netherlands

Ken M. Wallace, Engineering Design Centre, Department of Engineering, Cambridge University, Trumpington Street, Cambridge CB2 1PZ, UK

Brett Watson, Department of Mechanical Engineering, University of Queensland, St Lucia, Australia 4072

Masataka Yoshimura, Department of Precision Engineering, Kyoto University, Sakyo-ku, Kyoto 606-01, Japan

Frederick S. Ziegler, S.R.I. 333 Ravenswood Avenue, Menlo Park, California 94025, USA

This book is concerned with Design for X (DFX) - imperative practice in product development to achieve simultaneous improvements in products and processes. With DFX, quality, cost and speed are not compromised, but all improved to become more competitive. The book is designed for managers, practitioner and research engineers, academic and industrial consultants, and graduate engineers who are interested or involved in developing and implementing DFX, or anyone who wishes to know more about the subject. They will find this book a compass in the journey of searching for answers to the following questions:

1. What is DFX?
2. Which DFX tool should be used?
3. How does DFX work?
4. Why, where and when is DFX used?
5. Who uses DFX?
6. How to implement DFX ?
7. How to develop DFX ?
8. What is the latest development?

This book has brought together the expertise of practitioners and researchers from over ten countries in order to answer the above questions. Experience and good practice within both world-class and small-medium manufacturers are disseminated. Alternative approaches and common elements are examined. Latest developments are outlined. Emerging issues such as integration and tradeoff are explored.

This is the first comprehensive text on the subject of DFX. Twenty two chapters have been selected to systematically cover a wide range of major topics. The introductory chapter gives an overview of the subject in relation to all contributions included in this book. The chapters are logically grouped into four parts. The first part consists of six chapters to report on practical experience in developing and implementing DFX. In Chapter 1, Professor Boothroyd explains one of the best known Design for Assembly techniques and points out benefits achieved and lessons learnt by some of their successful clients. In Chapter 2, Dr Leaney investigates three well-known Design for Assembly tools using a retrospective industrial case study. Chapter 3 extends the industrial experience gained in applying Design for X techniques such as Design for Assembly and Manufacture into a relatively new area of Design for Environment. In Chapter 4, Professors Norell and Andersson report on the Swedish experience of developing and implementing DFX tools. Chapters 5 and 6 present relatively generic frameworks for developing and implementing DFX, respectively.

Nine chapters are included in Part Two, each presenting a DFX tool specific to a major life-cycle in product development from design through production to recycling. In Chapter 7,

Professor Doumeingts and the co-workers present GARI integrated methodology (GIM) and discuss its application in organising and rationalizing product design activities. In Chapter 8, Dr Leaney discusses the importance and techniques in managing dimensional variability in product design. Professor Remer and colleagues present a cost estimation tool specifically developed PCB (Printed Circuit Boards) assemblies. In Chapter 10, Professor Drury outlines a systematic Design for Inspectability procedure. Professor Gopalakrishnan and his colleagues explore a technique of Design for Effective Material Storage and Distribution in Chapter 11. A Design for Reliability technique under development at the Cambridge University Engineering Design Centre is outlined in Chapter 12. Chapter 13 presents findings from a major research project on Design for Electromagnetic Compatibility at the University of York. In Chapter 14, Professor Dewhurst leads the discussion on the latest development of their Design for Serviceability system. Chapter 15 deals with disassembly aspects in Design for Recycling with a case study on computer keyboards.

Part Three includes four chapters, dealing with DFX techniques for achieving corporate competitiveness. Multiple life cycles are usually considered and tradeoffs are carried out in this type of DFX. In Chapter 16, Professor Rovida and his colleague present a technique of Design for Quality by selecting best concepts from as many conceivable alternatives as possible. The issue of flexibility or modularity is addressed in Chapter 17. A methodology for optimising overall environmental impact of product designs is presented in Chapter 18. Chapter 19 introduces a number of concepts such as Activity-Based Cost and Action Charts which are invaluable for developing concurrent life-cycle design tools.

Three chapters are included in Part Four to investigate emerging issues such as integration and tradeoff analysis. Professor Yoshimura outlines mathematical models for optimal product life-cycle design in Chapter 20. Chapter 21 explores a meta-methodology of tradeoff among Design for X guidelines. Chapter 22 presents a method of Design for Technical Merit developed at the Cambridge University Engineering Design Centre.

The presentation of this book strives for a balance between modularity and integrity. Individual chapters are carefully structured in a self-contained fashion. Each starts with an overview of the technique and proceeds to outline the systematic procedure, followed by case studies to demonstrate its use and merits. Readers can choose the most relevant materials to achieve incremental understanding and implementation.

During the process of preparing this book, great help has been received from many people. I am most grateful to Professor B. Nnaji for his encouragement throughout this project. My sincere gratitude is also due to Professor R. W. Johnson, Head of School of Engineering, for his generous supports of the school facilities. Comments from the reviewers are greatly appreciated.

This book is never possible without the supports from enthusiastic and patient contributors. My sincere gratitude also extends to those whose proposed contributions were unfortunately not included because of the limited space in this volume.

Finally, I would like to thank my family, Peihua, my wife for sharing weal and woe with me, my parents for encouraging me to explore, and Margaret, my daughter for switching my computer off and on and for her "jigsaw puzzle" cover story.

George Huang

INTRODUCTION

George Q. Huang

The aim of this introductory chapter is to present an overview of the subject of Design for X (DFX) in relation to chapters included in this book. The following questions are discussed:

1. What is DFX?
2. How does DFX work?
3. Why is DFX used?
4. Which DFX, when and where?
5. Who is involved in DFX?
6. What is next?

WHAT IS DESIGN FOR X (DFX)?

Concurrent Engineering (CE) is an ideal environment for product development. Its objectives include improving quality, reducing costs, compressing cycle times, increasing flexibility, raising productivity and efficiency, and improving the social image. The means of achieving these objectives is through cooperative teamwork between multiple disciplinary functions to consider all interacting issues in designing products, processes and systems from conception through production to retirement.

Design for X (DFX) is one of the most effective approaches to implementing CE. It focuses on a limited number, say 7 ± 2, of vital elements at a time (Miller, 1956). This allows available resources to be put into best use. For example, Design for Assembly (DFA) focuses on the business process of "Assembly" which is part of the life cycle of "Production". DFA considers 5-9 primary factors related to the subject product, including part symmetry, size, weight, fits, orientation, form features, etc. It considers 5-9 primary factors related to the assembly process such as inserting, handling, gripping, orienting, special tooling and equipment, etc. Careful examination of these issues and their relationships results in better design decisions with respect of the ease of assembly. At the same time, an atmosphere of teamworking cooperation is created, thus assembly efficiency is improved.

With the success of DFA, other types of DFX tools can be introduced. Therefore, more life cycle issues and other factors are brought in for consideration. Better overall decisions are arrived at without losing the necessary focus and vision. Perhaps more importantly, a CE environment for product development is incrementally created and dynamically improved.

Thinking about manufacturing aspects when designing a product has always been laudable, though not practised enough or often omitted. Searching for early work on DFX is like mining for gold. *Engineering Design: A systematic approach* by Matousek (1957), *Designing for Production* by Niebel and Baldwin (1957), *Handbook of Parts, Forms, Processes, Materials in Design Engineering* by Everhart (1960), *Designing for Manufacturing* by Pech (1973) were among precious "ancient" texts on the subject. They covered a wide range of issues from basic drawing skills, design features, datums, metal cutting processes, casting processes, and assembly. These textbooks were derived from many years of research and practical experience before they were published. In fact, Ziemke and Spann (1993) told a few DFX and CE stories dating back to the World War II era.

During 1960s and 1970s, the subject of design for economic manufacture received noticeable attention from professional bodies. For example, the CIRP (College Internationale de Recherches Pour la Production) recognized the issue and called for systematic study (Gladman, 1968). As a matter of urgency, a working group "O" - the optimization subcommittee within the CIRP was established in 1970 (Chisholm, 1973). There were other professional activities. At a conference organised by PERA (Production Engineering Research Association) in 1965, some industrialists reported their experience with "Design for Mechanized Assembly" (Tipping, 1965). Another example is the workshop dedicated to "Design for Production" sponsored by the Ministry of Technology (UK) and the University of Strathclyde and took place at the Birniehill Institute in 1970 (anon., 1970). A number of standard institutions such as BSI (British Standard Institution) and VDI (Verein Deutscher Ingenieure) provided guidelines for design for economic manufacture (BSI PD6470 : 1981) in late 1970s and early 1980s. One recent event especially dedicated to DFX research was the WDK (Workshop Design-Konstruktion) DFX-Workshop organized by Professor Andreasen and his colleagues (1993).

As early as the 1960s several companies were developing guidelines for use during product design. One example is the Manufacturing Productivity Handbook compiled for internal use by General Electric in the USA. Manufacturing data were accumulated into one large reference volume and product designers could have the information necessary for efficient design.

However, significant benefits were not realized until systematic DFA were introduced in 1970s. One such early work was the Hitachi Assemblability Evaluation Method (Hitachi AEM) (Miyakawa, Ohashi and Iwata, 1990; Shimada, Miyakawa and Ohashi, 1992). Although the method was publicized in mid 1980s, its successful application in the development of an automatic assembly system for tape recorder mechanisms was awarded the Okochi Memorial Prize in 1980. Another early work started in 1970s by a group of researchers between Salford University in the UK and Massachusetts University in the USA. This work has resulted in two different commercial DFA tools: Boothroyd-Dewhurst DFA (Boothroyd and Dewhurst, 1983) and Lucas DFA (Swift, 1981; Miles, 1989).

These systematic DFA tools have revolutionized the thinking and practice in Design for Assembly. The breakthrough was largely due to the introduction of quantification, systematic procedure, comprehensive data and knowledge base in the form of handbooks or manuals, and well-structured worksheets. These features overcome limitations of design guidelines.

DFA was once pushed by automation technology. The Hitachi AEM was directed at simplifying automatic insertion of parts. The Boothroyd-Dewhurst DFA grew out of research on automatic feeding and automatic insertion. Products designed for manual assembly were found to require redesign for automatic assembly. However, DFA is now "pulled" by its ability to solve problems and achieve dramatic savings, not only in automated assembly but more astonishingly in manual assembly. Hundreds of successful applications with the Boothroyd-Dewhurst DFA and other DFA methods have been published.

Substantial benefits achieved by using DFA has been the locomotive engine pulling the recent development in several directions. First, more and "better" DFA tools have appeared. But most of them are research or teaching systems and their practicality is yet to be proved. Second, DFA has found itself more users although the number is still tiny in contrast to the entire engineering manufacturing industries. Third, new tools have penetrated into other life cycles such as manufacturing, service, recycling, etc. Fourth, new tools have appeared to cover important facets of competitiveness such as quality, costs, flexibility, time to market, environment, etc. Ultimately, new issues such as integration and tradeoff analysis between these tools in product development have emerged for further investigation.

Such proliferation and expansion have led to a string of new terms such as Design for Manufacturability, Design for Inspectability, Design for Environmentality, Design for Recyclibility, Design for Quality, Design for Reliability, etc. "Design for X" has been devised as an umbrella for these terms and DFX for their acronyms (Gatenby and Foo, 1990; Keys, 1990; Meerkamm, 1994). DFM (Design for Manufactrability) has been used for similar purposes (Stoll, 1988; Youssef, 1994; Dean and Salstrom, 1990).

X in DFX stands for manufacturability, inspectability, recyclability, etc. These words are made up of two parts: life cycle business process (x) and performance measures (*bility*), that is,

$$X = x + bility.$$

For example, "x = total" and "*bility* = quality" in "design for total quality"; "x = whole-life" and "*bility* = costs" in "design for whole-life costs"; "x = assembly" and "*bility* = cost" in "design for assembly cost" (or simply assemblability if other *bility* measures such as assembly times are used); and so on. If a DFX tool focuses on one life cycle process and uses more than one performance metrics, it is referred to as a tool of the "Design for the Life Cycle" type. Techniques included in Part Two of this book belong to this category. On the other hand, if a DFX tool focuses on one performance metric but covers a range of life cycle processes, it is then referred to as a tool of the "Design for the Competitiveness" type. Techniques included in Part Three of this book belong to this category.

Design in "Design for X" or *D* in DFX is interpreted as *product design* in the context of DFA, meaning the design of the product for the ease of assembly (Boothroyd, Dewhurst, and Knight, 1994). However, it can be seen from many successful DFA case studies that the assembly processes and systems are affected by the changes in the subject product. That is, the assembly processes and systems are often redesigned as a result of DFA analysis. For this reason, it is logical to interpret the *D* in DFX or *Design* in "Design for X" as *concurrent design* of products, and associated processes and systems. A generic definition can be given as *making decisions in product development* related to products, and processes and plants.

HOW DOES DFX WORK?

How successful Design for Assembly tools have worked is well understood. A question is whether they and other DFX tools follow a basic pattern. Olesen (1992) has explored for a generic DFX pattern. As a result, the Theory of Dispositions has been proposed (Andreasen and Olesen, 1990). The search continues.

The need for such a basic DFX pattern can be seen from the recent development and difficulties encountered:

1. A basic pattern would help understanding how DFX works and what DFX does. Much unnecessary confusion can be avoided.
2. A basic pattern would help selecting the most appropriate DFX tool for a problem at hand from a large toolbox.
3. A generic DFX model would speed up the development of specific DFX tools dramatically. This can be explained by the effect of the *learning curve factor* because different DFX tools share similar constructs which can be reused.
4. Learning curve factor can also be gained during DFX implementation if multiple DFX tools follow a general pattern. Once the team becomes familiar and experienced with one DFX tool, the members can easily adapt to new DFX tools which share a common basis.
5. A generic DFX model can provide a platform for integrating multiple DFX tools to facilitate the flow of data and decisions between them.
6. A generic DFX model can provide a common basis on which tradeoff can be carried out among competing issues when multiple DFX tools are used.
7. A generic DFX model can provide a platform for integrating a DFX tool with other decision support systems used in product development such as CAD/CAM (computer Aided Design and Manufacture), CAPP (Computer Aided Process Planning), and CAPM (Computer Aided Production Management), to facilitate the flow of data and decisions between them.

A basic DFX pattern is not only necessary, but also feasible. This can be seen from a number of observations obtained by examining existing successful DFX tools and proven product development models:

1. Most DFX tools are not usually considered as design systems. They do not make design decisions. Instead, they evaluate design decisions from specific points of view.
2. Main DFX functionality accomplished by DFX tools and their human users is summarized in Table 1. The first four functions are usually provided by DFX tools and the second five functions are carried out mainly by human users although a few research systems can achieve them to some extent.
3. Successful DFX tools rationalise product and process designs by assessing not only individual design decisions but also their interactions.
4. Successful DFX tools provide pragmatic product and process models which are familiar or easily become familiar to their users.
5. Successful DFX tools define clearly their specific areas of concern and thus provide the essential focus for the project team to make the best use of resources available to them.
6. Successful DFX tools focus on a few important aspects to evaluate the design decisions and their interactions. This allows the project team to view the subject problem from different perspectives without losing the necessary focus.

7. Successful DFX tools are equipped with logical worksheets, systematic procedures, and comprehensive data and knowledge bases, delivered as a complete package in the form of DFX handbooks, paper-based or computerised.
8. Successful DFX tools avoid unnecessary sophistication in modelling and measuring. Fabricated complexity is regarded as hindrance to communication and cooperation that DFX tools aim to achieve.
9. Successful DFX tools avoid requiring data which are too expensive to collect. They usually provide generic databases in the form of DFX manuals.
10. Successful DFX tools strike the balance between creativity and discipline, and the balance between structure and freedom.
11. Successful DFX tools are consistent and integrative with proven product development process models.

Table 1 What does a DFX tool do?

---

1. Gather and present facts about products and processes.
2. Clarify and analyze relationships between products and processes.
3. Measure performance.
4. Highlight strengths and weaknesses and compare alternatives.
5. Diagnose why an area is strong or weak.
6. Provide redesign advice on how a design can be improved.
7. Predict what-if effects.
8. Carry out improvements.
9. Allow iteration to take place.

Based on the above observations, a conceptual DFX model - PARIX can be proposed. Figure 1 shows the overview of the model. Main components are briefly explained as follows in relation to a number of concepts developed by others:

x - This variable represents business processes or organisational functions corresponding to life-cycles in product development. It is the prefix making up the words such as *produc*ibility, *manufactur*ability, and *inspect*ability. DFX may focus on one or more life cycle processes.

PAR - Duffey and Dixon (1992) consider the product realization process as a triple (P, A, R) of **P**roducts, **A**ctivities which realize products, and **R**esources which are available for realization. Customers and suppliers can also be included in this product realization model (Andreasen and Hein, 1987).

I - P, A, and R are interrelated to each other. *I*nteractions can be explained using the ABC (Activity-Based Costing) principle that products consume activities and activities consume resources (Johnson and Kaplan, 1987; Cooper, 1988; Brisom, 1991). Alternatively, the concept of dispositions (Andreasen and Olesen, 1990) can be used to describe interactions between decisions or decision activities in different functional areas. Finally, interactions can be mathematically represented by a constraint:

$$I_x(P, A, R) = 0$$

DFX functionality shown in Table 1 can be partially built into the PARIX model. The first function to introduce is probably the capability of measuring decisions related P, A, and R, and effects of their interactions. Appropriate competitiveness performance measures "*bility*" must be determined. This is the suffix for making up the words such as manufactura*bility*, produci*bility*, and inspecta*bility*. Performance can be measured using actual data or estimates. Competitiveness can be presented using empirical matrices as used in QFD (Quality Function Deployment) or mathematically represented by the following objective function:

$$\text{``bility''} = M_x(P, A, R)$$

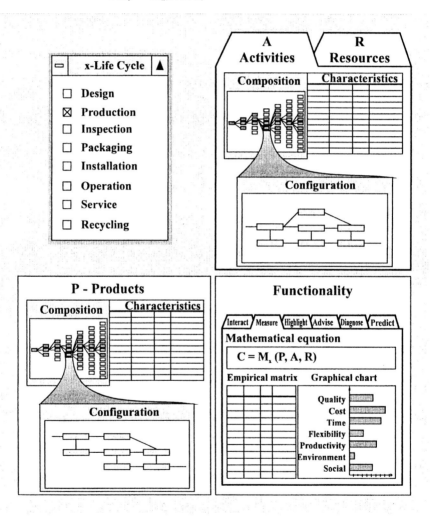

Figure 1 PARIX: A conceptual model of how DFX works.

WHY DFX?

The reason why DFX is used is simple: it works! It is limited here in space to enumerate all successful case studies. Benefits can be grouped into three categories. The Category 1 benefits are directly related to the competitiveness measures (Maskell, 1991), including improved quality, compressed cycle time, reduced life-cycle costs, increased flexibility, improved productivity, more satisfied customers, safer workplace and happier workforce, and lower adverse environment impact.

The second category of benefits include improved and rationalized decisions in designing products, processes and resources. The concept of cost drivers is used here to measure such achievements. Table 2 shows major items for measuring quality, costs and time. Table 3 lists typical cost drivers of major life cycles. It is relatively straightforward to explain improvements in Category 1 by relating these cost drivers to elements determining quality, costs, time, and flexibility. For example, significant reduction in part count has been reported as a result of DFA. This leads to a chain of not only direct savings but also overheads savings throughout the organization. Take the number of engineering changes as another example. The use of DFX tools is likely to increase the number of design changes at early stages but reduce the number of late design changes significantly. Because it is easy to change early than late, substantial savings can be achieved.

The third category of benefits of applying DFX is its far-reaching effect on operational efficiency in product development. In general, DFX leads to the rationalization of decision-making and realization activities in designing products, processes, and resources. For example, Chapter 7 extends the GRAI Integration Methodology (GIM) into a DFX method which can be loosely described as a "Design for Design" tool. Its use would lead to the reengineering of the product development and design process and improve its efficiency. As another example, the use of a "Design for material logistics" can not only improve product designs but also reengineer the "material logistics" business process. Let us return to Design for Assembly (DFA). The use of DFA would reduce the number of assembly operations and rationalize the remainder. These Category 3 improvements are fundamental to Category 2 and in turn to Category 1 benefits. Following is a list of typical Category 3 benefits:

1. Better communications and closer cooperation.
2. Concurrence and transparency.
3. Better job hang-over.
4. Improved customer and supplier involvement.
5. Easier project management.
6. Team-building in design work.
7. Rationalizing and structuring product development.
8. Promoting concurrent engineering practice.

A DFX tool does not work on its own, just like a hammer does not bang a nail by itself. Benefits are gained by using it, not by owning it. How beneficial a DFX project depends very much on how DFX is implemented. A comprehensive DFX tool is usually accompanied by a structured procedure which systematically describes instructions about its implementation, just like instructions for installing a software system on a computer. However, successful DFX implementation is much more complicated than this. Chapter 6 is prepared for those who want to investigate into aspects of DFX implementation.

Table 2 Sample performance characteristics

Quality	Costs	Time
Q1 Quality control	C1 Material	T1 Receive and administer customer orders
Q2 Market research	C2 Purchased parts	T2 Order administration
Q3 Design and development control	C3 Productive labour	T3 Product / process / production planning
Q4 Quality planning	C4 Subcontracts	T4 Order scheduling
Q5 Equipment maintenance and calibration	C5 Consumable tools and fixtures	T5 Order materials - Vendors' lead times
Q6 Supplier certification and accreditation	C6 Indirect labour	T6 Receive & inspect materials
Q7 Supplier quality assurance	C7 Depreciation	T7 Move to stock inventory
Q8 Training	C8 Repairs & Maintenance	T8 Release work order to manufacturing
Q9 Inspection and test, re-inspect and re-test	C9 Communication	T9 Order picking from stock
Q10 Material inspection and test	C10 Inventory storage	T10 Move materials to department
Q11 Scrap, rework and repair	C11 Design and development	T11 Perform operations
Q12 Trouble shooting and defect analysis	C12 Design modification	T12 Set-ups and changeovers
Q13 Modification permits and concessions	C13 Engineering data management	T13 Equipment maintenance and service
Q14 Handling complaints	C14 Production preparation	T14 Move to next work centre
Q15 Analysis and correlation of feedback data	C15 Assembly	T15 Store as WIP inventory
Q16 Warranty	C16 Materials management	T16 Inspection & test
Q17 Excessive stocks to buffer delivery failures	C17 Defects	T17 Packaging
Q18 Prepare specification	C18 Tooling	T18 Store in finished goods inventory
Q19 Administration	C19 Equipment	T19 Shipping

Table 3 Sample cost drivers

Life cycle	Products	Processes	Resources
Design & development	# of parts in a product % of common, standard, or unique parts Shape, depth and width of bills of materials # of new products introduced each year	# of review / check points # of design change notices % of back orders	# of pages of drawings (A4 equivalent) # of drawing errors
Purchasing	# of "buy" parts # of purchase orders # of raw materials	# of supply expedites # of inspections of incoming materials # of supplier visits	# of suppliers
Production planning & Shop-floor manufacture	# of manufacturing features # of "make" parts # of reworks # of scraps	# of operations # of processes # of production orders # of schedule changes # of engineering change requests # of setups # of changeovers	# of machines # of part movements # of tools and fixtures # of tool movements
Inspection and test	# of parts to be inspected # of key characteristics being inspected	# of inspections # of defects / faults	# of check points # of special test / inspection equipment # of inspectors
Storing & shipping	Product size, shape, weight etc.	# of movements distance of movements	# of tablets pallet space
Sales and marketing	# of product models # of product variants in a model # of cancellations # of defects recalled	# of order changes # of promotions and surveys # of customer order expedites	# of new customers # of customers withdrawn # of sales orders / invoices # of customer enquiries
Installation	# of components to be installed	# of installation operations	# of special tools
Use & operation	# of panel controls # of safety devices	# of operator movements	# of operators space requirements
Service & repair	# of customer complaints # of site service visits	# of service operations	# of tools # of special skills
Recycling & disposal	# of parts recyclable # of parts disassemble	# of disassembly operations	# of disassembly tools and equipment

WHICH ĐFX TO APPLY, WHEN AND WHERE?

The DFX toolbox has expanded rapidly from a few some fifteen years ago to many hundreds today, and the proliferation continues. This book adds over fifteen DFX tools into the DFX toolbox. Table 4 presents samples of DFX tools. From the table, it can be seen that almost all *X* areas have been dealt with to some extent.

It is ideal to apply multiple DFX tools to obtain overall optimal solutions. This is easily said but rarely done. Resources seem always limited to permit this. Usually, one DFX tool is applied at a time. A question arises: Which one? There are wide choices even with one type of DFX. For example, there have appeared dozens of Design for Assembly tools according to surveys carried out several years ago (Sackett and Holbrook, 1988; Carlsson and Egan, 1994; O'Grady and Oh, 1991). It is increasingly difficult, even confusing, to choose a DFX which is most appropriate for the problem at hand. There are many reasons, for example:

1. "Hammer or screwdriver?" What tool to use depends on what problem exists at hand. If the problem is a "nail", then use a hammer; if the problem is a "screw", use a screwdriver. The rule is simple. But practice is vague. Some DFX tools have been promoted as panaceas for curing all sorts of illness. Practitioners tend to be overwhelmed by the wide spectrum and the diverse nature. They have to spend more time and effort in evaluating question like "Do we use a hammer or a screwdriver?" than concentrating on actually identifying and solving their problems (Weber, 1994).
2. "Too many cooks spoil the broth." There has been elaboration in parallel to proliferation. No DFX tools are perfect and each suffers from shortcomings of one kind or another. Researchers have attempted to improve them by assuming the availability of required input data which may be expensive to collect and by introducing novel algorithms for data processing which may be hard for the user to comprehend. As a result, sophisticated systems may lose the advantage of being focused and pragmatic. Practitioners become increasingly sceptic and gradually lose their interests and commitments.

DFX applies when and where it helps - never too late, never too early. The point is when and where it helps most. Figure 2 shows an general applicability envelope of DFX tools (McGrath, Anthony and Shaprio, 1992). The "Where" axis corresponds to life cycles or business processes involved in product development. The "When" axis identifies different stages in product design. The shading of horizontal bars indicates the level of involvement each function has at the various points in product development. Following are a few general guidelines regarding when and where to apply what DFX:

1. A consensus view is that DFX should be used as early as feasible. The earlier, the greater the potential.
2. DFX such as Design for Assembly and Design for Variety should be used to rationalize product assortments and structures before other types of DFX tools.
3. What the problem is and where it lies determine what DFX to use.
4. Exactly which specific DFX tool should be used is affected by a number of factors such as availability, applicability, vendor experience, etc.

Table 4 Sample DFX tool kits

X	Design for X Examples	References
Design & development	*GRAI Integrated Methodology* Various approaches to improving product development	*Chapter 7*
Purchasing	Design for profits	Mughal and Osborne, 1995
Fabrication	*Design for dimension control* Hitachi MEM Design for manufacturing mfk For more DFM techniques	*Chapter 8* Arimoto *et al.*, 1993 Boothroyd *et al., 1994* Meerkamm, 1993 see Bralla, 1986
Assembly	*Boothroyd-Dewhurst DFA,* *Design for PCB assembly cost* Lucas DFA, Hitachi AEM For more DFA techniques, see Sacket and Holbrrok, 1988, O'Grady and Oh, 1992	*Chapter 1, also see Chapter 2* *Chapter 9* Chapter 2 for an overview Chapter 2 for an overview
Material logistics	Design for material logistics	Foo *et al.*, 1990
Material handling		
Inspection and test	*Design for inspectability* *Design for dimensional control*	*Chapter 10* *Chapter 8*
Storage / distribution	*Design for storability and distribution*	*Chapter 11*
Sales / marketing	Design for marketability Quality Function Deployment	Zaccai, 1994 Akao, 1991
Installation		
Use / operation	*Design for reliability* *Design for EMC* Design for safety Design for human factors	*Chapter 12* *Chapter 13* Wang and Ruxton, 1993 Tayyari, 1993
Service / repair	*Boothroyd-Dewhurst DFS* Design for serviceability Design for diagnosis Design-based diagnosis Design for reliability and maintainability	*Chapter 14* Gershenson and Ishii, 1991 Ruff and Paasch, 1993 Alexander *et al.*, 1993 Gardner and Sheldon, 1995
Recycling & disposal	*Design for disassembly for recycling* *Life-cycle design based on ABC* *Design for optimal environment impacts* Design for ease of recycling	*Chapter 15* *Chapter 19* *Chapter 18* Beitz, 1990
Quality	*Design for quality* Design for quality Quality Function Deployment Failure Modes and Effects Analysis	*Chapter 16* Morup, 1994 Akao, 1991 BSI, 1991
Cost	*Design for assembly cost* Design for whole life costs Cost information tools for designers	*Chapter 9* Sheldon *et al.*, 1990 Wierda, 1990
Flexibility	*Design for modularity* Variety reduction program Relationships between ...	*Chapter 17* Suzue and Kohdate, 1988 Andreasen and Ahm, 1986
Environment	*Design for environment* *Design for optimal environment impacts* *Design for life cycle* Life cycle design Design for environmentality	*Chapter 3* *Chapter 18* *Chapter 19* Alting, 1993 Navinchandra, 1991

Note: Those in italic are discussed in this book.

The final point is whether to "Make" or "Buy" a DFX tool. If there is a relevant DFX tool in the toolbox, then the decision is to "Buy" and to buy the most suitable tool. If no relevant DFX tool can be found, the decision is to "Make" a new DFX tool. Care should be taken with this "Make" decision because developing a DFX tool can be very time consuming and labour intensive. This is especially true when the developers have little DFX experience in development or implementation. According to Whitney (1994), one third of the Japanese companies he visited had developed their own DFX software in one way or another. Chapter 5 is prepared for those who are considering developing their own DFX tools.

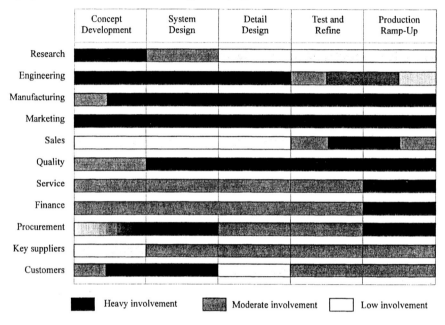

Figure 2 Involvement of different functions and applicability envelope of DFX.

WHO IS INVOLVED?

DFX has been used in manufacturing industries in engineering fields of mechanical, aerospace, automotive, electronic, electrical, etc. The size of companies ranges from multi-national giants to SME's (Small and Medium sized Enterprises). Subject products range from sophisticated aircraft to entertaining toys; from automobiles on the road to appliances in the kitchen; from as large as oil supertankers to as small as cut-off service fuses.

If anyone wants to benefit from a DFX project, then involvement and participation are necessary in exchange. DFX is primarily about improving a subject product. Therefore, design engineers are almost always involved. DFX is often concerned with improving a subject business process. Naturally, the subject business process should be represented in the DFX project. For example, if "assembly" is the subject process, then this function should be represented in the project. If "service" is the subject business process, then this function should be represented. If the subject process covers "whole life", then representatives from all functions should participate the DFX project in one way or another.

The second part of X in DFX is *"bility"*, i.e. specific definition of performance indicators. Therefore, a representative responsible for such performance indicator(s) should be represented in the DFX project. For example, if *"bility"* is defined by "quality" indicators, then QC/A (Quality Control / Assurance) should be represented. If *"bility"* is defined by "cost" indicators, then costing function should be represented. Because both quality and cost are of paramount importance their representation at some stage of DFX is imperative.

Above mentioned are prime DFX users who make direct contributions to the project by providing data and expertise for problem analysis. They constitute a so-called core team, in whichever form it may exist. Such a core team may be assisted by DFX facilitator(s) or consultant(s), especially in the beginning.

DFX tools are not usually developed for management although there have been a few managerial uses such as strategic product planning. Its involvement is essential to the success of the project. Opening speeches by chief executives and allocating funds which could be cut next year are not commitment. Commitment without direct participation and involvement is not enough. Leaderships must come from management, not someone who enjoys a word of appraisal nor someone who is made scapegoat blamed for everything.

Those who are affected by the DFX project should not be neglected, whether they are beneficiaries or "victims". They form what can be called an extended team that may work in a different way from the core team. Their involvement and participation in evaluating and implementing solutions can be decisive in making the benefits lasting and permanent.

A DFX project should be organised in a way that the benefits are maximized. Larger companies may choose to adopt a formal approach to organizing a DFX project. This usually includes (1) establishing a steering committee consisting of managers at varying levels and leading practitioners; and (2) a project core team led by the project leader and consisting of a variety of disciplines. These teams and committees hold regular working meetings.

To the other end, smaller companies may adopt an informal approach to organizing a DFX project. It is more economically viable for smaller companies to embed DFX activities into day-to-day operation in product development after receiving necessary training. A coordinator ensures proper DFX considerations and is responsible for convening meetings when necessary.

WHAT'S NEXT?

The current status of DFX practice is both encouraging and disappointing. It is encouraging in the sense that an ever-increasing number of companies are introducing DFX, with many more wishing to do so. Problems related to *x-bility* have been recognized as major cost drivers, time wasters, and quality barriers. This is confirmed by the findings from recent industrial surveys (Dean and Salstrom, 1990; Youssef, 1994; Sehdev and Fan *et al.*, 1995).

In sharp contrast, the number of companies who are using DFX is small relative to the manufacturing population and the pace at which companies are starting DFX is slow relative to the seriousness of *x-bility* problems. In November 1995, an E-Mail was broadcast from *Eric.Sleeckx@wtcm.kuleuven.ac.be* among Engineering Design MailBase expressing the disappointment that DFX has not been used widely enough in industry despite its great potentials. This question might have been so tough that very few replies were broadcast to the audience who received the question (Probably, the original broadcaster may have received more messages). One reply pointed out that this question has been puzzling researchers in the

Engineering Design arena for many years and many causes have been identified but only a few solutions have been proposed.

Typical barriers to halfway implementation are compiled from a number of surveys and are listed in a table towards the end of Chapter 6. Two thirds of these barriers are common to any improvement project, whether it is DFX, TQC/M (Total Quality Control / Management), JIT (Just In Time), or something else. Perhaps they should be dealt with seriously (Evans, 1993). Few than one third of these barriers are specific to DFX. One of the reasons listed in the survey reported by Sehdev and Fan *et al.* (1995) is that DFM (DFX) is not well understood. This is surprising because DFX and CE have been widely preached at conferences, workshops, seminars, and respectably lauded in journals and magazines. It is hoped that this book will help in this respect.

The future work should be centred around addressing the question why so few companies are using DFX tools. Progresses are expected in the following directions:

1.   More DFX users and more successes.
2.   Better DFX tools and better DFX results.
3.   More DFX tools focused on specific problems.
4.   Easier and more effective to use.
5.   Search for basic DFX model for development and implementation.
6.   Search for common basis for integration and tradeoff for overall optimum.
7.   Search for "plugs and sockets" with other decision support systems in product development such as CAD/CAM, CAPP, and CAPM.

SUMMARY

DFX is both a philosophy and a methodology that can help companies change the way that they manage product development and to become more competitive. This book is designed to answer the many questions companies may have, such as what is DFX? how does DFX work? what are the benefits, what are the techniques, how to implement DFX, which, when and where is DFX used, who is involved in DFX and what is good practice? It is hoped that this book will contribute to the understanding and the increased adoption of one of the most effective and exciting approaches available to manufacturing companies.

REFERENCES

Akao, Y. (1990) *Quality Function Deployment: Integrating Customer Requirements into Product Design*, Productivity Press, Cambridge, Massachusetts, USA.

Alexander, S.M., Lee, W.Y., Graham, J.H. (1993) Design-based diagnosis, *International Journal of Production Research*, 31 (9), 2087-2096.

Alting, L. (1993) Life cycle design of products: A new opportunity for manufacturing enterprises, In: *Concurrent Engineering: Automation, Tools, and Techniques*, Edited by A. Kusiak, John Wiley & Sons, Inc., New York, USA, Chapter 1, 1-18.

Andreasen, M.M., Ahm, T. (1986) The relationships between product design, production, layout and flexibility, In: *Proceedings of 7th International Conference on Assembly Automation*, Zurich, Switzerland, IFS (Publications) Ltd, UK, 161-172.

Andreasen, M.M., Hein, L. (1987) *Integrated Product Development*, IFS Publications / Springer-Verlag, London, UK.

Andreasen, M.M., Olesen, J. (1990) The concept of dispositions, *Journal of Engineering Design*, **1** (1), 17-36

Andreasen, M.M., Olesen, J. (1993) *Minutes of WDK Workshop on DFX Research*, Technical University of Denmark, Denmark.

anon. (1970) *Conference on Design for Production*, Birniehill Institute, Scotland.

Arimoto, S., Ohashi, T., Ikeda, M., Miyakawa, S. (1993) Development of Machining-Producibility Evaluation Method (MEM), *Annals of CIRP*, 42 (1), 119-122.

Beitz, W. (1990) Design for ease of recycling (Guidelines VDI 2243), In: *Proceedings of the ICED 90 Dubrovnik*, Heurista, Zurich.

Boothroyd, G., Dewhurst, P. (1990) *Product Design for Assembly*, Boothroyd Dewhurst, Inc., Wakefield, RI, USA, (First Edition 1983).

Boothroyd, G., Dewhurst, P., Knight, W. (1994) *Product design for manufacture and assembly*, Marcel Dekker Inc.

Bralla, J.G. (1986) *Handbook of Product Design for Manufacturing, A practical guide to low-cost production*, McGraw-Hill.

Brimson, J.A. (1991) *Activity accounting: An activity-based costing approach*, John Wiley & Sons, Boston, Mass.

British Standard PD 6470 : 1981, *The management of design for economic production*.

BSI (1991) *BS 5760 : Part 5 : 1991, Guide to Failure Modes, Effects and Criticality Analysis* (FMEA and FMECA).

Carlsson, M., Egan, M. (1994) *Design for manufacture and assembly - A missing link between quality and product development*, Technical Report MEC-Report 94-1, Department of Mechanical Engineering, College of Applied Engineering and Maritime Studies, Chalmers University of Technology, Sweden.

Chisholm, A.W.J. (1973) *Annuals of the CIRP*, **22** (2), 243-247.

Cohen, L. (1995) Quality Function Deployment: How to make QFD work for you, Engineering Process Improvement Series, Addison-Wesley Publishing Company, Reading, Massachusetts, USA.

Cooper, R. (1988) The rise of activity-based costing - Part one: What is an activity-based cost system? *Journal of Cost Management*, Summer, 1988, 45-54.

Dean, B.V., Salstrom, R.L. (1990) Utilization of Design for Manufacturing (DFM) techniques, In: *1990 IEEE International Engineering Management Conference*, 223-232.

Duffey, M.R., Dixon, J.R. (1993) Managing the product realization process: A model for aggregate cost and time-to-market evaluation, Concurrent Engineering: Research and Applications, **1**, 51-59.

Evans, S. (1993) Common failure modes and success factors, In: *Concurrent Engineering*, Edited by H.R. Parsaei and W.G. Sullivan, Chapman & Hall, 42-60.

Everhart, J.L. (1960) *Handbook of parts, forms, processes and materials in design engineering*, Van-Nostrand-Reinhold, New York.

Foo, G., Clancy, J.P., Kinney, L.E., Lindemudler, C.R. (1990) Design for material logistics, *AT & T Technical Journal*, **69** (3), 61-76.

Gardener, S., Sheldon, D.F. (1995) Maintainability as an issue for design, *Journal of Engineering Design*, **6** (2), 75-89.

Gatenby, D.A., Foo, G. (1990) Design for X (DFX): Key to competitive, profitable products, *AT & T Technical Journal*, May/June, 2-13.

Gershenson, J., Ishii, K. (1991) Life cycle serviceability design, In: *Proceedings of ASME Conference on Design Theory and Methodology*.

Gladman, C.A. (1968) *Annuals of the CIRP*, **16**, 3-10.

Holbrook, A.E.K., Sackett, P.J. (1988) Design for assembly - guidelines for product design, *Assembly Automation,* 210-212.

Johnson, H.T., Kaplan, R.S. (1987) *Relevance lost: The rise and fall of management accounting,* Harvard Business School Press, Boston, Mass.

Keys, L.K. (1990) System life cycle engineering and DF"X", *IEEE Transactions on Components, Hybrids, and Manufacturing Technology,* 13 (1), 83-93.

Maskell, B.H. (1991) Performance Measurement for World Class Manufacturing, Productivity Press.

Matousek, R. (1957) *Engineering Design - A Systematic Approach,* The German edition by Springer-Verlag, Berlin; The English edition translated by A.H. Burton and Edited by D.C. Johnson, Published by Lackie & Son Ltd, London, UK.

McGrath, M.E., Anthony, M.T., Shaprio, A.R. (1992) Product Development: Success through product and cycle time excellence, The Electronics Business Series, Butterworth-Heinemann, Stoneham, USA.

Meerkamm, H. (1993) Design system mfk - An important step towards an engineering workbench, *Proceedings of IMechE, Part B Journal of Engineering Manufacture,* 207 105-116.

Meerkamm, H. (1994) Design for X - A core area for design methodology, *Journal of Engineering Design,* 5 (2), 145-163.

Miles, B.L. (1989) Design for assembly - a key element within design for manufacture, *Proceedings of IMechE, Part D: Journal of Automobile Engineering,* 203, 29-38.

Miller, G.A. (1956) The magical number seven, plus or minus two: Some limits on our capacity for processing information, *Psychological Review,* 63, 81-97.

Miyakawa, S., Ohashi, T., Iwata, M. (1990) The Hitachi New Assemblability Evaluation Method, *Transactions of the North American Manufacturing Research,* Institution (NAMRI) of the SME, the NAMR Conference XVIII, May 23-25, 1990, Pennsylvania State University, Dearborn, USA.

Morup, M. (1994) Design for Quality, PhD Thesis, Institute of Engineering Design, Technical University of Denmark, Denmark.

Mughal, H., Osborne, R. (1995) Design for profit, *World-Class Design to Manufacture,* 2 (5), 16-26.

Navichandra, D. (1991) Design for environmentality, In: *Proceedings of ASME'91 conference on design theory and methodology,* New York, USA.

Niebel and Baldwin (1957) *Designing for Production,* Irwin.

O'Grady, P., Oh, J. (1991) A review of approaches to design for assembly, Concurrent Engineering, 1, 5-11.

Olesen, J. (1992) *Concurrent development in manufacturing - based on dispositional mechanisms,* PhD Thesis, Technical University of Denmark.

Peck, H. (1973) *Designing for Manufacture,* Topics in Engineering Design series, Pitman & Sons Ltd, London, UK.

Ruff, D.N., Paasch, R.K. (1993) Consideration of failure diagnosis in conceptual design of mechanical systems, In: *Proceedings of the 5th International Conference on Design Theory and Methodology,* ASME DE 53, 175-187.

Sackett, P., Holbrook, A. (1988) DFA as a primary process decreases design deficiencies, *Assembly Automation,* 12 (2), 15-16.

Sehdev, K., Fan, I.S., Cooper, S., Williams, G. (1995) Design for manufacture in the aerospace extended enterprise, *World-Class Design to Manufacture,* 2 (2), 28-33.

Sheldon, D.F., Perks, R., Jackson, M., Miles, B.L., Holland, J. (1990) Designing for whole-life costs at the concept stage, In: Proceedings of ICED'90, Heurista, Zurich.

Shimada, J., Miyakawa, S., Ohashi, T. (1992) Design for manufacture, tools and methods: - the Assemblability Evaluation Method (AEM), *FISITA'92 Congress*, London, 7-11 June, Paper C389/460, FISITA, SAE No. 925142, IMechE, 53-59.

Stoll, H.W. (1988) Design for manufacture, *Manufacturing Engineering*, January.

Suzue, T., Kohdate, A. (1988) *Variety Reduction Program: A production strategy for product diversification*, Productivity Press, Cambridge, Massachusetts, USA.

Swift, K.G. (1981) *Design for Assembly Handbook*, Salford University Industrial Centre Ltd., UK.

Tayyari, F. (1993) Design for human factors, In: *Concurrent Engineering*, Edited by H.R. Parsaei and W.G. Sullivan, Chapman & Hall, 297-325.

Tipping, W.V. (1965) Component and product design for mechanized assembly, In: *A PERA Conference and Exhibition*, Section 14.

Wang, J., Ruxton, T. (1993) Design for safety of make-to-order products, *Presented at the National Design Engineering Conference*, ASME, 93-DE-1

Weber, N.O. (1994) Flying high: Aircraft design takes off with DFMA, *Assembly*, September.

Whitney, D. (1994) Integrated design and manufacturing in Japan, *Target*, 10 (2), 14.

Wierda, L.S. (1990) *Cost Information Tools for Designers*, Delft University Press, the Netherlands.

Youssef, M.A. (1994) Design for Manufacturability and time to market - Part 1: Theoretical foundations, *International Journal of Operations and Production Management*, **14** (12), 6-21.

Zaccai, G. (1994) The new DFM: Design for Marketability, *World-Class Manufacture to Design*, 1 (6), 5-11.

Ziemke, M.C., Spann, M.S. (1993) Concurrent engineering roots in the World War II era, In: *Concurrent Engineering*, Edited by H.R. Parsaei and W.G. Sullivan, Chapman & Hall, Chapter 2, 24-41.

DESIGN FOR MANUFACTURE AND ASSEMBLY: THE BOOTHROYD-DEWHURST EXPERIENCE

Geoffrey Boothroyd

This chapter explains how the Boothroyd-Dewhurst (B&D) Design for Manufacture and Assembly (DFMA) works, discusses the experience and benefits of using DFMA by world-class manufacturers, and highlights implementation issues.

It has been estimated that, in the US, manufacturing contributes about 23% of the gross national product but, more importantly, about 70% of all wealth producing activities. Those who complacently say that the US is changing to a service economy might eventually find that they no longer have the means to purchase these services. The US has been losing $340 million per day to its foreign competitors and the national debt is now around $4 trillion!

Competitiveness has been lost in many areas, but most notably in automobile manufacture, as highlighted by the results of the $5 million world-wide study of this industry that was published in 1990 (Womack *et al., 1990*). The study, which showed that Japan has the most productive plants, attempted to explain the wide variations in auto assembly plant productivity throughout the world. It was found that automation could only account for one-third of the total difference in productivity between plants world-wide and that, at any level of automation, the difference between the most and least efficient plant is enormous.

Womack *et al.* (1990) concluded that no improvements in operation can make a plant fully competitive if the product design is defective. However, they failed to make a direct connection between product design and productivity. Whereas the author of this chapter believes that, and as this chapter will help to show, there is now overwhelming evidence to support the view that product design for manufacture and assembly can be the key to high productivity in all manufacturing industries.

1.1 DESIGN FOR MANUFACTURE AND ASSEMBLY

That designers should give attention to possible manufacturing problems associated with a design has been advocated for many years. Traditionally, the idea was that a competent designer should be familiar with manufacturing processes to avoid adding unnecessarily to manufacturing costs.

However, for reasons such as the increasingly complex technology incorporated within many products; the time pressures put on designers to get designs on to the shop floor; the "we design it, you manufacture it" attitude of designers; and the increasing sophistication of manufacturing techniques, this simple view of the product development process has become invalid.

It is, therefore, becoming recognized that more effort is required to take manufacturing and assembly into account early in the product design cycle. One way of achieving this is for manufacturing engineers to be part of a simultaneous or concurrent engineering design team.

Within this teamworking, design for manufacture and assembly (DFMA) analysis tools help in the evaluation of proposed designs. It is important that design teams have access to such tools in order to provide a focal point which helps identify problems from manufacturing and design perspectives. In terms of the 80/20 rule, teams spend 80% of the time on 20% of the problems, and DFMA helps the team identify the right 20% to work on.

DFMA is a systematic procedure that aims to help companies make the fullest use of the manufacturing processes that exist and keep the number of parts in an assembly to the minimum. It achieves this by enabling the analysis of design ideas. It is not a design system, and any innovation must come from the design team, but it does provide quantification to help decision-making at the early stages of design.

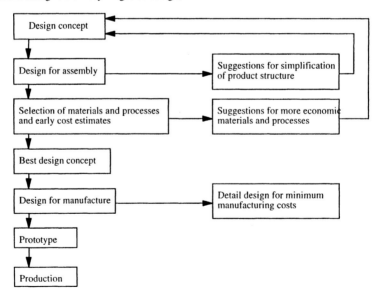

Figure 1.1 Typical steps taken in a simultaneous engineering study using DFMA.

Figure 1.1 summarizes the steps taken when using DFMA during design. The design-for-assembly (DFA) analysis is conducted first, leading to a simplification of the product structure. Then, early cost estimates for the parts are obtained for both the original design and the new design in order to make tradeoff decisions. During this process, the best materials and processes to be used for the various parts are considered. For example, would it be better to manufacture a cover from plastic or sheet metal? Once the materials and processes have been finally selected, a more thorough analysis for design for manufacture (DFM) can be carried out for the detail design of the parts.

It should be remembered that DFMA is the integration of the separate but interrelated design issues of assembly and manufacturing processes. Therefore, there are two fundamental aspects to producing efficient designs: DFA and the early implementation of DFM.

1.1.1 Boothroyd-Dewhurst DFA Method

Development of the Boothroyd Dewhurst DFA method started in 1977 with funding from the US National Science Foundation (Boothroyd and Dewhurst, 1983). It was first introduced in handbook form in 1980, with Salford University Industrial Centre producing a UK version of the handbook authored by K. G. Swift (1981). These handbooks included analysis methods and databases for both manual and high-speed automatic assembly. For each process, the handling of the parts and their insertion were considered separately. The original procedure for design for automatic assembly was the result of collaboration between the author and A. H. Redford and K. G. Swift in Salford.

Since the initial work, the author and his colleague P. Dewhurst have developed, in the US, a personal computer program for DFA which was introduced in 1982. In 1983, a new handbook, based on the lessons learned in implementing DFA in industry, was introduced and since then design for robot assembly and PCB assembly have been added (Boothroyd and Dewhurst, 1983).

With DFA, the greatest improvements tend to arise from simplification of the product by reducing the number of separate parts. In order to give guidance in reducing the part count, the Boothroyd Dewhurst DFA methodology provides three criteria against which each part must be examined as it is added to the product during assembly:

- During operation of the product, does the part move relative to all other parts already assembled?
- Must the part be of a different material than, or be isolated from, all other parts already assembled? Only fundamental reasons concerned with material properties are acceptable.
- Must the part be separate from all other parts already assembled because the necessary assembly or disassembly of other separate parts would otherwise be impossible?

If the answer is yes to any of these questions, then the part must be a separate item - a critical part. The number of critical items is regarded as the theoretical minimum number of parts for the design, since all the others can, in theory, be removed or merged with these critical parts. Therefore, the DFMA team must have a good reason for a part being included as a separate item in the design if it does not meet one of these criteria.

This assessment procedure leads to ideas as to how the product may be simplified. At this stage, these are not cost or analyzed and some may be impractical, but, from this, viable ideas come forward.

The next step is to estimate the assembly time for the product design, and establish its efficiency ratings in terms of difficulty of assembly.

Each part in the design is examined for two considerations: how the part is to be grasped, orientated and made ready for insertion, and how it is inserted and/or fastened into the product.

The difficulty of these operations is rated, and from this rating standard times are determined for all the operations necessary to assemble each part. The DFA time standard is a classification of design features which affect part assembly. It is a system for designers to use - similar to MTM (Methods-Time Measurement) standards for industrial engineers - which has been developed through years of experimentation. Usage has proved the data to be quite accurate for the overall times.

The total assembly time for the product can then be estimated and, using standard labour rates, so can assembly costs. Also the efficiency of a design from an ease of assembly point of view can be determined.

Based on the assumption that all of the critical parts could be made easy to assemble - requiring only three seconds each - the minimum assembly time (MAT) equals theoretical minimum number of parts times three. Assembly efficiency percentage equals MAT divided by the estimated total assembly time times 100.

At this stage, part manufacturing costs are not brought into the analysis, but the efficiency rating and estimated assembly times provide benchmarks against which further design iterations, previous estimates for an original product design or a competitor's product can be compared.

1.1.2 Boothroyd-Dewhurst Manufacture Analysis

After the DFA analysis and the simplification of the product structure, the next step is to analyze the manufacture of the individual parts. Few design engineers have detailed knowledge of all the major shapeforming processes and, consequently, they tend to design for the ones with which they are comfortable. The purpose of the DFM cost estimating process is to enable design teams to weigh alternative designs and production processes, quantify manufacturing costs, and make the necessary trade-off decisions between parts consolidation and increased material/manufacturing costs.

Table 1.1 DFM analysis of injection-moulded heater cover (Dewhurst, 1988)

	Old design	New design
Cost of one cavity and core	$ 8,032	$11,625
Cycle time (s)	42.8	13.3
Number of cavities required	6	2
Cost of production mould	$ 36,383	$ 22,925
Cost per part (inc. 5 cents for material)	25.1 cents	16.8 cents

The DFM system provides data, based on experimental work, for the cost estimation of a variety of processes. Although they may be rough estimates, they are ample for projecting costs at this stage of the design process. In fact, some companies have utilized this information for negotiating with vendors.

Since 1985, Boothroyd, Dewhurst and Knight have developed methods for designers to obtain cost estimates for parts and tooling during the early phases of design. Studies have been completed for machined parts (Boothroyd and Radovanovic, 1989), injection-moulded parts (Dewhurst, 1988), die-cast parts (Dewhurst and Blum, 1989), sheet-metal stampings

Zenger and Dewhurst, 1988) and powder-metal parts (Knight, 1991). The objective of these studies was to provide methods with which the designer or design team can quickly obtain information on costs before detailed design has taken place. For example, an analysis (Dewhurst, 1988) of an injection-moulded heater cover gave the results shown in Table 1.1. It was evident that certain wall thicknesses were too large, and that, through some fairly minor design changes, the processing cost could be reduced by 33%. If these studies had taken place at the early design stage, the designer could also have considered the cost for an equivalent sheet-metal part for example. In fact, the use of these analysis techniques is now allowing designers and purchasing managers to challenge suppliers' estimates. In one example, it has been reported that Polaroid Corporation has saved $16,000-20,000 on the cost of tooling for an injection-moulded part (Kirkland, 1992).

1.1.3 How DFMA Works

By way of example, Figure 1.2 shows the requirements of a motor-drive assembly that must be designed to sense and control its position on two steel guiderails. The motor must be fully enclosed for aesthetic reasons, and have a removable cover for access so that the position sensor can be adjusted. The principal requirements are a rigid base that is designed to slide up and down the guiderails, and that supports the motor and sensor. The motor and sensor have wires that connect them to a power supply and a control unit, respectively.

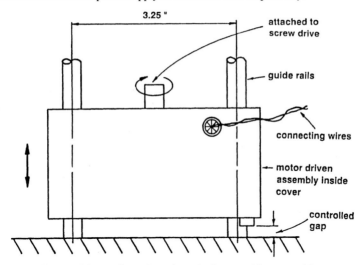

Figure 1.2 Configuration of required motor-drive assembly.

A proposed solution is shown in Figure 1.3. The base is provided with two bushes to provide suitable friction and wear characteristics. The motor is secured to the base with two motor screws, and a hole in the base accepts the cylindrical sensor, which is held in place with a set screw. To provide the required covers, an end plate is secured by two end-plate screws to two standoffs, which are, in turn, screwed into the base. This end plate is fitted with a plastic bush through which the connecting wires pass. Finally, a box-shaped cover slides over the whole assembly from below the base, and is held in place by four cover screws, two passing into the base, and two into the end cover.

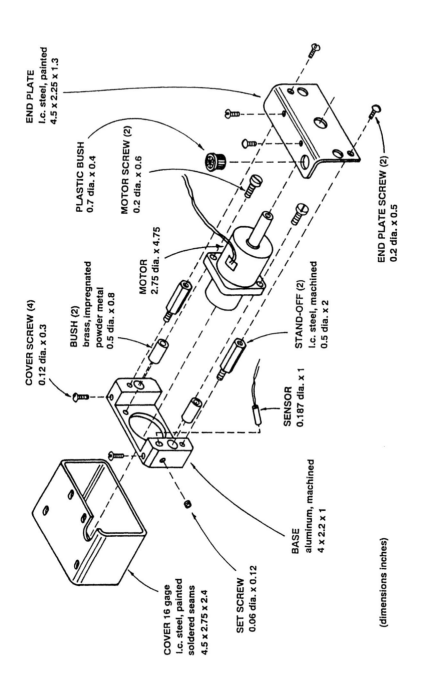

END PLATE
l.c. steel, painted
4.5 x 2.25 x 1.3

PLASTIC BUSH
0.7 dia. x 0.4

MOTOR SCREW (2)
0.2 dia. x 0.6

END PLATE SCREW (2)
0.2 dia. x 0.5

COVER SCREW (4)
0.12 dia. x 0.3

BUSH (2)
brass, impregnated
powder metal
0.5 dia. x 0.8

MOTOR
2.75 dia. x 4.75

STAND-OFF (2)
l.c. steel, machined
0.5 dia. x 2

SENSOR
0.187 dia. x 1

BASE
aluminum, machined
4 x 2.2 x 1

COVER 16 gage
l.c. steel, painted
soldered seams
4.5 x 2.75 x 2.4

SET SCREW
0.06 dia. x 0.12

(dimensions inches)

Figure 1.3 Initial design of motor-drive assembly.

Two subassemblies are required, the motor and the sensor, and, in this initial design, there are eight additional main parts, and nine screws, making a total of 19 items to be assembled.

The application of the minimum part criteria to the proposed design proceeds as follows:

- The base is assembled into a fixture, and, since there are no other parts with which to combine it, it is a theoretically necessary part.
- The two bushes do not satisfy the criteria, and can theoretically be integral with the base.
- The motor is a standard subassembly of parts which is a purchased item. Thus, the criteria cannot be applied unless the assembly of the motor itself is considered as part of the analysis. In this example, we assume that motor and sensor are not to be analyzed.
- Invariably, separate fasteners such as the two motor screws do not meet the criteria, because an integral fastening arrangement is always theoretically possible.
- The sensor is a purchased item
- The set screw is theoretically not necessary.
- The two standoffs do not meet the criteria; they could be incorporated into the base.
- The end plate must be separate for reasons of assembly.
- The two end-plate screws are theoretically not necessary.
- The plastic bush can be of the same material as, and therefore combined with, the end plate.
- The cover can also be combined with the end plate.
- Finally, the four cover screws are theoretically not necessary.

From this analysis, it can be seen that, if the motor and sensor subassemblies can be arranged to snap or screw in the base, and a plastic cover can be designed to snap on, only four separate items will be needed, instead of 19. These four items represent the theoretical minimum number needed to satisfy the constraints of the product design without consideration of the practical limitations.

It is now necessary for the designer or design team to justify the existence of those parts that have not satisfied the criteria. Justification may arise from practical, technical or economic considerations. In this example, it can be argued that two motor screws are needed to secure the motor, and one set screw is needed to hold the sensor, because any alternatives would be impractical for a low-volume product such as this.

It can be argued that the two powder metal bushes are unnecessary, because the base could be machined from an alternative material with the necessary frictional characteristics.

Finally, it is very difficult to justify the separate standoffs, end plate, cover, plastic bush and associated six screws.

Now, before an alternative design can be considered, it is necessary to have estimates of the assembly times and costs, so that any possible savings can be taken into account when considering design alternatives. Using DFMA time standards and knowledge bases, it is possible to make estimates of assembly costs, and then to estimate the cost of the parts and associated tooling, without having final detail drawings of the parts.

First, Table 1.2 shows the results of the DFA analysis; the total assembly time is estimated to be 160 s. It is also possible to obtain an absolute measure of the quality of the design for ease of assembly. The theoretical minimum number of parts is four, as explained above, and, if these parts were easy to assemble, they would take 3 s each to assemble on average. Thus, the theoretical minimum (or ideal) assembly time is 12 s, a figure which can be compared with the estimated time of 160 s, giving an assembly efficiency of 12/160, or 7.5%.

Table 1.2 Results of DFA analysis for initial design of motor-drive assembly

Item	Number	Theoretical part count	Assembly time (s)	Assembly cost (US cents)
Base	1	1	3.5	2.9
Bush	2	0	12.3	10.2
Motor subassembly	1	1	9.5	7.9
Motor screw	2	0	21.0	17.5
Sensor subassembly	1	1	8.5	7.1
Set screw	1	0	10.6	8.8
Standoff	2	0	16.0	13.3
End plate	1	1	8.4	7.0
End plate screw	2	0	16.6	13.8
Plastic bush	1	0	3.5	2.9
Thread lead	-	-	5.0	4.2
Reorient	-	-	4.5	3.8
Cover	1	0	9.4	7.9
Cover screw	4	0	34.2	26.0
Totals	19	4	160.0	133.0

[Design efficiency = 4 x 3 / 160 = 7.5%]

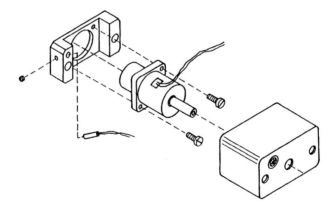

Figure 1.4 Redesign of motor-drive assembly following DFA analysis.

The elimination of parts not meeting the minimum part-count criteria, and which cannot be justified on practical grounds, results in the design concept shown in Figure 1.4. Here, the bushes are combined with the base, and the standoffs, end plate, cover, plastic bush and six associated screws are replaced by one snap-on plastic cover. The eliminated items entailed an assembly time of 97.4 s. The new cover takes only 4 s to assemble, and it avoids the need for a reorientation. In addition, screws with pilot points are used and the base is redesigned so that the motor is self-aligning. Table 1.3 presents the results of a DFA analysis of the redesigned assembly; the new assembly time is only 46 s, and the design efficiency has increased to 26%.

Finally, Table 1.4 compares the cost of the parts for the two designs. It can be seen that there is a saving of $13.71 in parts costs. However, the tooling for the new cover is estimated to be $5000 -- an investment that would have to be made at the outset. Thus, the outcome of this study is a second design concept that represents a total saving of $14.66, of which $0.95 represents the savings in assembly time.

Table 1.3 Results of DFA analysis for redesign of motor-drive assembly

Item	Number	Theoretical part count	Assembly time (s)	Assembly cost (US cents)
Base	1	1	3.5	2.9
Motor subassembly	1	1	4.5	3.8
Motor screw	2	0	12.0	10.0
Sensor subassembly	1	1	8.5	7.1
Set screw	1	0	8.5	7.1
Thread leads	-	-	5.0	4.2
Plastic cover	1	1	4.0	3.3
Totals	7	4	46.0	38.4

[Design efficiency = 4 x 3 / 46.0 = 26.0%]

Table 1.4 Comparison of part costs for motor-drive assembly design and redesign

Proposed design		Redesign	
Item	Cost $	Item	Cost $
Base (aluminium)	12.91	Base (nylon)	13.43
Bush (2)	2.40*	Motor screw (2)	0.20*
Motor screw (2)	0.20	Set screw	0.10*
Set screw	0.10*	Plastic cover include tooling	8.00
Standoff (2)	5.19		
End plate	5.89		
End plate screw (2)	0.20*		
Plastic bush	0.10*		
Cover	8.05		
Cover screw (4)	0.40*		
Totals	35.44		21.73

[* Purchased in quantity. Purchased motor and sensor subassemblies not included. Redesign: Tooling cost for plastic cover - $5,000]

1.2 RESULTS OF DFMA APPLICATIONS

DFMA provides a systematic procedure for analyzing proposed designs from the point of view of assembly and manufacture. This procedure results in simpler and more reliable products which are less expensive to assemble and manufacture. In addition, any reduction in the number of parts in an assembly produces a snowball effect on cost reduction, because of the drawings and specifications that are no longer needed, the vendors that are no longer needed and the inventory that is eliminated. All of these factors have an important effect on overheads, which, in many cases, form the largest proportion of the total product cost.

DFMA tools encourage dialogue between designers and the manufacturing engineers and any other individuals who play a part in determining final product costs during the early stages of design. This means that team working is encouraged, and the benefits of simultaneous or concurrent engineering can be achieved. The following selection of published case studies illustrates the results of DFMA applications.

GE Automotive

Sorge (1994) reported that, around 1992, GE Automotive created two kinds of joint, cross-functional teams. Productivity teams work on short term solutions while design for manufacture and assembly (DFMA) teams are charged with getting long-term results. Their job is to cut costs, improve efficiencies, add capacity, create new business, and produce better quality. Simply put, the challenge is to "minimize the agony and maximize the ecstasy of reaching those goals" says A. J. Febbo, GE Vice President, Auto Industry.

Consisting of ten to fifteen members, the DFMA teams are cross-functional and often include representatives from two or three companies plus a facilitator from GE. In early 1993, GE invested $200,000 in a DFMA centre which houses the necessary software and an area where vehicles can be dismantled.

When the DFMA team process works, spectacular results can be achieved, says GE. For example, DFMA studies done in 1992 and 1993 showed the following:

- In a headlamp assembly project, the number of parts dropped from 67 to 42; the assembly cost fell from $11.81 to $6.96, and the total assembly cost was reduced from $19.79 to $13.90. These figures are for each headlamp.
- In a structural instrument panel, the number of parts was whittled down from 178 to 107; the number of assembly operations declined from 245 to 172; and the total assembly cost dropped from $13.51 to $9.46.
- The number of parts in a front door fell from 327 to 307, while the number of operations plunged from 696 to 522, and the total assembly cost shrank from $38.44 to $27.21.
- In an accelerator pedal, the number of parts dropped from 13 to 2 while the number of assembly operations plunged from 24 to 2, a whopping 92% decline. Total assembly cost went down 93% to 9 cents from $1.28.

Those are just a few examples. In 1993, the DFMA teams had 21 projects, 14 still under way, and another 7 are complete for a three-year projected saving of twenty million dollars. Another 10 pending projects could save about thirty millions dollars. The average saving per project is about $500,000 a year says Mr. Isaac.

Parts reduction alone can create substantial savings over time. Just keeping the drawings for a specific part costs about $300 a year, says Mr. Isaac.

Brown & Sharpe

The need for a low-cost, high-accuracy coordinate-measuring machine (CMM) was the impetus behind the development of the MicroVal personal CMM by Brown & Sharpe (McCabe, 1988). The primary design consideration was to produce a CMM which would sell for one-half of the price of the existing product. The CMM was to compete with low-priced imports which had penetrated the CMM market to an even greater extent than imports had in the automotive industry. Since the CMM customer is not driven by price alone, the new

CMM would have to be more accurate than the current design, while also being easier to install, use, maintain and repair.

Brown & Sharpe started with a clean sheet of paper. Instead of designing the basic elements of the machine and then adding on parts which would perform specific functions required for the operation of the machine, it was decided to build as many functions into the required elements as was feasible. This concept was called integrated construction. However, until the DFA methodology was applied, the cost objectives could not be met with the original design proposal. After DFA, for example, the shape of the Z rail was changed to an elongated hexagon, thus providing the necessary anti-rotation function. As a result, the number of parts required to provide the anti-rotation function was reduced from 57 to four. In addition, the time required to assemble and align the anti-rotation rail was eliminated. Similar savings were made in other areas, such as the linear-displacement measuring system and the Z-rail counterbalance system. On its introduction at the Quality Show in Chicago, IL, USA, in 1988, the machine became an instant success, setting new industry standards for price and ease of operation. The product has proved popular not only in the USA and Europe, but also in Japan.

NCR

Following a year-long competition for the USA's "outstanding example of applied assembly technology and thinking", Assembly Engineering magazine selected Bill Sprague of NCR Corporation, Cambridge, OH, USA, as the PAT (Productivity Through Technology) recipient. Sprague, a senior advanced-manufacturing engineer, was recognized for his contribution in designing a new point-of-sale terminal called the NCR 2760. The DFA methodology, used in conjunction with solid modelling, assisted NCR engineers in making significant changes from the previous design. Those changes translated into dramatic reductions and savings, as follows (Kirkland, 1988).

- 65% fewer suppliers
- 75% less assembly time
- 100% reduction in number of assembly tools
- a total lifetime manufacturing cost reduction of 44% (translating into savings of millions of US dollars).

Indeed, Sprague estimated that the removal of one single screw from the original design would reduce lifetime product costs by as much as $12,500.

Digital Equipment

A multifunctional design team at Digital Equipment Corporation redesigned the company's computer mouse (Digital, 1990). They began with the competitive benchmarking of Digital's products and mice made by other companies. They used DFMA software to compare such factors as assembly times, part counts, assembly operations, labour costs, and total costs of the products. They also consulted with hourly-paid people who actually assembled the mice. Gordon Lewis, the DFMA coordinator and team leader, stated that DFMA gives the design team a "focal point so that [they] can go in and pinpoint the problems from a manufacturing perspective and a design perspective." "It's the 80/20 rule", said Mr. Lewis. "You spend 80% of your time on 20% of your problems." DFMA is one of the tools that helps design teams identify the right 20% of the problems to work on," he said.

Figure 1.5 shows the old and new mice. In the new DFMA design, 130 s of assembly for a ball-cage device has been reduced to 15 s for the device that has replaced it. Other changes to the product structure have also brought cost savings. For instance, the average of seven screws in the original mouse has been reduced to zero with snap fits. The new mouse also requires no assembly adjustments, whereas the average number for previous designs was eight. The total number of assembly operations has decreased from 83 in the old product to 54 in the new mouse. All these improvements add up to a mouse that is assembled in 277 s, rather than 592 s for the conventional one. Cycle time, too, has been reduced by DFMA. A second development project that adhered to the new methodology was finished in 18 weeks, including the hard-tooling cycle. "That's unbelievable", admitted Mr. Lewis. "Normally it takes 18 weeks to do hard tooling alone."

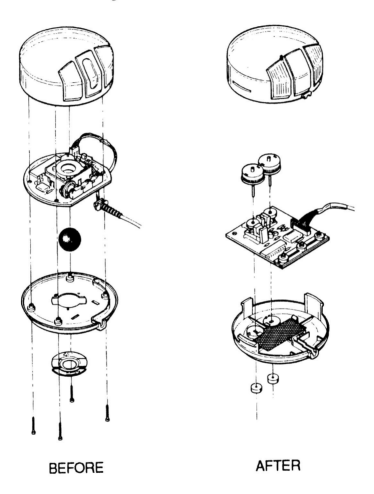

BEFORE   AFTER

Figure 1.5 Old and new designs of Digital mouse (Digital, 1990).

Motorola

DFMA methods have been used at Motorola to simplify products and reduce assembly costs. As part of the commitment to total customer satisfaction, Motorola has embraced the six-sigma philosophy for product design and manufacturing. It seemed obvious that simpler assembly should result in improved assembly quality. With these precepts in mind, they set about designing the new generation of vehicular adapters (Branan, 1991).

The portable-products division of Motorola designs and manufactures portable 2-way Handi-TalkieTM radios for the landmobile-radio market. This includes such users as police, firemen and other public-safety services, in addition to the construction and utility fields. These radios are battery-operated, and are carried about by the user.

Table 1.5 Redesign of vehicular adaptor - Motorola (Burke and Carlson, 1990)

	Old design	New design	Improvement %
DFA assembly efficiency, %	4	36	800
Assembly time (seconds)	2742	354	87
Assembly count	217	47	78
Fasteners	72	0	100

The design team embraced the idea that designing a product with a high assembly efficiency would result in lower manufacturing costs, and the provision of the high assembly quality desired. They also considered that an important part of any design was to benchmark competitors' products as well as their own. At the time, Motorola produced two types of vehicular adapter called Convert-a-ComTM (CVC) for different radio products. Several of their competitors also offered similar units for their radio products. The results of the redesign efforts were so encouraging (Table 1.5) that Motorola surveyed several products which had been designed using the DFA methodology to see if there might be a general correlation of assembly efficiency with manufacturing quality. Figure 1.6 shows what they found. The defect levels are reported as defects per million parts assembled, which allows a quality evaluation to be made that is independent of the number of parts in the assembly. Motorola's six-sigma quality goal is 3.4 defects per million parts assembled. Each result in Figure 1.6 represents a product with an analyzed assembly efficiency and a reported quality level.

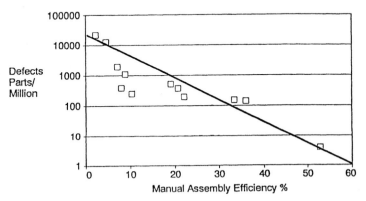

Figure 1.6 Product assembly efficiency correlation - Motorola.

Ford Motor Company

Ford leads the field as an aggressive user of DFMA tools. To date, they have trained thousands of engineers in the DFA methodology, and they have contributed heavily to new research programs, and to expanding the existing DFMA tools. Ford is now even requiring its vendors to conduct DFA analysis prior to submitting bids on subcontracted products.

James Cnossen, Ford's manager of manufacturing systems and operations research, has concluded that "it's part of the very fabric of Ford Motor Co." This is not surprising, when Ford reports savings of over $1000M annually as a result of applying DFMA to the Taurus line of cars.

DFMA has become part of the simultaneous-engineering environment, which supports Ford's "Concept to Customer" theme. Using the DFMA software, teams made up from product design, manufacturing, suppliers and other representatives regularly meet to review not only the conceptual design of their future products, but also the products that are currently being manufactured. Gains in productivity are shown not only in reduced manufacturing costs, but also in the design lead-time required to bring new products to market. The adoption of these types of engineering tool is allowing Ford to reap tremendous benefits in both quality and customer satisfaction.

The Transmission and Chassis (T&C) Division of Ford is responsible for the design and manufacture of automatic transmissions of Ford vehicles. The transmission is a complex product, with approximately 500 parts and 15 model variations. The steps in the introduction and implementation of DFA in the Transmission and Chassis Division (Burke and Carlson, 1990) are as follows:

- Provide DFA overview for senior management.
- Choose DFA champion/coordinator.
- Define objectives.
- Choose pilot program.
- Choose test case.
- Identify team structure.
- Identify team members.
- Coordinate training.
- Have first workshop.

During the workshop:

- Review the parts list and processes.
- Break up into teams.
- Analyze the existing design for manual assembly.
- Analyze the teams' redesigns for manual assembly.
- Teams present results of original design analysis versus redesign analysis.
- Prioritize redesign ideas: A, B, C, etc.
- Incorporate all the A and B ideas into one analysis.
- Assign responsibilities and timing.

The combined results of all of the workshops held in the T&C Division of Ford indicated potential total assembly labour savings of 29%, a reduction in part count of 20%, and a reduction in the number of operations of 23%.

The cost benefits that have been gained since the introduction of the DFA methodology in the T&C Division are nothing less than staggering. Even more importantly, the changes resulting from DFA have brought substantial quality improvements. Moreover, the design lead-time has been reduced by one-half, and is expected to be halved again. Reduced cost and improved manufacturability was reflected in Ford's profits for 1988.

General Motors

A few years ago, General Motors (GM) made comparisons between its assembly plant for the Pontiac at Fairfax, KS, USA, and Ford's assembly plant for its Taurus and Mercury Sable models near Atlanta, GA, USA. GM found that there was a large productivity gap between its plant and the Ford plant. GM concluded that 41% of the producibility gap could be traced to the manufacturability of the two designs. For example, the Ford car had many fewer parts (ten in its front bumper compared with 100 in the GM Pontiac), and the Ford parts fitted together more easily. The GM study found that the level of automation, which was actually much higher in the GM plant, was not a factor in explaining the productivity gap.

Kobe (1992) explains that the result of the application of DFMA can be seen in selected areas of the 1992 Cadillac Seville and Eldorado. For example, the new bumper system reduces part count by half over the previous generation, and assembly time is about 19 minutes less than the pre-DFMA design. A further example is the Cadillac full console. In this case a reduction of 40% in assembly time and a 33% reduction in part count was achieved by employing DFMA from the concept stage, capitalizing on the real benefits of the methodology by improving on the concept itself.

Hewlett-Packard

It was reported by Colucci (1994) that Hewlett-Packard's Loveland, Colorado division implemented a concurrent engineering program to produce its 34401A multimeter, which reportedly has the performance of a $3-5,000 instrument at a $1,000 price. The implementation program used DFMA software to encourage team input and quantified results as the development process gradually evolved. Every part of the 34401A was analyzed using DFMA. The most significant results: a complete redesign of the input connection scheme and a front panel design that assembles with no screws.

Robert Williams, Manufacturing R & D engineer at HP, admits that many of the ideas for these changes were conceived before the bulk of the concurrent engineering team met, but he still attributes the success of the project to the team effort. "It took the efforts of the cross-functional design teams to identify producible designs, materials, and the correct suppliers to make the ideas work," he says. "The key deliverable of any DFMA effort is a significantly reduced part count. The lower part count we achieved allowed us the freedom to try new manufacturing processes."

The finished 34401A multimeter has only 18 parts, compared to 45 parts for the previous model. It can be assembled manually by one person in just over six minutes; much less than the twenty minutes required for the unit it replaces. Says Williams, "the key point is the part count drives virtually all downstream processing in manufacturing. Without development tools, particularly DFMA, these competitive advantages could not be realized."

McDonnell Douglas Corporation

Weber (1994) explains that like many other companies, McDonnell Douglas Corporation (MDC) has realized that to stay competitive it must reduce costs without compromising product quality. This requires the careful consideration of manufacturing and assembly costs during product design.

MDC has found that applying DFA reduces parts and fastenings, which in turn reduces the opportunities for defects. Additionally, applying DFM to structure design further reduces defects during production.

For fighter aircraft, MDC applies DFMA primarily to structure design done mostly in-house. Secondarily, DFMA is applied to system design -- landing gear systems, controls, electronics/electric, hydraulics, and environmental control systems.

Aircraft structure is very complex, typically requiring large quantities of parts and fasteners. Because many components are used, assembly is labour intensive. Fighter planes may require more than 100,000 structural fasteners, while large commercial aircraft may use more than one million. The MD-11 wide body commercial aircraft, for example, has 1.3 million fasteners, 184,000 other parts, one hundred miles of electrical wiring (50,000 segments), 5,200 feet of hydraulic pipe with 2,765 joints and 400 control cable segments.

MDC has applied DFA to reduce parts and defects on a wide variety of fighter and commercial aircraft. They have found that DFA benefits include:

- Fewer parts -- lower inventory, and lower assembly costs
- Fewer fasteners -- high speed machining and high speed machining techniques are replacing many traditional riveted sheet metal assemblies
- Reduced weight -- very critical to aircraft design
- Fewer opportunities for defects -- a very significant benefit due to the large number of fasteners in aircraft assemblies
- Improved reliability -- using fewer parts and fasteners enhances reliability
- Less maintenance -- improves mean time between failures
- Fewer manufacturing operations -- assembling fewer parts/fasteners cuts manufacturing operations
- Less tooling -- reduces tool design, fabrication, and maintenance. Important savings when aircraft volume production is low
- Less analysis work -- strengths, loads, materials
- Fewer CAD models/drawings -- parts/fastener reduction means fewer CAD models/drawings

According to Nelson Weber, too much time, two years, was spent investigating and evaluating DFMA, instead of implementing DFMA. Such questions as "Does it really work?" and "Is it really applicable to the aerospace industry?" had to be answered before DFMA could be implemented. Hindsight shows we should have used it, instead of questioning it.

The primary DFMA application for large commercial transport aircraft was systems and structure. Applying DFA reduced part count by 37 percent and fastener count by 46 percent on average. DFMA is now being applied to new aircraft designs, and to selected existing designs as resources allow.

Hasbro

According to Kirkland (1995), toy manufacturers today must comply with some of the most demanding time-to-market schedules of any industry on the planet. With an average product life cycle of only one year, toys are serious business for the development teams in the promotional division of the largest toy company in the world, Hasbro, Inc. (Pawtucket, Rhode Island).

Hasbro uses DFMA to identify design and cost improvements at the earliest concept stages of design. "Working for a toy company is a lot of fun," says Jim Tout, Hasbro's director of design engineering. Toy retailers want products to reach their shelves right at the time consumers are going to buy them adds Tout. The retailers do not want to carry inventories. Because timing is so critical to Hasbro's success, the emphasis is on getting products shipped on schedule. "DFMA is a big part of this movement, because it helps eliminate problems in the debug production startup process by analyzing part counts, assembly times, and material costs before a design concept is locked in and changes become too time consuming to implement."

Hasbro's Tout can cite a number of cases where DFMA software has cut redesign time and cost. One is the Talk n' Play Fire Truck, the most successful fire truck of the 1993 Christmas season. A product of Hasbro's Tonka line, this fire truck, like other Tonka products, was traditionally made of metal. After a DFMA analysis had been performed, it was evident that there were significant opportunities for cost reduction if the product was redesigned in plastic. "The team justified the changes by looking at assembly times, metal vs. plastic," Tout says.

The original ladder assembly was composed of 33 total parts and subassemblies, with an assembly time of 198 s. The redesigned ladder brought the number of parts down to its theoretical minimum of only five parts -- all plastic -- with an assembly time of just 22 s. "It looks as nice as the metal assembly and it performs the same functions," boasts Egan. "Plus, it's more reliable when subjected to abuse testing."

Hasbro is expecting to get a strong second year out of the product -- a remarkable accomplishment in this industry. "If we had stalled on this project, we probably would have missed our retailing window," Tout adds. "DFMA enabled us to come up with trade-off information up front, so we could develop a high-quality, profitable product, and still fall within our aggressive schedule requirements." Hasbro also has found that DFMA provides a nonthreatening way to get team members talking about a design without anyone feeling as though others are encroaching on his or her territory. And it allows Hasbro's tooling and manufacturing engineers to get involved at the concept stage, eliminating any surprises.

In addition, DFMA helps teams quantify their design decisions, which can be beneficial in getting changes actually implemented. After analysis, a product component not only can be simplified or consolidated, but engineers can examine how that change will impact, say, assembly time vs. a possible part cost increase, in dollars and cents. It can be done up front, in about an hour.

1.3 ROADBLOCKS IN IMPLEMENTATION OF DFMA

As to the implementation of DFMA, the format for success varies from company to company, but some major points stand out. Firstly, DFMA is a team tool and should be utilized as such.

Training is important. Today, most DFMA implementation efforts employ the software system, and for this reason some companies believe it is, for example, like using Lotus 123. This is not the case. It is important to train people in a workshop environment - a team using the system on an on-going project with the company's "champion" or an outside system consultant providing help. In this way, one or two days provides useful training plus, often as not, real results.

Finally, it is important to remember that it is often not the target, but the journey through the systematic procedure that matters. Experience has shown that there are many barriers to the implementation of DFMA.

Within many companies, reasons for resisting the implementation of DFMA are put forward, but all can be effectively argued against:

No Time

The most common complaint among designers is that they are not allowed sufficient time to carry out their work. Designers are usually constrained by the urgent need to minimize the design-to-manufacture time for a new product. However, more time spent in the initial stages of design will reap benefits later in terms of reduced engineering changes after the design has been released to manufacturing. Company executives and managers must be made to realize that the early stages of design are critical in determining not only manufacturing costs, but also the overall design-to-manufacturing cycle time.

Not invented here

Enormous resistance can be encountered when new techniques are proposed to designers. Ideally, any proposal to implement DFMA should come from the designers themselves. However, more frequently it is the managers or executives who have heard of the successes resulting from DFMA and wish that their own designers would implement the philosophy. Under these circumstances, great care must be taken to involve the designers in the decision to implement these new techniques. Only then will the designers feel that they 'invented' or 'thought of' the idea of applying DFMA.

The ugly baby syndrome

Even greater difficulties exist when an outside group or a separate group within the company undertakes to analyze existing designs for ease of manufacture and assembly. Commonly, this group will find that significant improvements could be made to the original design and, when these improvements are brought to the attention of those who produced the design, this can result in extreme resistance. Telling a designer that this designs could be improved is much like telling a mother that her baby is ugly!

It is important, therefore, to involve the designers in the analysis and provide them with the incentive to produce better designs. If they perform the analysis, they are less likely to take any problems that may be highlighted as criticism.

Low assembly costs

The first step in the application of DFMA is a DFA analysis of the product or sub-assembly. Quite frequently, it will be suggested that since assembly costs for a particular product form only a small proportion of the total manufacturing costs, there is no point in performing a DFA analysis. However, a DFA analysis might suggest the replacement of a complete assembly with, say, a machined casting and might reduce total manufacturing costs by over 50%.

Lower volume

The view is often expressed that DFMA is only worthwhile when the product is manufactured in large quantities. It could be argued, though, that use of the DFMA philosophy is even more important when the production quantities are small. This is commonly because reconsideration of an initial design is usually not carried out for low volume production. Applying the philosophy "do it right the first time" becomes even more important, therefore, when the production quantities are small. In fact, the opportunities for part consolidation are usually greater under these circumstances because it is not usually a consideration during design.

The database doesn't apply to our product

Everyone seems to think that their own company is unique and, therefore, in need of unique databases rather than the ones incorporated within the DFMA system. However, when one design is rated better than another using the DFA database, it would almost certainly be rated in the same way using a customized database. Remembering that there is a need to apply DFMA at the early design stage before detailed design has taken place, there is a need for a generalized database for this purpose. Later when more accurate estimates are desired, the user can employ a customized database if necessary.

We've been doing it for years

When the claim, "We've been doing it for years" is made, it usually means that some procedure for "design for producibility" has been in use in the company. However, design for producibility usually means detailed design of the individual parts of an assembly for ease of manufacture. It was made clear earlier that such a process should only occur at the end of the design cycle; it can be regarded as a "fine tuning" of the design. The important decisions affecting total manufacturing costs will already have been made. In fact, there is a great danger in implementing design for producibility in this way.

It has been found that the design of individual parts for ease of manufacture can mean, for example, limiting the number of bends in a sheet metal part. This invariably results in a more expensive assembly where several simple parts are fastened together, rather than a single, more complicated part. Again, experience has shown that it is important to combine as many features in one part as possible. In this way, full use is made of the abilities of the various manufacturing processes. Therefore, when the claim is made that the company has been implementing DFMA for some time, this should be taken with a very large pinch of salt.

It's only value analysis

It is true that the objectives of DFMA and value analysis (VA) are the same. However, it should be realized that DFMA is meant to be applied early in the design cycle, and that value analysis does not give proper attention to the structure of the product and its possible simplification. DFMA has the advantage that it is a systematic step-by-step procedure, which can be applied at all stages of design and challenges the designer or design team to justify the existence of all the parts and consider alternative designs. VA, on the other hand, only looks at major points; it is often the screws and washers - often not shown on drawings - that impose the difficulty during assembly.

Experience has shown that DFMA can still make significant improvements even after value analysis has been carried out.

DFMA is only one among many techniques

Since the introduction of DFMA, many other acronyms have been proposed, for example, design for quality (DFQ), design for competitiveness (DFC), design for reliability, etc. Some have referred to this proliferation of acronyms as alphabet soup! Many have even suggested that design for performance is just as important as DFMA. One cannot argue with this. However, DFMA is the subject that has been neglected over the years while adequate consideration has always been given to the design of a product for performance, appearance, etc. The other factors, such as quality, reliability, etc., will follow when proper consideration is given to the manufacture and assembly of the product. The earlier example from Motorola (Figure 1.6) illustrates how DFMA can lead to higher product quality.

DFMA leads to products which are more difficult to service

It has been claimed that DFMA leads to products which are more difficult to service. This is absolute nonsense. Experience shows that a product that is easy to assemble is usually easier to disassemble and reassemble. In fact, those products that need continuous servicing, involving the removal of inspection covers and the replacement of various items, should have DFMA applied even more rigorously during the design stage. How many times have we seen an inspection cover fitted with numerous screws only to find that after the first inspection only two are replaced?

I prefer design rules

There is a danger in using design rules, because they can guide the designer in the wrong direction. Generally, rules attempt to force the designer to think of simpler-shaped parts which are easier to manufacture. In an earlier example, it was pointed out that this can lead to more complicated product structures and a resulting increase in total product costs. In addition, in considering novel designs of parts which perform several functions, the designer needs to know the penalties when the rules are not followed. For these reasons, the systematic procedures used in DFMA, which guide the designer to simpler product structures and provide quantitative data on the effect of any design changes or suggestions, are found to be the best approach.

I refuse to use DFMA

Although a designer may not say out loud that he refuses to use DFMA, if he does not have the incentive to adopt this philosophy and use the tools available, then no matter how useful the tools or how simple they are to apply, he will see to it that they do not work.

Therefore, it is imperative that the designer or the design team is given the incentive and the necessary facilities to incorporate considerations of assembly and manufacture during design.

The main argument, however, against any reservations about adopting DFMA are the savings in manufacturing costs obtained by the hundreds of companies world-wide which have adopted the system. Some examples of these were described earlier.

1.4 SUMMARY

DFMA provides a systematic procedure for analyzing proposed designs from the point of view of assembly and manufacture. It encourages teamwork and a dialogue between designers and the manufacturing engineers, and any other individuals who play a part in determining final product costs during the early stages of design.

This DFMA procedure often produces a considerable reduction in part count, resulting in simpler and more reliable products which are less expensive to assemble and manufacture. In addition, any reduction in the number of parts in an assembly produces a snowball effect on cost reduction because of the drawings and specifications that are no longer needed, the vendors that are no longer needed and the inventory that is eliminated. All of these factors have an important effect on overheads which, in many cases, form the largest proportion of the total product cost.

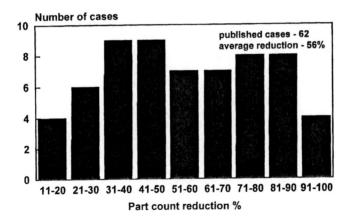

Figure 1.7 Part count reductions when Boothroyd Dewhurst DFMA methods were used.

Table 1.6 Improvements due to DFMA applications

Category	Number of cases	Average reduction (%)
Part count	61	56
Assembly time	38	62
Product cost	21	50
Assembly cost	17	45
Assembly operations	14	57
Separate fasteners	12	72
Labour costs	8	42
Manufacturing cycle	6	58
Weight	6	31
Assembly tools	5	69
Part cost	3	56
Unique parts	3	57
Material cost	3	37
Manufacturing process steps	3	45
Number of suppliers	3	55
Assembly defects	3	68
Cost savings per year	6	$1,283,000

As we saw earlier, there are many widely publicized DFMA case studies to illustrate these claims. By way of a summary, Figure 1.7 shows the effect of DFA on part count reduction from published case studies and Table 1.6 presents details of other improvements from the same case studies.

In spite of all the success stories, the major barrier to DFMA implementation continues to be human nature. People resist new ideas and unfamiliar tools, or claim that they have always taken manufacturing into consideration during design. The DFMA methodology challenges the conventional product design hierarchy. It re-orders the implementation sequence of other valuable manufacturing tools, such as SPC (Statistical Process Control) and Taguchi methods.

Designers are traditionally under great pressure to produce results as quickly as possible and often perceive DFMA as yet another time delay. In fact, as numerous case studies have shown, the overall design development cycle is shortened through use of early manufacturing analysis tools, because designers can receive rapid feedback on the consequences of their design decisions where it counts - at the conceptual stage.

Overall, the facts are that DFMA is a subject that has been neglected over the years while adequate consideration has always been given to the design of a product for performance, appearance, etc. The other factors such as quality, reliability, etc. will follow when proper consideration is given to the manufacture and assembly of the product. In order to remain competitive in the future, every manufacturing organization will have to adopt the DFMA philosophy and apply cost quantification tools at the early stages of product design.

REFERENCES

Boothroyd, G., Dewhurst, P. (1990) *Product Design for Assembly*, Boothroyd Dewhurst, Inc., Wakefield, RI, USA, (First Edition 1983).

Boothroyd, G., Radovanovic, P. (1989) Estimating the cost of machined components during the conceptual design of a product, *Annuals of CIRP*, **38** (1), 157.

Branan, B. (1991) DFA cuts assembly defects by 80%, *Appliance Manufacture,* November.

Burke, G.J., Carlson, J.B. (1990) DFA at Ford Motor Company, *DFMA Insight*, **1** (4), Boothroyd Dewhurst Inc.

Colucci, D. (1994) DFMA Helps Companies Keep Competitive, *Design News*, November, 21.

Dewhurst, P. (1988) Computer-aided assessment of injection moulding cost - a tool for DFA analyses, *Report 24*, Department of Industrial and Manufacturing Engineering, University of Rhode Island, USA.

Dewhurst, P. (1988) Cutting assembly costs with moulded parts, *Machine Design*, July 21.

Dewhurst, P., Blum, C. (1989) Supporting analyses for the economic assessment of due casting in product design, *Annuals of CIRP*, **28** (1), 161.

Digital (1990) Digital builds a better mouse, *Industrial Week,* April 16, 50-58.

Kirkland, C. (1988) Meet two architects of design - integrated manufacturing, *Plastics World,* Dec.

Kirkland, C. (1992) Design Watch, *Plastics World*, March.

Kirkland, C. (1995) Hasbro Doesn't Toy with Time to Market, *Injection Moulding*, Feb., 34.

Knight, W.A. (1991) Design for manufacture analysis: early estimates of tool costs for sintered parts, *Annuals of CIRP*, **40** (1), 131.

Kobe, G. (1992) DFMA at Cadillac, *Automotive Industries*, May, 43.

McCabe, W.J. (1988) Maximizing design efficiencies for a coordinate measuring machine, *DFMA Insight*, **1** (1), Boothroyd Dewhurst Inc.

Sorge, M. (1994) GE's ongoing mission: cut costs, *Ward's Auto World*, February, 43.

Swift, K.G. (1981) *Design for Assembly Handbook*, Salford University Industrial Centre, UK.

Weber, N.O. (1994) Flying High: Aircraft Design Takes Off with DFMA, *Assembly*, Sept.

Womack, J.P., Jones, D.T., Roos, D. (1990) *The Machine that Changed the World*, Macmillan, USA.

Zenger, D., Dewhurst, P. (1988) Early assessment of tooling costs in the design of sheet metal parts, *Report 29*, Department of Industrial and Manufacturing Engineering, University of Rhode Island, USA.

CASE EXPERIENCE WITH HITACHI, LUCAS AND BOOTHROYD-DEWHURST DFA METHODS

Paul G. Leaney

This chapter presents a case for the importance of DFA and its relevance within a structured product development framework based on concurrent engineering, provides an insight into three DFA evaluation methods, namely Hitachi, Lucas and Boothroyd-Dewhurst, and provides advice on good practice.

2.1 ROLE OF DFA

The aim of this section is to emphasise the importance of DFA, especially for manufactured products. Its relevance within the context of concurrent engineering is also highlighted.

2.1.1 DFA - A Manufacturing Perspective

Let us start with the concept of lean production (Womack *et al., 1*990). This concept is built on the Toyota's just-in-time (JIT) approach, endeavouring to achieve the efficiencies of mass production for a market place demanding more product variety and forcing a manufacturing strategy based on batch production. Just as one may visualise continuous flow of material through transfer lines in a mass production situation then the JIT philosophy is based on the idea of achieving continuous material flow (ideally down to batches of one) through a batch manufacturing facility. The aim is to minimise non-value added operations. This is lean production because material in buffers and storage are non-value added and should be minimised.

Without buffers quality problems become evident immediately since if material stops moving in one place it stops all along the line. In this situation quality is promoted to the number one concern and 'make it right first time' becomes the imperative. The required manufacturing response needs to be built on providing the necessary manufacturing base (that is the production facilities and the empowered workers in a team based culture) to provide the volume leaving managers to chase quality. This is the Japanese lesson in manufacturing. Quality defects on the shop floor are interpreted as flags that highlight problems with the process of making things. The aim is not to inspect for defective products but to control the manufacturing process so that a 'wrong one' is never made in the first place. It is the process that needs controlling. This is why process monitoring and techniques such as SPC (statistical process control) are important.

The challenge for us is to recognise the advantages that this way of systems thinking can have when applied to the engineering function of the business enterprise. Whereas the manufacturing function is concerned with enabling the value added processing of materials without bottlenecks, delays or errors then engineering should be concerned with the added value processing of information and ideas. Meeting market demand with appropriate products becomes a matter of timeliness. Poor judgement or simple oversights at the design stage will have consequences, i.e. time lost and costs incurred, that ripple and grow throughout the organisation. Design it 'right first time' becomes the imperative.

This is easier said than done but steps can be taken. Simultaneous engineering might provide a managerial structure, and concurrent engineering with CAD (Computer Aided Design) the technological base, but at the end of the day people in teams need to focus on a problem or a goal. At this level particular tools precipitate thinking and can provide measures against which management can set goals and monitor progress. It is in this context that a number of team driven approaches, e.g. DFA, DFM, FMEA (Failure Modes and Effects Analysis), QFD (Quality Function Deployment), CPI (Continuous Process Improvement), Taguchi and Robust Design are promoted in relevance. The mechanics of the design process itself is now under more scrutiny.

"Design right first time" means that product development teams need guidance and support. Some broad based guidelines for design for assembly are listed in Section 2.4. These guidelines are not new or revolutionary but they are now taking on a particular relevance and importance. The difficulty in ensuring the application of these types of guidelines is starting to be overcome with the development of recent DFA and DFM evaluation methods. Underlying all of these methods is the ability to quantify penalties or costs and giving designers direct feedback on manufacturability and assemblability. These measures also provide managers with a means of setting objectives and measuring progress.

In addition, as managers of the leaner enterprise scrutinise their product development process it is becoming clear that DFA and DFM technique(s) can be used to lubricate some of the changes necessary with people. An example is provided by a retrospective look at the lessons of the endeavours made during the late seventies and early eighties when visions of advanced manufacturing technology (AMT) and flexible manufacturing systems (FMS) conjured up ideas of the lights out factory. Assembly automation and robotics attracted a quantum leap in interest. People did not want to 'miss the boat'.

When current product lines were considered for automatic assembly it started to dawn on people that the current designs were not sympathetic to automation. Design for assembly (DFA) grew in prominence as new products were developed. Following DFA studies the new products, in many cases, would take a lot less labour to assemble and for this to be done more reliably and consistently. The reduced labour content then made the original automation even more difficult to economically justify. Shop floor workers are still quick and flexible at assembly processes and errors can be minimised through sensible product design and production aids. The lesson from all of this is that there is something inherently useful about DFA whether or not automation is actually used.

One of the early DFA techniques (that of Hitachi) did not make explicit the distinction between automatic and manual assembly in its evaluation procedure. This did not appear to weaken the attraction of the Hitachi assemblability evaluation method to Japanese, and some US, companies. These companies seemed to build the technique into the way things were naturally undertaken. Other up and coming DFA techniques such as Boothroyd-Dewhurst from the US and Lucas from the UK (the Lucas method is now part of CSC TeamSET™) were, arguably, promoted on the crest of the automation wave sweeping industry at the time and, consequently, they do explicitly provide the mechanisms for evaluating product designs for automatic assembly.

The full title for the Hitachi method is Assemblability Evaluation Method (AEM). Its underlying methodology was first developed in the late 1970's. The term 'design for assembly' (DFA) was introduced later (circa, 1980) to describe the methodology and associated databases developed by Geoffrey Boothroyd at the University of Massachusetts in the 1970's, otherwise known as the UMass system. The term 'design for manufacture and assembly' (DFMA) was introduced a little later to cover the continued work of Professors Boothroyd and Dewhurst at the University of Rhode Island, with their design for manufacture modules (machining, sheet metalwork, injection moulding etc.). The terms DFA and DFMA have now been widely adopted and are often used as generic terms just about everywhere. However Boothroyd-Dewhurst Inc. retains a trademark on DFMA™ when referring to their software suite of programs.

The real achievement of DFA methods is their ability to provide measurements of assemblability which allows an objective criteria to apply in a team based situation. Section 2.4 outlines one checklist of good DFA practice and what the methods of Hitachi, Lucas and Boothroyd-Dewhurst now give us is the means of promoting this good design practice at the earliest stages in design as well as during detailed design.

The other real benefit of DFA is that it centres attention on the complete product (or sub-assembly) as a whole and then promotes the ideas of parts reduction, standardised parts and product modularisation. In this way it acts as the driver for DFM. DFA thus plays an integrative role as a DFM strategy based totally on 'design for process' is in danger of becoming too piece part oriented. The process by which piece parts are individually manufactured is only one aspect of the total scene which encompasses the whole product and includes such things as production control and material flow logistics, assembly, test and quality. Product designs subject to DFA were not only becoming more sympathetic to assembly automation they were also becoming JIT friendly.

Clearly product functionality is uppermost in the designers mind but the customer expects more than this. The customer expects value for money, good service and quality. The need is to design for whole life cost. This idea is based on the fact that the cost of a product to a customer is the purchase price plus the cost of waiting for delivery plus the cost of ownership and, with due regard for the environment, cost of disposal. Engineering needs to work in partnership with marketing in assessing the needs of customers but engineering must also work with manufacturing in satisfying these needs. Engineering plays a central and pivotal role. To achieve successful customer satisfaction and competitiveness it is necessary to pursue the means of synchronising product development strategies with manufacturing strategies. DFA provides one such bridging mechanism.

In the 'lean production' paradigm of manufacturing any quality problem on the shop floor is seen as an 'error flag' that highlights something wrong with the process of making things. The time, effort and cost of putting right these errors immediately is allowed to outweigh the direct and evident costs caused by the error as these might not, at the time, seem significant. This is because correcting errors at their source is seen to save magnified costs that would otherwise emerge later. The danger is that if corrective action is not taken immediately then the corrective action taken later would only focus on mitigating the symptoms and not on eliminating the cause. In an analogous way the DFA method can be used to 'flag' a problem with a product design. In other words, a product design that is reflected badly in a DFA evaluation should be flagged as a poor design and that the efforts of rectification may well be allowed to outweigh the direct savings anticipated in assembly. Often the largest savings to be made are in materials and overheads, just as they are with JIT and lean production. Indirect costs are notoriously difficult to predict so that any indication of direct cost savings highlighted by DFA (and DFM) assessments should be interpreted as opportunities being 'flagged'. As with JIT and lean production the real cost savings always emerge in retrospect.

In summary, DFA acts as a driver for Concurrent Engineering, it acts as a flag for poor designs, it can be used to direct the effort of teams, and it provides a metric for managerial control. However the full benefit of DFA comes out of the context in which it is pursued and this often occurs within a broader product development process or strategy centred on concurrent engineering.

2.1.2 Design, Concurrent Engineering and DFA

The aim of this section is to promote the consideration of DFA (and other formal methods) within the context of concurrent engineering which, in-turn, embodies the design activity in the manufacturing business enterprise. One well known and accepted definition of concurrent engineering has been documented by an Institute for Defence Analysis (IDA) report (Winner *et al.*, 1988): "A systematic approach to the integrated, concurrent design of products and their related processes, including manufacture and support. This approach is intended to cause the developers, from the outset, to consider all elements of the product life cycle from conception through disposal, including quality, cost, schedule and user requirements".

A well considered exposition of this definition is presented by Keys (1992), who states that the implementation of concurrent engineering takes a variety of forms. However, he identifies three generic elements:

- Reliance on multi-functional teams to integrate the designs of a product and its manufacturing and support processes.
- Use of CAD/CAE/CAM to support design integration through shared product and process models and databases.
- Use of a variety of formal evaluation methods to optimise a product's design and its manufacturing and support processes, *e.g.* FMEA, QFD, DFA, DFM, SPC.

This presents concurrent engineering as being more than CAD/CAE/CAM and that team working and formal methods provide equally important support.

In a later paper, Dowlatshahi (1994) identifies 5 forms of successful concurrent engineering approaches by categorising them in a way that reflects their philosophy of integration:

- Information systems, software design and artificial intelligence.
- CAD/CAM.

- Life cycle engineering.
- Design for manufacture and assembly.
- Organisational and cultural changes.

An analysis of the concurrent engineering definition (Winner, 1988) and the explanations of Keys (1992) and Dowlatshahi (1994) reveal perspectives that are different but complementary. Keys identifies the generic elements necessary to facilitate concurrent engineering that, in turn, enables engineering work to be done effectively. Dowlatshahi demonstrates that the particular approach to concurrent engineering depends on perspective. Dowlatshahi adds two possible things to the argument:

- That effective concurrent engineering is based on systems thinking. The various approaches he outlines relate to different ways of rationalising the system (or process) by addressing particular inputs, outputs and interactions.
- That organisational and cultural changes can drive changes in the way engineering work is carried out.

Keys (1992) and Dowlatshahi (1994) provide considered views of concurrent engineering. However the pragmatism of engineers working in industry, and their equally rational managers, channels them into the perception that concurrent engineering means implementing CAD/CAE/CAM integrated systems and then addressing (or ignoring) the concomitant 'problems' of team working and formal methods. In this way concurrent engineering does not necessarily have the required impact on the design process. This results is a number of areas of potential weakness, for example:

- Poor integration of formal methods (FMEA, QFD, DFA, DFM, Concept Evaluation and Convergence, Requirements Capture and Analysis, etc.) into the design process.
- Poor management of technical requirements versus business requirements versus customer requirements.
- Lack of methods for negotiating and resolving design conflicts.

Efforts to document procedures through BS EN ISO 9000, for example, sometimes highlights these issues but the use of quality standards does not address the problem head on. There is, however, an increasing need for the development of design standards (e.g. BS 7000).

An analysis of the definition of concurrent engineering, reproduced earlier, shows it to contain and cover a wide range of topics that allows it to be an equally good definition for 'design integration' across the marketing, engineering and manufacturing functions of the business enterprise. Topics like quality, cost, user requirements, manufacture and support (including acquisition and logistics) will involve people from a range of disciplines and professions (e.g. finance, management, marketing, manufacture, design, engineering). However, it may seem reasonable for the phrase concurrent engineering to refer to the engineering aspects of the topics listed in the definition.

Unfortunately that interpretation is to allow a 'divide to conquer' mentality. This worked well for Henry Ford who developed the techniques of mass production for his assembly line by breaking tasks down. Since then, however, even the Ford Motor Company has modified its approach in the light of the Toyota Production System (or lean production) which advocates an emphasis on throughput rather than utilisation and on shop floor teamwork in tackling more broadly defined work tasks. These developments come out of the re-evaluation of the 'system' or 'process' being addressed. The concepts underlying 'continuous process improvement' and 'business process re-engineering' are providing the necessary reorientation in business thinking. The underlying concept of systems thinking is giving perceptive insight for seeking improvements. A re-evaluation of the process and role of concurrent engineering might lead to broader opportunities.

By using the phrase concurrent engineering to solely refer to the engineering aspects of the topics listed in the definition used, is to draw the engineers away from a truly 'systems thinking' approach. Or rather, it constrains the 'systems thinking' approach to the sub-system levels within the engineering function of the manufacturing enterprise. The resultant 'engineering thinking' approach acknowledges the existence of the marketing, manufacturing and commercial functions of a company but would rather interface than integrate. This 'engineering thinking' leads to the idea that an integrated approach means the integration of the engineering aspects. It leads to the idea that the development of mechatronic products will drive design integration because electrical and electronic engineers need to work in multi-disciplinary teams with software and mechanical engineers. In reality the opportunity for true 'design integration' is much broader and techniques such as DFA, DFM, QFD, etc underpin the broader opportunity.

At the engineering / manufacturing interface the communication is mainly between engineers, i.e. design engineers and manufacturing engineers, and concurrent engineering can be useful for providing the basis for that communication. However, co-operative working and the necessary communication needs actively managing by the various functional managers. For example, a particular danger recognised by the manufacturing fraternity (who operate in the real world and deal with variation on a daily basis) is that those in product engineering can become increasingly sucked into their virtual (and rather perfect) world rather than deal with the real life problems of manufacture. The means of communication and the means for conflict resolution are major managerial problems that need to be addressed outside the engineering remit of concurrent engineering. It is in this arena that DFA methods, and the like, can start to have a real impact.

At the other interface, i.e. the engineering / marketing interface, the communication will be, predominantly, between engineer and non-engineer. Here industrial product designers have a particular responsibility dealing on the one hand with customer perceptions (aesthetics) and on the other with engineering aspects (of materials, for example) and engineering designers. Bridging this professional divide could present an even bigger managerial problem. Techniques such as QFD and the product simplification element of DFA (developed by Boothroyd-Dewhurst) has a role here. The Lucas DFA method within CSC's TeamSET™ (Tibbetts, 1995) is presented within a suite of computerised formal methods to specifically capitalise on the need for precipitative tools for teams.

This discussion draws out a number of points:

- Design integration is consistent with a systems thinking approach.
- The system, or process, being addressed is the product development process, i.e. the process which provides products that meet or exceed customers' expectations.
- This process involves all three major functions of the business enterprise (namely marketing/commercial, engineering and manufacturing).
- The design activity can be used as the basis for integration.
- Some key issues include concurrent engineering, role of formal methods, organisation and cultural change, teamwork, design management, design standards, design in the

extended enterprise, global operations, involving suppliers, design as a business language.

- DFA supports the design activity from concept to customer.
- DFA presents a product based view.
- DFA drives product simplification.
- DFA can be used to cut across functional barriers - and can precipitate contributions from a wide range of people.
- DFA is one key component of a successful concurrent engineering strategy.

2.2 DFA METHODS

This review of DFA evaluation methods is restricted to three methods, namely Hitachi, Lucas and Boothroyd-Dewhurst. This particular choice shares two important characteristics in common. Firstly, they enjoy an industrially based pedigree and continued industrial support. Secondly, they are commercially available. In this way they distinguish themselves from a raft of other DFA methods largely in the research domain with many described as 'knowledge based'.

An insight is provided into the three leading DFA evaluation methods in a comparative way. It complements other papers in the literature on Hitachi method (Miyakawa at al., 1990; Shimada *et al.*, 1992), on the Lucas method (D'Cruz, 1992; Miles, 1989) and on the Boothroyd-Dewhurst method (Boothroyd *et al.*, 1994). Leaney and Wittenberg (1992) have already provided a comparative view that is supplemented here with case study material. Clearly each of the DFA methods is based on their own synthetic assembly data, which is not in the public domain, but scrutiny can be applied to the different types of data and the way in which it is manipulated and interpreted. This gives perceptive insight into the underlying principles of DFA independent of particular methods. Brief descriptive narratives are complemented by the consideration of a simple case study subject to the three DFA evaluation mechanisms. These DFA methods are focused on mechanical based assemblies of a size that could be conveniently assembled at a desk top. Typical assemblies would be tape recorders, video recorders or many car sub-assemblies such as alternators, water pumps or pedal boxes. The procedures are not applicable to products of the size of, say, a complete car or vehicle. For this size of product the size and weight of component parts, and the need for the assembly worker to walk about, means that the DFA synthetic data is not applicable. Other problem products include wiring and wiring harnesses. However Boothroyd-Dewhurst continues to develop a range of software modules that address DFA for large parts, wiring harness assembly, design for service, design for disassembly, etc.

In the early days of the DFA methods, paper based versions existed. Although driving a paper based version provides useful insight into the workings of the DFA evaluation mechanisms, the overall advantages of using computerised (PC based) versions have grown and now dominate. The advantage of computer support is that it aids the DFA evaluation procedure by prompting the user, providing help screens in context and by conveniently documenting the analyses. The user can quickly analyse the effect of a proposed design change by editing a current analysis. Although computer support is excellent in 'what if' and on going studies it is generally useful to drive the paper based method in pilot studies and training in early stages of DFA adoption. Driving the paper based version in the initial stages deepens the DFA understanding of the user. Unfortunately the paper based versions are dropping out of use and are not generally supported for understandable reasons. Developments continue at a pace with the computerised versions but the underlying DFA principles remain largely intact. The case study presented later (Section 2.3) is based on the paper based versions for the purpose of conciseness, clarity, and ability to focus on the underlying principles of assemblability.

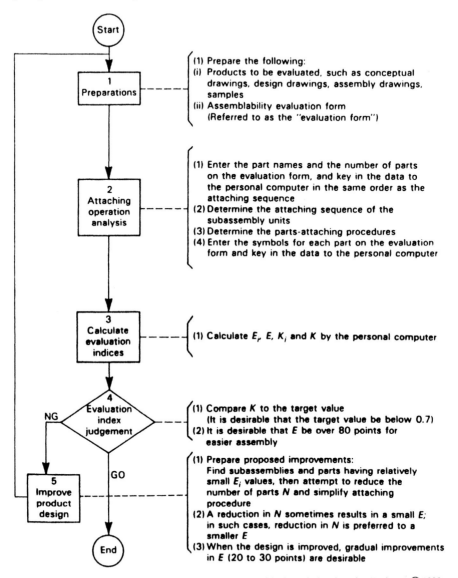

Source: NAMRI/SME Technical paper. Reproduced courtesy of Society of Manufacturing Engineers © 1990

Figure 2.1 Hitachi assemblability evaluation procedure.

Step / Examples	Product structure and assembly operations		E_i Part assemblability evaluation score	E: Assemblability evaluation score	K Assembly cost ratio	Part to be improved
Structure 1 (before improvement)	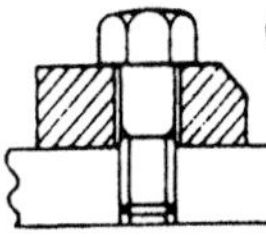 C(↓◯) B(↓···) A(↓−)	1. Set chassis A	100	73	0.1	B
		2. Bring down block B and hold it to maintain its orientation	50			
		3. Fasten screw C	65			
Structure 2	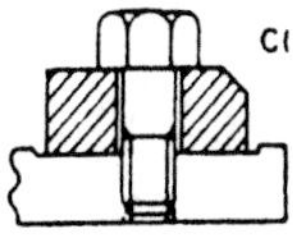 C(↓◯) B(↓) A(↓−)	1. Set chassis A	100	88	Approx. 0.8	C
		2. Bring down block B (orientation is maintained by spot-facing)	100			
		3. Fasten screw C	65			
Structure 3	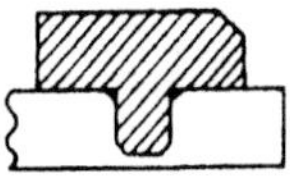 B(↓···) A(↓−)	1. Set chassis A	100	89	Approx. 0.5	B
		2. Bring down and pressfit block B	80			

Source: NAMRI/SME Technical paper. Reproduced courtesy of Society of Manufacturing Engineers © 1990

Figure 2.2 Hitachi assemblability evaluation and improvement examples.

2.2.1 Hitachi AEM

The Hitachi (AEM) method was first developed in 1976 (Miyakawa and Ohashi, 1990). After ten years of use the need to improve the methodology was evident and changes were made. One requirement was for it to be made compatible with its sister method the Hitachi Machinability Evaluation Method, (MEM). The 'New AEM' has other refinements and particularly in relation to the assembly operation cost of individual parts.

The New AEM endeavours to assess the assemblability of a product design by making use of two indices: (i) the assemblability evaluation score, E, which is used to assess design quality or the difficulty of assembly operations and (ii) the estimated assembly cost ratio, K, used to estimate assembly cost improvements. The term assemblability is interpreted as meaning - 'assembly producibility'. The implication of this is that the assemblability evaluation is built around the assessment of what are called assembly operations. These assembly operations particularly relate to the insertion (and fixing) processes. No direct analysis is available for part feeding and orientation. It is for this reason that 'design for automatic assembly' is not explicitly available. Nevertheless it is covered in so far as the estimated operation time obtained by AEM includes time for feeding and orienting parts. The argument being that these are sensitive to part configurations and are rather difficult to handle precisely at early design stages. These aspects would be dealt with at later design stages.

The AEM procedure is illustrated in Figure 2.1. Stages 1 and 2 in this figure are predominantly preparatory stages prior to evaluating the indices at stage 3. Some of the concepts behind the Hitachi method are illustrated in Figure 2.2 which shows a simplified assemblability evaluation with examples of design improvements given that the requirement is for block B to be located and fixed with respect to chassis A.

The procedure starts with defining the motions and operations necessary to insert each part of the product. Penalty points are assigned to every motion or operation that is different from a simple downward motion. A simple downward motion is regarded the fastest and easiest assembly operation for a human or machine to perform. This is the base motion onto which additional motions or processes accumulate penalty points. The AEM uses symbols to represent specific motions and processes (collectively termed operations). There is a choice from approximately twenty symbols covering such things as part insertion motions (e.g. down in a straight line, horizontal in a straight line), fixturing (e.g. holding, steadying or securing unstable parts), forming, rotating, and joining.

The evaluation procedure is based around the filling of a form in the same order as the anticipated assembly sequence. Each row occupies a part and intersecting columns will, variously, contain information relating to that part such as part description and symbol(s) that represent specific motions and processes (called elemental operations) of attaching that part. Each elemental operation is provided with a penalty score from their own synthetic assembly data. The basic elemental operation, i.e. simple downward motion, has a penalty score of zero. The penalty scores are manipulated to give an assemblability for each part (Ei - for part 'i') and then all the Ei values are combined with N (the total number of parts) to produce the total assemblability evaluation score, E. If each of the parts were to be assembled with a simple downward motion only (being the fastest and easiest assembly operation), each Ei would have a value of 100 and the total E would be 100. The score of 100 represents the ideal situation.

For ease of interpretation the E score may be thought of as an assemblability design efficiency. An efficiency of 100% would then indicate that all the assembly operations necessary were the best possible, i.e. with a simple downward motion only. The guidance given is that an E score of 80 or more is desirable. The higher the E score the lower are the manual assembly costs and the greater the ease of assembly automation. The general advice is that products with an E score of over 80 can be assembled automatically.

What the E score does not do, in itself, is to provide feedback on the advantages of parts reduction and for that the assembly cost ratio K is used. The cost ratio K can be interpreted as total assembly operation cost of new product design divided by the total assembly operation cost of the previous (or standard or basic or old) product design. The method for determining assembly costs includes a mechanism for calibrating estimated costs with historical actual costs. This is done by allocating a time (and cost) to the basic elemental operation or the simple downward motion. Calculation of K depends on the earlier calculations for E. The design target suggested is to achieve a K value of 0.7 or less. That is a cost saving of 30% or more. This can be achieved by reducing the number of parts in the redesign and/or making the assembly operations easier. The AEM analysis will help the designer focus in on problem areas in the design in endeavouring to achieve target values of E and K.

2.2.2 Lucas DFA

The Lucas DFA method came out of collaborative work between Lucas Engineering and Systems with the University of Hull, England. The first commercial computer version was launched in October 1989 following a period of successful application of the paper based version. The method revolves around the need to complete a form called the assembly flowchart. In 1995 Lucas Engineering and Systems was taken over by CSC which is a large

IT services corporation. Through this take-over CSC now has a computer based product called TeamSET™ (Tibbetts, 1995) which is an integrated suite of formal methods that is presented as a constituent part of any well balanced concurrent engineering implementation strategy. The case for the relevance of formal methods (including DFA) was made earlier in Section 2.1.2. The TeamSET™ software accesses a common (relational) database in supporting the following methods: (i) Design for Assembly; (ii) Manufacturing Analysis; (iii) FMEA; (iv) Concept Convergence; (v) QFD; (vi) Design to Target Cost. Here we centre on the DFA method only, which is otherwise referred to as the Lucas DFA method.

The method involves assigning and summing penalty factors associated with potential design problems in a way that is reminiscent of Hitachi AEM although the Lucas method includes as assessment for handling (or feeding) as well as insertion (or fitting). The penalty factors are manipulated into three assemblability indices called design efficiency, feeding ratio and fitting ratio. These indices are compared against thresholds or values established for previous designs. The DFA evaluation is not based on monetary costs and in this respect differs from both Hitachi and Boothroyd-Dewhurst. The Lucas penalty factors and indices give a relative measure of assemblability difficulty. The penalty factors are established for the feeding of each part and for the subsequent fitting operations. The feeding and fitting analyses are preceded by a functional analysis (described later) and all the information is entered onto the assembly flowchart.

The assembly flowchart comprises of a component description in the first column followed by columns containing the component number, a functional analysis and feeding analysis. The fitting analysis, that comprises the assembly operations, is built up on the rest of the form using different shaped symbols for different assembly operations.

The Lucas method distinguishes between manual and automatic assembly but it does not distinguish different types of automatic assembly. In this sense it takes an approach that lies somewhere between Hitachi (no explicit consideration given to automation) and Boothroyd-Dewhurst (which has a comprehensive approach to assembly automation). The Lucas method uses the term handling when components are handled manually and it uses the term feeding when components are handled by automation. In the feeding analysis the types of questions to be answered for the automation analysis is more extensive and quite different to the questions to be answered for the manual analysis. By and large the kinds of questions that need answering are similar to the kinds of questions that need answering in the Boothroyd-Dewhurst method although not to such detail. The fitting analysis is much the same for manual or automatic fitting. The questions are much the same in both cases but the differences come mainly in the penalty indices allocated. A more detailed look at functional analysis, feeding (or handling) analysis and fitting analysis will now be given.

The functional analysis comprises of addressing each component in turn and establishing whether or not the part exists for fundamental reasons. Each part is established as either an essential part (called an A part) or non-essential (called a B part). These values are entered into the assembly flowchart. A design efficiency is then defined as essential parts divided by all parts, i.e. A/(A+B). Essential and non-essential parts are evaluated in a way that is analogous to Boothroyd-Dewhurst's method except in one important way. The Lucas method refers the user to the requirements of the product design specification (PDS). This imposes the worthwhile discipline of developing the product design within the mantle of the PDS. The other advantage of organising the evaluation this way, i.e. performing a functional analysis on all parts before undertaking the feeding and fitting analyses, is that if the efficiency is low then a redesign might be prompted before a more detailed analysis proceeds. The suggested design efficiency threshold is 60% but a practical working target is often taken as 45%. It should be noted from experience that values of 3% to 12% are not untypical for products before redesign. It is interesting to note that automotive products are found to fare better than aerospace products on initial assessment. This is seen to reflect the differences in the product development strategies of the two industries rather than any inherent difference in the products. The aerospace industry is following the automotive industry in becoming increasingly conscious of manufacturing and assembling costs through poor design.

The feeding analysis comprises of answering questions about each part in turn to identify a feeding (or handling) index. For the automatic handling, i.e. feeding, analysis the Lucas method will give the user guidance towards the appropriate feeding technology as either:

MT - mechanical tooling (e.g. bowl feeding using external part features).
LT - laser tooling (e.g. laser training using internal part features).
RO - retained orientation (e.g. in a magazine or roll).
M - manual orientation (when all else fails).

The minimum feeding index is 1. The suggested threshold is 1.5 which means that if a component part attracts a feeding (or handling) index of greater than 1.5 then the designer's attention is drawn to the possibility of improvements in part design for feeding. A very high index value sometimes occurs due to an accumulation of penalty features (e.g. it might be abrasive and have a tendency to nest).

After the feeding (or handling) analysis the user will engage in a fitting analysis. The fitting analysis is used to identify values for every possible operation during assembly. These are then entered in the assembly flowchart. The processes covered include:

(i)   Inserting and fixing (via riveting, screwing, bending, etc)
(ii)   Non-assembly operations (e.g. adjustments) or re-orientations (e.g. turnover).
(iii) Work holding (e.g. placing a temporary part to act as a guide to insertion).
(iv)   Gripping (for automation analysis only as it is not a problem in manual assembly).

Fitting indices have a suggested threshold of 1.5 apart from the gripping index (in the automation analysis) which has a threshold of 0. Any operation (or process) attracting values above these thresholds will also attract the attention of the designer who would be seeking improvements. Alternatively the overall results could be assessed by perusing the design efficiency (already explained), the feeding ratio and the fitting ratio where:

Feeding ratio = Feeding index total/No.of essential parts (Threshold 2.5)
Fitting ratio = Fitting index total/No. of essential parts(Threshold 2.5)

These measures of performance can be used to indicate the product 'state of health' with regard to assembly. The feeding ratio threshold (2.5) happens to be numerically equal to all feeding indices at 1.5 (the threshold) for a design efficiency at 60% (the threshold). The implication of having the fitting ratio threshold at 2.5 implies that the average fitting index should be below 1.5.

2.2.3 Boothroyd-Dewhurst DFMA™

The Boothroyd-Dewhurst DFA method is documented in a handbook now in its third edition (1989) being an updated and expanded version of the original document first published in 1980. Originally the handbook was available separate from the DFMA™ software but now the handbooks are only available to those who hold the software licence. The handbook allows a paper based DFA evaluation to be carried out. The first stage in the method is to establish whether the anticipated assembly system for the product will be: (i) by manual assembly, (ii) by high speed automatic assembly or (iii) by robotic assembly. This selection is based upon an analysis of anticipated annual production volume, payback period, number of parts in the assembly and, in the software package, on equipment costs. Clearly the higher the equipment costs in relation to labour costs the less viable automation becomes.

The particular DFA evaluation mechanism undertaken then depends on which of the three assembly systems is anticipated. High speed automatic assembly will be centred on an indexing or free transfer machine and is only feasible for very high volumes. Manual assembly is feasible for low volume and robotic assembly holds the middle ground. Boothroyd distinguishes three robotic assembly systems listed here in order of reducing tool change requirements: single station single robot, single station with two robots and multi-station robot line.

A recently appended extension to the Boothroyd-Dewhurst method allows the assembly cost of printed circuit boards (or products containing PCBs) to be evaluated and DFA applied to large parts.

Whether or not a design is to be evaluated for manual, high speed automatic or robotic assembly the first thrust is seen to be parts reduction. The opportunity for parts reduction is identified by evaluating each part of the assembly in turn and determining whether that part exists as a separate part for fundamental reasons. Boothroyd-Dewhurst suggests that there are only three fundamental reasons:

- The part moves relative to all other parts already assembled.
- The part is of a different material to those already assembled.
- The part is separate to allow assembly or disassembly of parts already assembled.

If the existence of a part cannot be justified by at least one of these reasons then it earns a theoretical minimum part value of 0. If the part does exist for fundamental reasons it earns a value of 1. This information is used in establishing the design efficiency as can be seen later.

All of the Boothroyd-Dewhurst's evaluation mechanisms are centred on establishing the cost of handling and inserting component parts whether this is done manually or by machines. The design for robotic assembly evaluation technique may be regarded as an extension of the same approach adopted for manual and high speed automatic assembly. Any additional complication in designing for robotic assembly is associated with a need to account for the three robot assembly systems identified earlier. In addition there is a need to account for general purpose equipment cost (e.g. the robots) and special purpose equipment costs (e.g. the tooling) as well as including time penalties for gripper changes.

The three paper based DFA evaluation techniques (for manual, high speed automatic and robotic assembly) all depend on the filling in of a worksheet with each individual component part of the assembly occupying a row. As you progress along the row the handling and insertion difficulties are accounted for, resulting in an operation cost per part. The total cost of handling and inserting all the parts then represents the total assembly cost for the product. If the product is redesigned the total assembly cost can be re-evaluated. Although this cost is expressed in monetary terms care must be exercised in interpreting the value in an absolute sense. First and foremost the value should be used as a comparative means of evaluating whether a design change is good or bad and whether or not it is worth implementing. If the decision is in the balance and there is a need to know a good estimate of the true cost of assembly, further thought can be given to the calibration constants in the calculation. These calibration figures relate to such things as the actual costs of shop floor wages, automation equipment, payback period required and accurate forecasts of required production volume.

The DFA (for manual assembly) procedure involves answering questions about potential handling difficulties, size, weight and amount of orienting necessary. This is necessary to extract a handling time from a chart of synthetic generalised assembly data built up over years of observation and research by Boothroyd and co-workers. After establishing the handling time the same procedure is applied to the insertion operation. Questions are asked about insertion restrictions such as access, vision, resistance to insertion etc. From this an insertion time is identified from a chart of synthetic data. The total operation time for that part is then the sum of handling and insertion time multiplied by the number of occurrences of that part. The operation cost is the time multiplied by the wage rate. It is recommended that the wage rate includes at least some component of overhead. Evaluation of the theoretical minimum number of parts is undertaken as explained earlier. A design efficiency (or index) is defined by the ideal assembly time divided by the estimated assembly time. The estimated assembly time is the sum of the operation times for all the component parts and the ideal assembly time is given by 3NM where NM represents the total theoretical minimum number of parts. The number 3 comes from the assumption that an ideal component part takes 1.5 seconds to handle and 1.5 seconds to insert, i.e. 3 seconds operation time.

The assumption in design for manual assembly evaluation is that the equipment costs are small and do not significantly affect the assembly cost. The opposite is true in design for special purpose assembly. A different worksheet is used for automatic assembly but the format is similar to that used for manual assembly. Further questions are asked when automatic feeding is considered. The further questions account for the extra difficulty in using machines to automatically feed one component part from bulk and to present the part in the right position and orientation for the workhead (insertion) mechanism. From synthetic data charts information is extracted which is related to the orienting efficiency and relative feeder costs. This is used to calculate the cost of handling per component. This cost is essentially established by amortising the equipment costs against the total number of components to be handled in the payback period. A similar exercise is carried out for insertion.

When the worksheet is complete the total handling and insertion cost per assembly is the sum of all the component part operation costs. If the true full cost of automatic assembly is required then account must be taken of the base machine cost (which may be an indexing or free transfer machine) and personnel (operator) costs. These calculations are undertaken separately from the worksheet and are then added to the handling and insertion costs established in the worksheet.

The DFA knowledge in the Boothroyd-Dewhurst method is twofold - (i) that which is embodied in the questions asked in identifying handling and insertion codes and (ii) the synthetic data used in the charts. The synthetic data for the manual assembly evaluation is embodied in handling and insertion times which relate to monetary costs by the wage rate which can be calibrated to provide for realistic absolute costs including overheads. The synthetic data for the automatic assembly evaluation is embodied in the orienting efficiency and relative feeder costs established for each part. The relative feeder cost is an index which provides a measure of handling, i.e. feeding, difficulty. If the index value is 1 then this corresponds to the basic feeder. The capabilities of the basic feeder are defined. If the relative feeder cost index is greater than 1 then feeding difficulties have been identified and quantified. If necessary an extra index is also used, called the additional feeder cost. This accounts for special factors which might exist such as a tendency to tangle or nest etc. The actual costs of assembly are established by defining the actual cost of the basic feeder as this is the way in which the calculated costs are calibrated with the real costs in the factory.

Experience has shown that Boothroyd-Dewhurst's evaluation of product designs for automatic assembly has real value in anticipating difficulties that would otherwise emerge much later on the shop floor. In other words the Boothroyd-Dewhurst method asks the right kind of questions and the penalty figures subsequently attributed give valuable insights.

2.3 DFA CASE STUDY

For comparative purposes the case study targets the three DFA methods on the same high volume product (a house service cut-out fuse) which has potential for automated assembly. The product design and process design have undergone a number of developments over the last number of years. The three DFA methods are applied retrospectively to the product design developments (which took place without the help of these formal methods) so that the results can be compared with actual experience. This proves to be instructive in illustrating:

- the advantages of DFA methods applied to this product design and the relationship with process design;
- the specific ways in which the three methods highlight potential problems with the earlier product designs.

As mentioned earlier the case study presented is based on the paper based DFA evaluation so as to illustrate the underlying principles and to be concise. The aim of this section is, therefore, to address the following points:

- use the same product (house service cut-out fuse) with each evaluation method;
- identify the way in which the potential design problems are highlighted by each evaluation method;
- provide an insight into the similarities and differences of the methods;
- relate the assemblability results to the case history, i.e. the developments in the product and process design, of the house service cut-out fuse;
- to draw out broad conclusions.

The way in which these points are to be addressed is to look, retrospectively, on the development of the house service cut-out fuse. The product was developed without the use of any formal or proprietary assemblability evaluation method. Nevertheless it is instructive to reflect on how the three evaluation methods might have highlighted certain aspects of the design if they had been used. For this reason it is appropriate to introduce the company, the product and to summarise its case history.

2.3.1 The House Service Cut-Out Fuse

The house service cut-out fuse is a set of products from a fusegear company that employs around 200 people and has a turnover in excess of £8M. The company manufacture 10 or so basic ranges of fuses with a total fuse variety count of approximately 9000 on their books. Their fuses range from a rating of 2 amps up to 2000 amps and cover a spread of voltages from low (240 volts) to high (72 kV). Although some ceramic and pressed metal parts are manufactured at the plant the main shop floor activity is assembly. The majority of the assembly is manual although assisted by a wide variety of fixtures and powered tooling. The company is showing an increasing interest in the role of automation and the present level of implementation is not insignificant. This aspect will be reflected in the case history of the house service cut-out fuse. The high variety of products, produced by the company, means that there is a high number of projects going through the Engineering Office at any one time. The case history of the house service cut-out fuse summarises the results of a number of smaller projects over the past number of years.

The house service cut-out fuse, as the name suggests, is the one fuse through which all the current to one house must pass. The fuse comes in 2 body sizes, the Type 2B has a bigger body diameter than the Type 2A. The full current rating range available is 5 amps to 100 amps with the Type 2B being used to accommodate more than one fuse element for the higher ratings. Around 1 million house service cut-out fuses are made per annum. It is a high volume product and the market is very cost conscious. Savings of pennies in the cost of a house service cut-out fuse can result in significant commercial advantage.

This thinking led the company, a number of years ago, to review the product and process design of their house service cut-out fuses.

2.3.2 The Case History

The original design of the Type 2A fuse is illustrated in Figure 2.3. The tree structure for the original double cap design is shown in Figure 2.4. It consists of 8 parts (plus the sand to fill the void around the fuse element) and is referred to as the double cap design due to the use of two types of end cap - inner and outer. The reason for the design being in this form was so that the fuse element tabs could be easily soldered to the inner end caps. Soldering of element tabs to caps is necessary for reliable electrical conductivity. Reliable electrical conductivity through to outer end caps is secured by the press fit between the inner and outer end caps and the large contacting area. The card discs were initially seen to have two functions. Firstly to achieve the required overall length of the product and secondly as an energy absorbing mechanism during fuse operation.

Development work led to the first certified redesign labelled 'single cap redesign 1' in Figure 2.3. This redesign now has 4 parts (plus sand). In the assembly the small hole in the end caps would line up with the end tabs of the element so that solder could be applied afterwards. The material cost of the single cap fuse was the same as the double cap fuse. The material cost savings in reducing the number of end caps and eliminating the card discs was offset by the extra cost of the ceramic in the longer body and the extra silver and copper in the longer element. Any overall savings would have to come from reduced assembly costs and since this was realised from the outset the aim for this redesign was for it to be assembled automatically.

The costing of the double cap fuse is broken down as follows: materials (45%), labour (20%), overheads (35%). For management accounting purposes the overhead costs are attached to labour. This practice is not unusual but it does mean any savings in labour will then be significantly enhanced by the concomitant savings in overhead. The automatic machine that was designed and built to automatically assemble single cap redesign 1 did work but it never produced a high enough yield for it to go on line on the shop floor.

The main problem with the automatic machine was that it did not provide the company with enough confidence (for the production environment) in aligning the hole in the cap with the tab of the fuse element. If the two were not accurately aligned the solder would leak into the fuse (and not perform adequate soldering) and the sand would leak out. Another problem was aesthetic as the solder blobs on the outside of the caps were unseemly although functionally quite acceptable.

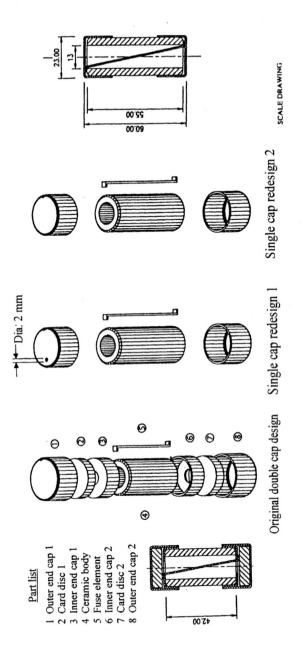

Figure 2.3 House service cut-out fuse (Type 2A).

Part list

1 Outer end cap 1
2 Card disc 1
3 Inner end cap 1
4 Ceramic body
5 Fuse element
6 Inner end cap 2
7 Card disc 2
8 Outer end cap 2

Original double cap design   Single cap redesign 1   Single cap redesign 2

Dia: 2 mm

SCALE DRAWING

23.00
13
55.00
60.00
42.00

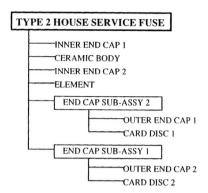

Figure 2.4 Structure tree for original (double cap) design of house service cut-out fuse.

Further development work led to single cap redesign 2 achieving design certification. This redesign had eliminated the small hole in the end cap and soldering was done blind. The solder was originally applied to the inside of the heated end caps which were then cooled prior to assembling. After the fuse was assembled it would be heated up (by induction) and the solder would reflow between the element tab and the end cap.

On paper this total process was suitable for automation. In practice the solder did not always cover the entire base of the end cap due to variations in the surface treatment (tin plating) of the end caps supplied. It was necessary for the solder to cover the entire base of the end cap so that it would contact the tab of the element whichever (beta) orientation the end cap was inserted. Another process problem existed with the control of the flux in the reflow soldering of the completed assembly - too little flux causes improper soldering and too much flux causes the end caps to rise and pop-off due to the expanding gases.

A small number of 'single cap redesign 2' Type 2A fuses were delivered to customers but these were all hand assembled for reliability. Automation was never used in production and the single cap design was soon abandoned. Production of the double cap Type 2A fuse continued.

However developments in solder pastes (no doubt driven by the burgeoning demands in electronics manufacture and increasing use of SMT) has led to a low temperature, low flux content paste that is suitable for the single cap design. Following some confirmation testing the decision was made to resurrect the blind solder single cap design. Solder paste is now syringed directly on to the tab of the fuse element - thus limiting the amount of solder material used and placing it where it is actually needed. The process of assembly has now been made semi-automatic with manual intervention between the automated assembling and automated soldering operations. Manual handling is reserved for loading and unloading and the dextrous handling of the fragile fuse element. Any fuse that is assembled is soldered directly and immediately. This is necessary as the very low resistance of the solder paste makes it very difficult to tell from external tests whether the soldering process has been completed. If the assembling and soldering operations were physically separated then there is a danger, however good the production control and material tracking system is, that fuses might escape the factory unsoldered.

2.3.3 Basis for DFA Case Study

Results of the assemblability evaluation analysis of the original (double cap) design of house service cut-out fuse will be presented as they relate to the three DFA methods. Some general comments need to be made:

1. Since the DFA methods are first used in the early stages of product design then the detailed process design has yet to be considered. Thus, early on a sensible assembly sequence can be assumed from the current state of the product's design and this is used as the basis for assemblability evaluation. The assumed assembly sequence for the various house service cut-out fuse designs, illustrated in Figure 2.3, appears in Tables 2.1-3.

2. Often the first phase of design evaluation is based on the assumed manual operations necessary. For this reason the assemblability evaluation analyses presented in this chapter relate to the assumed manual assembly operations as listed below. However, some features of the house service cut-out fuse design that may affect the efficiency of assembly automation will be highlighted and discussed.

3. The fuse contains sand for functional reasons. For clarity the sand filling operation is omitted from the assemblability evaluation analyses but the operation does not escape comment against each of the DFA methods. In practice the sand filling operation is automatic as it shaken with ultrasonics - the operation takes about 4 seconds.

Table 2.1 Assembly sequence for original double cap design

1	Place inner end cap into fixture.
2	Place ceramic body into inner end cap.
3	Bring down manually operated press.
4	Place other end cap on top of ceramic body.
5	Press.
6	Drop element through body so that element protrudes at the bottom and tag rests on inner end cap at top.
7	Place forefinger on top tag and pick up body and element so that protruding part of the element can be bent into a tag at other hand. The element now has a 'Z' shape.
8	Replace into fixture so that bottom tag is held against inner end cap by fixture bottom.
9	Solder tag onto inner end cap.
10	Reorientate 180 degrees in fixture.
11	Solder.
12	Place card disc into outer end cap 1.
13	Place card disc into outer end cap 2.
14	Place outer end cap 1 onto body sub-assembly (which is still in fixture).
15	Press
16	Reorientate 180 degrees in fixture.
17	Fill with sand (automatic).
18	Place outer end cap 2 on to body sub-assembly.
19	Press
20	Remove completed product from fixture.

Table 2.2 Assembly sequence for single cap redesign 1

1	Place ceramic body into fixture.
2	Drop element through body so that element protrudes at the bottom and tag rests on top of ceramic body.
3	Place forefinger on top tag and pick up body and element so that protruding part of the element can be bent into a tag at other hand. The element now has a 'Z' shape.
4	Replace into fixture so that bottom tag is held against inner end cap by fixture bottom.
5	Place end cap 1 onto body sub-assembly (which is in fixture).
6	Press and solder (solder applied at small hole in end cap).
7	Reorientate 180 degrees in fixture.
8	Fill with sand (automatic).
9	Place end cap 2 on to body sub-assembly.
10	Press and solder.
11	Remove completed product from fixture.

Table 2.3 Assembly sequence for single cap redesign 2

	In this sequence the end caps are supplied with solder lining the inside base.
1-5	Same as those for redesign 1 in Table 2.2.
6	Press.
7	Reorientate 180 degrees in fixture.
8	Fill with sand (automatic).
9	Place end cap 2 on to body sub-assembly.
10	Press.
11	Heat end caps simultaneously (heating tool part of fixture).
12	Remove completed product from fixture.

2.3.4 Hitachi AEM Evaluation

The Hitachi AEM method is enhanced by the additional consideration given to all parts in identifying candidates for elimination (CFE). This particular enhancement is based on the Boothroyd-Dewhurst approach and was first introduced by the General Electric Company in the US. This is the version used here. The Hitachi method has (and is) being continually refined so certain differences continue to exist in the evolving versions.

The evaluation procedure is based on completing the evaluation sheet in the same order as the envisaged assembly sequence. It is important to create a product structure tree to clarify the number of parts, sub-assemblies and the possible assembly sequences. The evaluation sheet consists of nine main column headings which, when completed for each part, will lead to the total assembly analysis. There are five scoring headings:

1. *Assembly time (AT)*. *AT* is measured in *T*-downs. One *T*-down is the time taken for one downward movement with a part. The assigned *T*-down value can be worked out for a specific factory, to give a true assembly cost. Alternatively it can be used as a relative measure.
2. *Assemblability (E)*. *E* is a number in the range 0-100. It is a relative measure of how producible a design will be in production such that:
   - $E = 0$ means an infinitely hard assembly.

- $E = 30$ then hard work to assemble.
- $E = 80$ then easy to assemble.
- $E = 100$ is ideal assembly as AT = one T-down per part.
3. *Assembly cost ratio (K)*. The ratio of how a design change has reduced the time and cost of assembly from the original design. Example: if $K = 0.74$ then the new design will cost 74% of the original design to assemble.
4. *Part count design efficiency (PCDE)*. A measure of the design efficiency of an assembly by justifying the existence of parts in terms of motion, material and service.
5. *Simplicity factor (SF)*. The overall efficiency of an assembly. Since E can be artificially elevated by adding parts and a good *PCDE* may not mean an assembly is easy then there is a need for one overall measure of a design which is a combination of E and *PCDE* scores, i.e. $SF = E * PCDE$.

The completed evaluation sheet is illustrated in Figure 2.5. The results for the original double cap design can be summarised as follows:

Assembly time	AT	= 20 T-downs.
Assemblability	E	= 40
Part count design efficiency	$PCDE$	= 0.5
Simplicity factor	SF	= 20

The assemblability (E) score of 40 indicates that the design is reasonably difficult to assemble. However this does not give an overall picture of the design since adding more parts could artificially increase the E score. The SF gives a more reliable impression of the efficiency of the design as it includes the *PCDE*. The SF score of 20 indicates a poor design. The *PCDE* of 0.5 means that only half of the parts used are there for fundamental reasons. These scores indicate that there are areas for improvement and these can be seen from the evaluation sheet:

- There are 4 candidates for elimination (*CFE*), 2 end caps and 2 card discs.
- The most difficult part to assemble is the fuse element. However this is a necessary part.

The sand filling operation is not included in the evaluation sheet. Sand could be regarded as another part. It would not be a candidate for elimination because it is necessary for functional reasons and is a different material to all other parts. In addition the author is not clear on how it could be included in the analysis without some confusion arising due to the lack of an obvious 'operation process' to choose from the list. The best guess would be that it is a part that is 'moved in a downward straight line', i.e. one T-down.
The evaluation of the two single cap designs (see Figure 2.3) reveal the following results:

	Single cap redesign 1	Single cap redesign 2
AT:	10.5 T-downs.	9.7 T-downs.
E:	38	41
PCDE:	1	1
SF	38	41
K:	0.525	0.49

Design For Assembly Evaluation

Attachment sequence	Part Name / Number	Quantity	Operation element symbol	m	Summation method Σn	100+Σn	T-n [x100-Σn]	T (n)	Motion	Material	Service	CFE	CFE x n
1	INNER CAP 1	1	F—F	2	80	180	207	207	2	7	2	2	0
2	CERAMIC BODY	1	↓C	2	20	120	138	138	2	7	2	2	0
3	INNER CAP 2	1	R↓C	3	60	160	184	184	2	7	2	2	0
4	ELEMENT	1	✓FRP↓↓SR↓↓S	11	270	370	925	925	2	7	2	2	0
5	END CAP 1	1	—↓C	2	20	120	138	138	2	2	2	4	1
6	CARD DISC 1	1	—↓	1	0	100	100	100	2	2	2	4	—
7	END CAP 2	1	—R↓C	3	60	160	208	208	2	2	2	4	—
8	CARD DISC 2	1	↓	1	0	100	100	100	2	2	2	4	—

$N = \Sigma n = 8$ Σ[T(n)] = 2000 Σ[CFE x (n)] = 4

Number of operational elements	m	0	1	2	3	4	5	6	7	8	9	10	m >10
ALPHA	α	1	1	1.15	1.30	1.45	1.60	1.75	1.90	2.05	2.20	2.35	0.85 + 0.15m

$K = \dfrac{AT}{AT\ (ORIGINAL)} = /$

$E = \dfrac{N(100)}{AT} = \dfrac{800}{20} = 40$

$AT = \dfrac{\Sigma[T(n)]}{100} = 20\ T_L$

$M = N - \Sigma[CFE \times (n)] = 8 - 4 = 4$

$PCDE = \dfrac{M}{N} = \dfrac{4}{8} = 0.5$

Simplicity Factor $= E \times PCDE = 40 \times 0.5 = 20$

ASSEMBLY: HOUSE SUPPLY FUSE Performed By: JULIE Date: 31/4/93

Figure 2.5 Hitachi evaluation sheet for original double cap design of Type 2A house service cut-out fuse.

The two sets of results are very similar. The slight differences account for the two ways the fuse element is soldered. No particular penalties were attributed to the extra requirement for orientation of the end caps with the small holes in single cap redesign 1. This is because the Hitachi method assumes that parts are located in a suitable position for attachment. Both designs have a part count of 4 (ignoring the sand as explained earlier) and both show an assembly time (*AT*) score of around half that value established for the double cap design. It is interesting to note that the *E* (assemblability) score for 'single cap redesign 1' is less than the double cap design. This has occurred because the parts eliminated from the design were easier to assemble than those not eliminated.

2.3.5 Lucas DFA Evaluation

This method encompasses a functional analysis, a handling (or feeding) analysis and a fitting analysis and the resulting penalty factors are entered into an evaluation sheet called an assembly flowchart. The penalty factors are manipulated into three assemblability scores. These scores are compared to thresholds or values established for previous designs.

1. *Design efficiency.* Design efficiency is defined as the number of essential parts divided by the total number of parts in a product. The suggested threshold is 60%.
2. *Feeding/handling ratio.* The feeding/handling analysis consists of answering questions about each part in order to determine a penalty index. The minimum feeding index is 1 and the suggested threshold is 1.5. A very high feeding index is sometimes due to a combination of penalty features, e.g. abrasive and has a tendency to nest. The feeding/handling ratio is the total of the feeding indices divided by the number of essential parts. The suggested threshold is 2.5.
3. *Fitting ratio.* The fitting analysis is used to determine penalty values for each operation during assembling (called fitting in the Lucas method). These values are entered in the assembly flowchart. Fitting indices have a threshold of 1.5. The fitting ratio is defined as the total of the fitting indices divided by the number of essential parts. Suggested threshold is 2.5.

The original (double cap) design is illustrated in Figure 2.3 and the completed evaluation sheet is illustrated in Figure 2.6. The results for the original double cap design can be summarised as follows:

Essential parts	= 4	
Design efficiency	= 50%	(threshold 60%)
Handling ratio	= 2.2	(threshold 2.5)
Fitting ratio	= 7.2	(threshold 2.5)

These results show that the areas to be addressed in the Type 2A fuse redesign should be in reducing the number of parts and making fitting easier. Most of the difficulty with fitting comes from the forming and soldering of the fuse element. The handling of parts does not appear to present any problems with the current design.

The sand filling operation was not included in the evaluation but the synthetic data allows for a filling operation (of fluids) with a penalty index of 5. Thus the sand filling operation could be easily included although there is no clear way to modify the index to account for simple or complex filling operations.

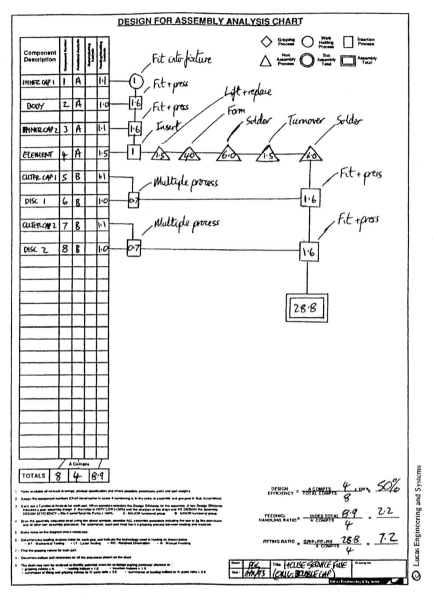

Figure 2.6 Lucas evaluation sheet for original double cap design of Type 2A house service cut-out fuse.

The evaluation of the two single cap designs reveal the following results:

	Single cap redesign 1	Single cap redesign 2
Design efficiency:	100%	100%
Handling ratio:	1.4	1.2
Fitting ratio:	6.0	4.5

The redesign 2 shows an improvement (in handling and fitting) over redesign 1 and this is due to the removal of the small hole in the end caps. Redesign 1 shows an overall improvement over the original design and this is mainly due to the reduction in the number of parts. The redesign 2 fitting ratio (of 4.5) is still a little high and is due to the forming and soldering of the element.

Assessment of the fuse element for automatic assembly would have highlighted further problems with feeding and gripping although indicating the feasibility of its automation. The small hole in the end cap (of redesign 1) requires the cap to be rotationally oriented but the feature is very small. This can be picked up by the Lucas method which would then suggest that manual orientation is necessary, i.e. that it is not feasible with automation. The difficulty is that a 'small geometric feature' is not clearly defined.

2.3.6 Boothroyd-Dewhurst DFA Evaluation

The design for (manual) assembly worksheet is used to establish the handling and insertion times of all the parts. A easy part to assemble will take no more than 1.5 seconds to handle and 1.5 seconds to insert. Any values greater than 1.5 seconds indicate some penalty. The worksheet identifies a number of scores that relate to assemblability:

1. *Assembly time (TM)*. This is the total handling and insertion times for all the parts.
2. *Assembly cost (CM)*. This is proportional to *TM* by a factor encompassing wage rate and overheads.
3. *Theoretical minimum number of parts (NM)*. The number of parts, in the design, that exist as a separate part for fundamental reasons. This aspect of the Boothroyd-Dewhurst method has been emulated in other DFA evaluation methods.
4. *Design efficiency (or index)*. This is defined as the theoretical ideal minimum time for assembly divided by the actual estimated time of assembly.

The completed evaluation sheet (for the original design) appears in Figure 2.7. The results for the original double cap design may be summarised as follows:

Assembly time	*TM*	= 71.7 seconds.
Min. no. of parts	*NM*	= 4
Design efficiency		= 17%
Theoretical assembly time		= *NM* * 3 = 4 seconds * 3 = 12 seconds

The overall assembly time is estimated to be 71.7 seconds. Of this 31.38 seconds is attributable to the fuse element. The opportunity for redesign with reduced number of parts is flagged by *NM* = 4. The handling of parts does not present any particular difficulty as only one part has a handling time of greater than 1.5 seconds. The element has a handling time of 2.25 seconds - this reflects a small penalty only. However 5 of the 8 parts had insertion times greater than 1.5 seconds. This indicates some complication with the insertion operations and these may be addressed by product redesign.

DESIGN FOR MANUAL ASSEMBLY WORKSHEET

part ID No (1)	number of times the operation is carried out consecutively (2)	two-digit manual handling code (3)	manual handling time per part (4)	two-digit manual insertion code (5)	manual insertion time per part (6)	operation time, seconds [(2) × [(4) + (6)]] (7)	operation cost, cents 0·4 × (7) (8)	figures for estimation of theoretical minimum parts (9)	Name of Assembly
									HOUSE SERVICE FUSE – KR (ORIGINAL DOUBLE CAP DESIGN)
1	1	10	1·5	00	1·5	3·0		1 (FIRST PART)	INNER END CAP 1
2	1	00	1·13	31	5·0	6·13		1 (DIFF MATL)	CERAMIC BODY (+press)
3	1	00	1·13	31	5·0	6·13		1 (DIFF MATL)	FUNNEL END CAP 2
4	1	15	2·25	00	1·5	3·75		1 (DIFF MATL + ASSY)	ELEMENT (any beta orientation)
-	1	-	-	98	9·0	9·0		-	Bend element (separate operation)
-	2	-	-	95	8·0	16·0		-	Solder (jist + seam)
-	1	00	1·13	00	1·5	2·63		-	Reorientate (code as HII)
5+7	2	10	1·5	00	1·5	(6)4·0		0	OUTER END CAPS {Two handed operation}
6+8	2	00	1·13	03	3·5	(4·26)6·17		0	CAPS ∴ × ²⁄₃
9+12	2	00	1·13	31	3·0	12·26		-	Sub-assy 1+2 (end caps and cords)
-	1	00	1·13	00	1·5	2·63		-	Reorientate
						71·7		4	
						TM	CM	NM	

©1982, 1985, 1989 Boothroyd Dewhurst, Inc.

$$\text{design efficiency} = \frac{3 \times NM}{TM} = 17\%$$

Figure 2.7 Boothroyd-Dewhurst evaluation sheet for original double cap design of Type 2A house service cut-out fuse.

The sand filling operation could have been included in the worksheet as a separate operation. From the synthetic data sheet the closest choice would come under non-fastening process, e.g. liquid insertion with an attributable time of 12 seconds. However since the synthetic data is not in terms of penalty indices then it is an easy matter to substitute the 12 seconds with a value that is better known. In this case a value of 4 seconds would be appropriate. Thus the sand filling operation is easy to accommodate in the Boothroyd-Dewhurst method.

The evaluation of the two single cap redesigns reveal the following:

	Single cap redesign 1	*Single cap redesign 2*
Assembly time:	57.39 secs.	38.01 secs.
Design efficiency:	21%.	32%.

It is interesting to note that the improvement from original design to single cap redesign 1 totals about 14 seconds assembly time. However the improvement from redesign 1 to redesign 2 is a further 19 seconds, i.e. even better. This is due to the reduced requirement for orienting the element and end cap together prior to their soldering through the small hole.

Further assessment by the Boothroyd-Dewhurst method highlights difficulties with the automatic assembly of the fuse element due to its delicate nature and its tendency to overlap in the feeder. In addition the end cap of redesign 1, which contains the small hole, is immediately and clearly flagged up as being inappropriate for automation and manual handling is required. The Boothroyd-Dewhurst method makes the definition of the small feature clear as being less than a tenth of the diameter of the rotational part. There is little possibility for confusion in interpretation here.

2.3.7 Discussion of the DFA Case Study Evaluations

Assemblability is a measure of how easy or difficult it is to assemble a product. The better the assemblability the higher the product quality in terms of fewer parts and simpler assembly operations. Fewer parts lead to less breakdowns, fewer workstations, less time to assemble and less overheads. Simpler assembly operations imply that the product fits together easier, leading to shorter lead times and less rework. It may even become easy enough for machines to assemble them.

The case study here has demonstrated how the various DFA evaluation methods act as a guide to the evaluation of assemblability. Clearly the results are subject to interpretation but the simple act of using the methods will promote assemblability as both a goal and a process that should be designed into the product from the start.

The drive for parts reduction clearly emerges out of the three DFA methods used in the case study. The approach used is very similar in each DFA method as they are all based on the initial ideas of Boothroyd-Dewhurst and their concept of the theoretical minimum number of parts for a product.

The sand filling operation, in the assembly of the fuse, would be interpreted as a simple assembly operation in the Hitachi method. In the Lucas method it was identified as a large (unchangeable) penalty. The Boothroyd-Dewhurst method identifies it as a penalty, but it can be modified by the user in the light of better knowledge than the generic synthetic data. The example shows that DFA methods can vary in their interpretation in certain assembly areas.

All three DFA evaluation methods identified the assembling of the fuse element to be the most costly part of the assembling process.

The case history of the Type 2A house service cut-out fuse demonstrated particular difficulties with the handling and orientation of the element and end caps in single cap redesign 1. This was not picked up explicitly by the Hitachi evaluation (which does not address the handling operations directly) but it was picked up well with the Lucas and Boothroyd-Dewhurst evaluations. In particular the DFA evaluations for assembly automation of these parts was flagged as being very difficult or impossible.

The Hitachi method demonstrated little assemblability advantage of single cap redesign 2 over redesign 1. The Lucas method demonstrated some advantage of redesign 2 over redesign 1. The B&D method demonstrated a large advantage of redesign 2 over redesign 1.

The end caps, in single cap redesign 2, were assumed to be supplied with solder already added and this would increase their cost. This would need to be taken into account before deciding in favour of redesign 2 over redesign 1. The Hitachi and Boothroyd-Dewhurst methods can and do relate to monetary costs. The Lucas method does not tie itself directly to monetary cost so any decision based on costs would involve further considerations. One way forward might be to consider the application of the solder to the end caps as an additional assembly process. An alternative process plan might be to assume that solder is added to the element tags instead. Different process plans could be easily and quickly assessed by the DFA evaluation methods.

In the end the real assembling cost savings came from a combination of product redesign and process design and development. It can be seen that the DFA methods could have aided not only product design for assembly but also the process design by quickly and easily evaluating alternatives. In the event the particular process plan for the single cap redesign 2 was achievable due to developments in the soldering process, i.e. application of solder paste.

The Hitachi method centres on insertion operations of parts and does not explicitly deal with automation. The Boothroyd method centres on the handling and insertion of parts with detailed consideration given to automation. The Lucas method adopts aspects of both by considering handling and insertion with some consideration of automation and some emphasis on the fitting (insertion) processes. Arguably the Hitachi and Lucas methods give a better process view of the assembly sequence and insertion operations as each fitting process is clearly documented. Boothroyd tends to have a more component oriented view. Although the handling and insertion processes are considered in detail by Boothroyd they are tagged to components. The Boothroyd method centres around the filling in and subsequent interpretation of worksheets. Nevertheless the Boothroyd software does retain all information entered in and this can be presented in other output formats.

The design efficiency of the Lucas method is based solely on the opportunity to reduce the number of parts in the product design. The design efficiency of the Boothroyd method reflects the opportunity for parts reduction plus the opportunity to improve the handling and insertion (manual) processes. The Hitachi E score (referred to here as a design efficiency) measure the efficiency of the insertion processes only. On this last point when General Electric Co. in the US adopted the AEM in the early 1980s they proceeded with a modification by adding the Boothroyd criteria for minimum part count.

DFA evaluation techniques are seen to provide a systematic and disciplined way of promoting the importance of assembly in the mind of the designer. Assembly is the point where piece part manufacture comes together and thus provides the ideal basis from which to develop an integrative view of design and manufacture in the product development process. DFA evaluation techniques can be seen to have an important role in facilitating concurrent engineering and the success of which is evidenced in the wide appeal of the Hitachi, Boothroyd and Lucas methods.

Table 2.4 General Producibility Checklist

1 Aim for simplicity.
  - Simplicity leads to lower costs and more reliability via fewer parts, fewer adjustments, simple shape, shortest manufacturing sequence, ease of component handling and insertion with foolproof assembly etc.
2 Use standard materials and components for a product.
  - Off the shelf components attracts the benefits of mass production to low unit quantity products.
  - Standardised components lead to less complications in inventory management, purchasing, tooling and manufacture.
3 Rationalise product design across modules and product families.
  - Same materials, parts and sub-assemblies in product families provides economy of scale for component production, simplifies process control, reduces tooling and equipment costs.
  - Modularise design and allow for product variants to be produced as late in assembly sequence as possible since controlled variation fits into JIT production.
4 Use appropriate tolerances.
  - The extra cost of tight tolerances stem from extra operations, higher tooling costs, longer processing times, higher scrap and rework, need for more skilled labour, higher material costs, and higher investments tied up in precision equipment.
5 Choose material for function and product process.
  - The challenge here is that the most economic choice of material is not necessarily the cheapest material that will satisfy the functional requirements. It must also account for the production process (yield and reliability) and subsequent product reliability which in turn affects warranty cost, service charges and product image.
6 Avoid non value added operations.
  - Time and cost can be added to a product's manufacture by such operations e.g.. deburring, inspection, finishing, heat treatment and materials handling.
7 Design for process.
  - The design should take advantage of process capabilities e.g. designing surface finish into injection moulded plastic parts or adopting the porous nature of sintered parts allowing lubrication retention that obviates the need for separate bushes.
  - Process limitations should be designed around e.g. inclusion of non-functional features on components to aid automatic feeding and orienting for assembly automation.
  - Avoid process restrictiveness, e.g. on part drawings. Specify only characteristics needed, allowing some flexibility for the manufacturing department in their process planning activity.
8 Adopt teamwork.
  - Simultaneous engineering including concurrent design of product and process.
  - Product or project based development organisation involving a formalised teamwork structure across functional activities or departments.
  - Success is seen to depend on (i) developing an 'open door' culture and removing the hierarchical view of working relationships, (ii) strategic commitment by senior management, (iii) formalised teamwork structure, (iv) training for all and (v) ongoing communications and continuous improvement.

Table 2.5 Design for assembly guidelines

1	Reduce part count and types.
2	Modularise the design.
3	Strive to eliminate adjustments (esp. blind adjustments).
4	Design parts for ease of feeding or handling (from bulk).
5	Design parts to be self aligning and locating.
6	Ensure adequate access and unrestricted vision.
7	Design parts that cannot be installed incorrectly.
8	Use efficient fastening or fixing techniques.
9	Minimise handling and reorientations.
10	Maximise part symmetry.
11	Good detail design for assembly.
12	Use gravity.

2.4 SUMMARY

There is much public domain literature on successful DFA and DFM case studies. Some generic advice from each specific case study is sometimes provided. Fewer articles are available that solely aim to provide this advice (Holbrook and Sackett, 1988; Huthwaite, 1990). The focus of this chapter has been on providing an insight into the systematic assemblability assessment of product design through three DFA methods. It is often found that after the successful introduction of DFA, practitioners then lead themselves into considering a more systematic approach (through DFMA techniques) to the expansive topic of producibility. Tables 2.4 and 2.5 present checklists of good practice. They provide some generic advice on achieving 'producibility' and 'assemblability' in a product's design. The particular advantage of DFM and DFA techniques is that they provide a systematic evaluation of a product's design that will inherently reflect this good practice.

REFERENCES

Boothroyd, G., Dewhurst, P., Knight, W. (1994) *Product design for manufacture and assembly*, Marcel Dekker Inc.

D'Cruz, A. (1992) Optimum efficiency, *Manufacturing Breakthrough*, **1** (2), 95-99.

Dowlatshahi, S. (1994) A comparison of approaches to concurrent engineering, *International Journal of Advanced Manufacturing Technology*, **9**, 106-113.

Holbrook, A.E.K, Sackett, P.J. (1988) Design for assembly - guidelines for product design, *Assembly Automation,* 210-212.

Huthwaite, B. (1990) Checklist for DFM, *Machine Design*, January.

Keys, L.K. (1992) Concurrent engineering for consumer, industrial products, and government systems, *IEEE Transactions on Components, Hybrids and Manufacturing Technology*, **15** (3).

Leaney, P., Wittenberg, G. (1992) Design for assembling: the evaluation methods of Hitachi, Boothroyd and Lucas, *Assembly Automation*, **12** (2), 8-17.

Miles, B.L. (1989) Design for assembly - a key element within design for manufacture, *Proceedings of IMechE, Part D: Journal of Automobile Engineering*, **203**, 29-38.

Miyakawa, S., Ohashi, T., Iwata, M. (1990) The Hitachi New Assemblability Evaluation Method, *Transactions of the North American Manufacturing Research*, Institution (NAMRI) of the SME, the NAMR Conference XVIII, May 23-25, 1990, Pennsylvania State University, Dearborn, USA.

Shimada, J., Miyakawa, S., Ohashi, T. (1992) Design for manufacture, tools and methods: - the Assemblability Evaluation Method (AEM), *FISITA'92 Congress*, London, 7-11 June, Paper C389/460, FISITA, SAE No. 925142, IMechE, 53-59.

Tibbetts, K. (1995) *An introduction to TeamSET^{TM}*, CSC Manufacturing, Computer Sciences Ltd, Dog Kennel Lane, Shirley, Birmingham, England.

Winner, R.I., *et al.* (1988), The role of concurrent engineering in weapons systems acquisition, *IDA Report R-338*, Institute for Defence Analysis, Alexandria, VA.

Womack, J.P., Jones, D.T., Roos, D. (1990) *The machine that changed the world*, Rawson Associates - Macmillan.

APPLYING "DESIGN FOR X" EXPERIENCE IN DESIGN FOR ENVIRONMENT

Carolien G. van Hemel; Troels Keldmann

This chapter is concerned with applying 'Design for X' (DFX) approaches and experiences to improve implementation of 'Design for Environment' (DFE). It first examines essentials of DFX in general, and Design for Manufacture and Assembly (DFMA), Design for Quality (DFQ), and Design for Costs (DFC) in particular. Approaches and experiences of developing and implementing these tools are then highlighted in order to search for some guidance for DFE. Difficulties in DFE implementation are outlined and counter measures are proposed.

DFE addresses environmental concerns in all stages of product development - production, transport, consumption, maintenance and repair, recovery and disposal. The aim of DFE is to minimize the environmental impact of products from their production through use to retirement. Environmental considerations can be taken at two levels. One is at the strategic level of making product policy decisions. This is primarily the domain of management. The other is at the operational level. This is referred to as the domain of product designers.

Companies can achieve competitive advantages by taking proactive actions in DFE. For example, environment-consciously designed products have less environmental load, lower energy consumption, lower life cycle costs, lower costs to comply with environmental legislation, innovative re-thinking, better social image, etc. On the one hand, companies must build up their internal DFE competence. On the other hand, the experiences of proactive companies can inspire others to follow to integrate DFE into product development.

However, only a few DFE projects can be identified in the industrial context and many of them are heavily subsidized by governmental grants. Rogers (1987) describes companies now implementing DFE as 'innovators' or 'early adopters'. Our experience shows that it is not easy to convince the 'majority' of companies of the need and benefits of practising DFE.

There are many other terms used alternatively with Design for Environment, for example 'sustainable development', 'ecodesign', 'green design', 'lifecycle design', 'environmental product design'. These terms may have their particular emphasis on some aspects. But they share similar goals. 'Green Products by Design' (1992) from the US Office for Technology Assessment provides a more detailed description of the various DFE perspectives. In this chapter, the term Design for Environment is used to cover all other expressions.

3.1 DFE AS A MEMBER OF THE DFX FAMILY

DFE is a recent development and therefore a younger descendant in the DFX family with respect to DFM, DFA, DFC and DFQ. Application of a certain DFX approach means adapting the product development process in order to improve the product with a certain focus and target. This chapter argues that according to this reasoning, DFE can indeed be considered as a new DFX type, since it strives for influencing the development process to create products with better environmental performance.

3.1.1 Virtues and lifephases are the foci of DFX

Olesen (1992) presents an overview of various DFX types and the attention the approaches have received from different academics, see Tables 3.1 and 3.2. The amount of DFX types is still growing.

According to Mørup (1994), the DFXs can be distinguished in two groups, related to their specific improvement character: DFX_{Virtue} and $DFX_{Lifephase}$. A certain DFX belongs to the DFX_{Virtue} group if the product is optimised according to a certain *virtue* in all its lifephases, like DFCost and DFQuality. A DFX is of the $DFX_{Lifephase}$ type when the product is optimised with respect to a certain *phase of its life*, like DFManufacturing or DFAssembly.

DFXs belonging to this $DFX_{Lifephase}$ group seem to be more widely implemented than DFX_{Virtue} tools, probably because the latter are (perceived as) more complex. The reason why application of DFX_{Virtue} tools is more difficult can be that they take all lifephases simultaneously into account. This leads to many trade-offs and thus a complex decision process in which many people are involved. $DFX_{Lifephase}$ types however focus on just one lifephase, so it is clear on which topic time and money will be concentrated. Moreover, $DFX_{Lifephase}$ tools often boil down to computer programs which are more concrete and applicable than the relatively abstract DFX_{Virtue} tools.

DFE is clearly of the DFX_{Virtue} type. That is, the environmental load of the product system should be as low as possible and all lifephases should be taken into account. On the other hand, Design for Recycling or Disassembly is of the less complex $DFX_{Lifephase}$ type and is just one element of DFE.

Statements

- $DFX_{Lifephase}$ techniques are easier to interpret and result in fewer tradeoffs than DFX_{Virtue} techniques.
- In short, DFX_{Virtue} tools tend to be translated to a $DFX_{Lifephase}$ variant, because the latter is more concrete and easier to handle.
- Design for Environment (DFX_{Virtue}) is often not distinguished in industry from Design for Disassembly or Design for Recycling ($DFX_{Lifephase}$). This implies that in many product branches lifecycle thinking is not yet there.

Table 3.1 Attention to DFX$_{Virtue}$ types (Olesen, 1992)

	Cost	Quality	Lead time	Efficiency	Flexibility	Risk	Environment
Analysis and diagnosis	Sheldon et al 90		Stalk & Hout 90		Eversheim 91	The Design Council 92	
Advising	Ehrlenspiel 85 Pahl & Beelich 87 Jorden 88 Kunne 88 etc.	Morup & Pihl 90 Hubka 92 etc.					Beitz 90 Jorden & Gehrmann 90
Computer-based	Ehrlenspiel 88 Dewhurst 88						

Table 3.2 Attention to DFX$_{lifephase}$ types (Olesen, 1992)

	Design for production	Design for assembly	Design for distribution	Design for service	Design for recycling
Analysis and diagnosis		Poli & Graves 85			Navinchandra 91
Advising	Pahl & Beitz 84 Sant 77 Pighini 89 Ruiz & Koeningsberger 70	Boothroyd & Dewhurst 86 Seliger et al 87 Bassler 88 Boothroyd 87 Andreasen et al 87 Andreasen & Ahm 88			Beitz 90 Beitz & Meyer 82
Computer-based	Meerkamm et al 89	Boothroyd & Dewhurst 88 Miles & Swift 92		Gershenson & Ishii 91	

3.1.2 Elements of DFX

DFX is an 'umbrella phrase', representing elements such as a specific mindset, procedures, models and tools. These elements are the means which facilitate focused improvement of the product design when a certain DFX focus is chosen.

Gatenby and Foo (1990) state that elements such as technical core (knowledge base, development processes and information systems), education, training and managerial considerations are necessary for supporting DFX. In the definition of each product development project the DFX related tasks should be stated, by selecting the project related DFX rules, checklists and targets, by defining the team design and redesign activities, team inspections and reviews of designs and at last tracking DFX performance.

Mørup (1994) discusses the preconditions for and main elements of Design for Quality, represented in Figure 3.1. This framework, strongly overlapping Gatenby's view, seems to be valid for all DFXs.

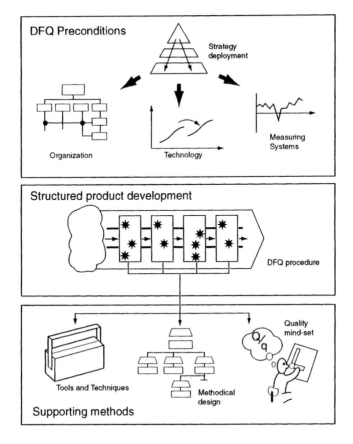

Figure 3.1 Preconditions for and elements of Design For Quality (Mørup, 1994).

DFE is an overall expression as well. The various elements, in Figure 3.1 called 'Supporting Methods', are:

1)   DFE mindsets

A general DFE mindset ('environmental demands should be integrated into the product development process') should be disseminated in the company from strategic to operational level and thus belong to the domain of the company's management and the members of the R&D function. A distinction between a managerial mindset and a designer mindset should be made.

The managerial DFE mindset concerns how to develop DFE strategies which fit in the general company policy, how to determine the level of emphasis, and how to establish the right conditions and measuring systems. The product development DFE mindset is more concrete. It concerns the understanding of ecosystems, type and severity of environmental problems, the role of the company and its products in this picture, and their own role in relation to the company, the product and the stakeholders.

2) DFE procedures

DFE procedures, like the US EPA 'Lifecycle Design Guidance Manual' (1993), the Dutch NOTA 'PROMISE Manual for Environmental Product Development' (1994), the US OTA 'Green Products by Design' (1992), are methods to structure and support the development process as a whole. They should assist product planners in the set-up and control of a product development project and give the product developers an overview of the path to follow.

3) DFE tools

DFE tools are meant to support elements or phases of the product development process. So far, they are mostly computer tools. Notorious are the various computerised Life Cycle Analysis programs. Many researchers are also working on computer programs to optimise product disassembly procedures.

A question arises which DFE elements are needed in the product development processes. Each of the elements has its specific value and consequences. The following questions can be used to identify the most appropriate DFE element.

1) What is the purpose of a certain DFE element? To get started, to organise the project, for decision support, to predict financial/environmental outcomes, for design evaluation, to supply information, to visualise, communicate, document decisions, to convince, for design education?
2) Which phases of the product development process does it cover?
3) How does it fit in the general product development process?
4) Who is actually going to learn and handle the DFE element? Governmental normalisation institutes, branch organisations, knowledge institutes, environmental specialists, company management, marketeers, R&D, designers, design students?
5) What is required for its application? Knowledge, time, money, organisational changes, a minimal level of company ambition?
6) Can the element 'work on its own' or does it need support from other elements?

Answering these questions is very important, but goes beyond the scope of this chapter.

Statements

- Like any DFX, DFE is an umbrella phrase covering a number of elements, such as mindsets, procedures and tools.
- DFX mindsets should belong to the domain of every product developer. When relevant, he or she should be able to focus on specific areas. DFX procedures are used by product managers and designers. DFX (computer)tools are mostly used by specialised designers or environmental experts.
- The lack of declarations for DFE elements on their purpose, requirements for application and consequences causes confusion.

3.1.3 DFX provides focused loops in the development process

In general, a DFX can be defined by its aim and result of its application: optimising the fit between the product design and the specific systems it will meet in all phases of its 'product life'. DFXs can be deployed at different stages of the product development process to facilitate continuous improvements of the engineering solutions. This is illustrated by Olesen (1992), Figure 3.2, after Meerkamm's improvement iteration model (1990).

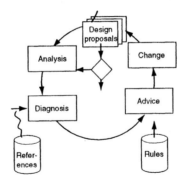

Figure 3.2 The X-loop (Olesen, 1992), after Meerkamm (1990).

The model shows that the DFXs are to be integrated in the various stages in the development process. They are not representing each a design procedure which makes the general product development process obsolete.

In this sense DFE is seen as a new member of the DFX family. It aims to generate solutions with better environmental performance. DFE does not change the general approach to product development; it is intended to be incorporated in the general process, both at strategic and operational level.

3.1.4 The level of ambition is reflected in the DFX application

Companies are often conscious about what is realistic to expect from focused improvement activities in product development. If improvement activities are carried out *on-line*, thus in a running project with all 'usual' requirements to take care of, then it is realistic to have limited expectations of the results (Olesen, 1993).

On the contrary, it would be possible to make larger improvements if a project only focuses on e.g. a certain environmental problem in a certain product type. The efforts can be concentrated on the environmental issue. The results of this type of project can then be applied, as ready knowledge or technology, in an 'ordinary' product development project. This type of application of improvement tools is called *off-line* DFX, see Figure 3.3

Concerning DFE, most efforts today are off-line activities. An example is the Dutch EcoDesign programme (1994), covering 13 demonstration projects in industry. The follow-up of this programme, also called EcoDesign, started in 1995. It 'awakens' small- and medium-sized companies by advising them how to operationalise DFE in their business.

The on-line activities so far are restricted to the more traditional topic of energy reduction during product use and to activities in the area of Design for Recycling or Disassembly. There are just a few companies who are going beyond this, and even try to consider all DFE demands in relation to each other.

Statements

- Off-line DFE application has only been seen in a few large companies and in government supported industry projects.
- On-line DFE is, so far, mostly restricted to the traditional energy reduction and to DFR.

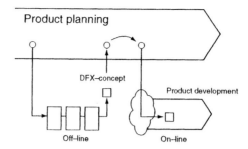

Figure 3.3 Off-line and on-line application of DFX (Olesen, 1993).

3.2 DFE ADDS NEW ASPECTS TO DFX

In addition to general DFX features that characterise DFE, it adds new aspects which are discussed in this section. We do not claim that these are unique for DFE. What is unique for DFE is that they all come together, thus impeding its autonomous diffusion in design practice.

3.2.1 No DFE without moral and ethics

DFE implies considerations of specific ethics and moral issues which are not evident regarding other DFXs. It is not easy to motivate companies to integrate environmental demands. The environmental impact of DFE actions is difficult to trace in short-term. Moreover, a so-called 'social dilemma' occurs. A company may deny the responsibility for environmental problems influencing the well-being of others, but not directly affecting its own business, and therefore take no action.

This all means that DFE has unfamiliar implications and can conflict with other sets of values. Therefore the environmental decisions should be made consciously, in relation to the sets of values of customers and other stakeholders.

What the aim of DFE application is, and not is, has been subject of much discussion. We see DFE as a means for improving the environmental performance of a product or service, contributing to a change towards sustainability.

But the use of DFE may also yield other benefits, like direct financial benefits which often are in focus for the companies involved. These reasons can be expressed in financial terms, showing that the bottom line in many companies refers to financial, and not to other results (however, this is slowly changing).

3.2.2 The mindset is the fundament

DFE requires a good understanding of a problem field. In order to make decisions on environmental issues from customers', company's and other stakeholders' perspectives, the product developers involved in design decisions with environmental implications must develop a new mindset. If the company does not assist them in developing this mindset, they will create it themselves - which means that newspapers and other media are going to form their references.

If the company has defined a clear DFE strategy, it can pass it's views to the product developers, thus ensuring that the product developer's mindset is in line with the company's mindset. This company environmental mindset should be close to the set of values of the product developers, keeping them motivated for DFE.

3.2.3 Setting the right goals is complicated

DFXs other than DFE have relatively clear targets: make the product better suitable for the life phases it will meet, or increase the general virtues of the product in order to improve product performance. The ultimate goal is to create financial benefits and ensure the company's continuity. Even when we think of DFQuality or DFSafety, results can -maybe on long term - be measured by an increase or decrease in the amount of sold products.

The ultimate goal of DFE however, is not to make more money or increase selling rates, but to contribute to sustainable development. None of us can predict whether this goal is or even can be reached. Even more confusing, environmental demands are often in contradiction with the demand for profit. This makes the implementation of DFE a complex issue.

As Walley and Whitehead (1994) argue, stating that DFE does always result in financial benefits seems hypocritical or at least naive.

3.2.4 DFE results are difficult to measure and communicate

Often it will be insufficient to measure the contribution of DFE to e.g. increased market share or decreased costs. In cases of compliance to laws or standards there is no 'choice' for the company, and the only reason for them to measure costs is to compare themselves with others: how effective is the organization in dealing with environmental requirements.

Defining the parameters to measure the product's or company's environmental performance is very difficult. Expressing the environmental performance of a product in absolute terms is simply not possible. In case of a product *re*design the environmental improvements can be indicated by constructing an environmental profile of the redesign. This profile can be compared with the environmental profile of a reference product, e.g. the original design or a competitor's product.

The profiles can be constructed by using a qualitative or quantitative form of the well-known lifecycle analysis (LCA) methods. Though, the value of the results of any LCA approach is still discussed, and it is not suitable to express the environmental improvements to external stakeholders, e.g. for competitive reasons. However, one can take the not-objective measures from an LCA and couple these to sets of values of the company's stakeholders. This can give an indication of the dimensions that are possible and relevant to measure, at least for in-company use.

In relation to this measuring problem, it is difficult to visualize and communicate the environmental results to the internal and external stakeholders. Added to the problems are fear of prosecution, media exposition and revealing problems, not all results are communicated.

3.2.5 Both product and lifecycle are synthesized

Product development has always involved more than just the design of the three dimensional product. Systems like production, marketing and maintenance should be developed simultaneously. This is illustrated in Figure 3.4.

Especially in DFE it is essential to approach the product from a broad perspective, with a holistic view. Instead of talking about the environmental performance of a product, we want to speak about the environmental performance of the *Product Lifecycle System*, see Figure 3.5.

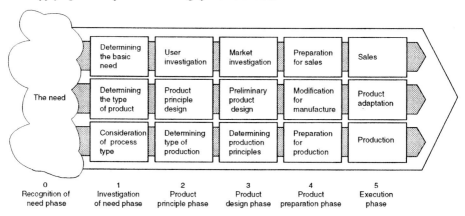

Figure 3.4 Integrated Product Development is simultaneous determination of the sales and marketing system, the product and the production system (Andreassen and Hein, 1987).

The term product life cycle system expresses:

1) **Inclusion of the functional product system.** Not only the product itself should be considered, but the additional products necessary for its functioning too. An example is a coffee machine. In DFE not only the materials of the machine itself should be studied, but additional materials like coffee and filters, and energy consumption too.

2) **Inclusion of the product lifecycle.** Not just one phase of the product life must be considered (like mostly the use phase) but all life phases from production to disposal.

   The three dimensional product is surrounded by necessary systems like production, distribution and maintenance systems, which all have their environmental burden and are subjects for improvement. In DFE even a new system must be developed, namely the End-of-Life system, in which is determined if and how the product can be reused, remanufactured or recycled and how it will ultimately be discarded.

The system thinking means that it is complicated to ensure the right environmental solutions, and therefore it will be necessary to have models and tools for describing environmental properties and possible life paths of the products.

The synthesis of the lifepath is evident when the company chooses to extend its control over maintenance, service and repair. Designing these systems to an optimal fit with the products may yield a better environmental performance and a better business performance. The potential in acquiring this control over the lifecycle is only realized by a few companies.

3.2.6 External relations are essential in DFE

The environmental decisions made in product development may have far reaching consequences, as these decisions will determine the conditions for environmental considerations and performance through the life path of the product. This means that relations to suppliers and other business partners will be affected by the ambitions, requirements and decisions with respect to environmental issues. Compared to some of the other Xs, the E reaches beyond the company both upstream and downstream from production. Where other Xs may focus on internal aspects of efficiency (DFA, DFM, DFC), then DFE addresses both internal and external relations and performance.

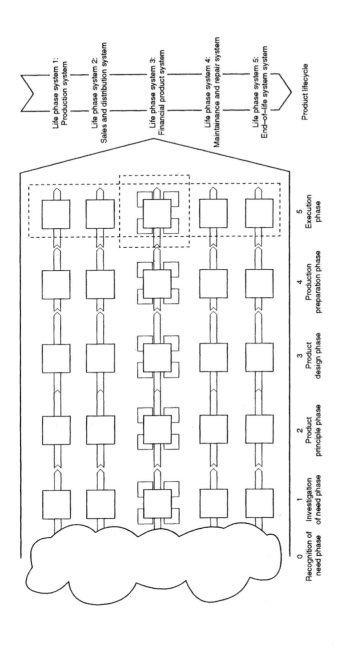

Figure 3.5 Synthesis of the Product Lifecycle System, including the functional product system and the product life cycle (consisting of various life phases systems), adapted from Andreassen and Hein (1987).

3.2.7 The stakeholder gallery is expanded

Not only the above mentioned relatively known external relations should be involved in DFE, also those external stakeholders who formerly were supposed to have no or minor relation to the company should be listened to. Not only customers, authorities and employees are involved in DFE, other stakeholders (like the recycling industry and consumer associations) all along the products' path of life too. Deciding which voices to listen to and which to satisfy are important considerations in DFE.

3.2.8 Legislation and regulation play important roles

A major stimulus for DFE is legislation. Since it is not yet clear in what directions policy and legislation will be developed on the long term by various governments, companies feel insecure about what long term product planning suits the future governmental demands best. Because of this insecurity, they will hesitate taking investment decisions, and therefore keep away from long term and more substantial environmental product improvements.

Statements

- DFE requires specific moral considerations, previously not known in DFX.
- Provide the product developers with a DFE mindset - or they will develop their own, which may not be in line with the company's environmental mindset.
- Radical DFE work addresses the core of a company's business and may therefore be perceived as threatening.
- Lifecycle thinking must be a vital part of the designers' environmental mindset.
- The potential, in terms of better environmental performance and business performance, in acquiring control over the lifecycle is realized by only a few companies.
- The complexity of measuring and communicating the environmental improvements - both internally and externally - is one of DFE's specialities.
- Where other Xs may focus mostly on internal aspects of efficiency (DFM, DFC), DFE addresses both internal and external relations and performances, reaching beyond the company both up- and downstream.
- DFE implies listening to stakeholders whose voice has never been interesting in other DFXs.

3.3 DFE IMPLEMENTATION HAS NOT YET HAD ITS BREAKTHROUGH

In 3.1 we stated that to a certain extent DFE is a descendant of the DFX family. That this descendant has a combination of specific characteristics, causing some problems in its maturation, is discussed in 3.2.

This section states more in detail what are bottlenecks which impede DFE to diffuse autonomously in every day design practice. The chapter starts with some remarks on the implementation of DFX in general, to refer to when discussing DFE implementation issues in 3.3.2 to 3.3.5.

3.3.1 DFX implementation has come a long way

To structure the discussion on the implementation of DFX tools, it can be useful to make a couple of distinctions.

Section 3.1.2 concerns the distinction between the various elements of DFX. The *DFX computer tools* like Finite Element Method (DFReliability) and Assembly Optimization

(DFAssembly) are chosen at the start of a project, specified as such in the contract and mostly used by specialists. On the contrary, *DFX mindsets and procedures* like FMEA belong in the toolbox of each product developer. The decision to focus on a certain DFX depends on management considerations concerning the problem definition. However, a focus can come from the R&D or marketing function too. The specific experience and interests of the product developer will reinforce or reduce this focus and determine the amount of time which is spent on the subject.

A next distinction is between the product type, the company or consultancy type and size which we want to adopt the DFX tools. It is clear that DFX tools all have their own field of application. Products with many components need DFAssembly tools. Highly cost-competitive products need DFCost tools. Highly stressed products with the need for high reliability need tools like Mould Flow Analysis and the Finite Element Method. Products with relatively new, but still implicit product attributes need a tool like the QFD House of Quality, to investigate the consumers' demands. A characteristic of DFE tools is that they are relevant for all product branches, and not have a branch related field of application, like most other DFX tools have.

Successful traditional companies who produce the same products for many years, don't feel the urge to adopt new tools. Interest in DFX tools often pops up when a company is trying to pass through a crises. Then it suddenly seems worthwhile to invest in tools to increase efficiency, decrease costs or find new product-market-technology combinations. Companies who are pro-active don't apply DFX tools to solve crises, but exploit them in a more aggressive way to generate competitive advantages and unique selling points.

Many small- or medium-sized companies, at least in The Netherlands, don't have their own department for product development and contract out-door design consultancies to do this job. To give a picture, the number of employees in a small design consultancies may be around 3; some large firms may employ up to 30 persons.

For small design consultancies, working for various clients with diverging problems, and for small companies, it doesn't make sense to invest in adoption of specific DFX tools. Small design consultancies and companies contract out-door specialists to apply certain DFX tools. However, some small design consultancies do invest in a certain DFX tool, and become specialists in that area. The larger design consultancies have specialists who know how to work with the tools and who are called in a project if there is agreement on applying a specific DFX tool.

This leads to the conclusion that though all companies and design consultancies should develop their own DFE mindset and procedures, we can expect only the larger ones to adopt DFE computer tools. Some small design offices do apply DFE tools and sell their DFE competence as a unique selling point.

Statements

- We cannot expect all small companies and design consultancies to master one or more DFX tools. Most small design consultancies adopting a certain DFX tool want to become specialists and concentrate on exploiting this knowledge.
- Proactive companies see new chances in exploiting DFX tools for creating competitive advantages.
- Reactive companies mostly start adopting DFX tools to pass through an internal crisis. Since environmental demands are often perceived as an external force, these companies will not adopt DFE principles without further stimulation.

Table 3.3 Company related obstacles for DFE

A   Bottlenecks which inhibit starting DFE:

a.1   Lack of vision. Management is not aware of the dispositional power and effect of decisions in product development;

a.2   Lack of motivation. Nor management, R&D or marketing shows interest in DFE since they don't see the benefit of it, though they are aware of the company's impact on the environment;

a.3   Insecurity. Management is insecure about regulatory initiatives and commercial effects, since there is hardly any DFE tradition;

a.4   Complexity of getting started
- The company doesn't have a systematic approach to product development in general, so it doesn't know how to integrate DFE in a structured way;
- There was no fixed procedure assisting companies in setting the stages for DFE (arranging preconditions, organizing the team, determining lacks of knowledge);
- The company is discouraged by the cost of acquiring the environmental 'start up' information;

a.5   Other priorities
- The company gives priority to investments in other new activities;
- The company gives priority to environmental work elsewhere in the company;

a.6   Unawareness. The company has never thought about its relation to environmental problems.

B   Bottlenecks which inhibit proceeding with DFE:

b.1   Lack of support. Though the R&D function started with DFE, their mindset was not in line with the managerial mindset. Therefore management did not support their work;

b.2   Complexity
- It was too complicated to balance DFE with other efforts;
- There was no access to assistance in critical steps;
- The designers did not know what to manipulate, because of lack of understanding of options and their effects;
- Lack of a structure to group suggestions for improvements;
- Goal-setting and measuring DFE results is complex, therefore there were no clear targets in mind for all participants, leading to demotivation;

b.3   Resistance. General resistance to technological or organizational change;

b.4   Opportunistic attitude. Only ad hoc decisions with short-term financial benefits.

3.3.2 The reasons for DFE's problems are known

Crul (1994) gives a comprehensive overview of problems concerning the implementation of DFE in practice. He has studied eight Dutch ECOdesign demonstration projects (1994), which were coordinated by TNO Product Centre and the Delft University of Technology, and funded by the Dutch government. He has made a distinction between problems which are familiar to innovation processes in general and problems which are specific for DFE.

Based on this research and our own experiences, the following summarises various attitudes of companies towards integrating DFE in their product development process:

A.   No interest. The company has not considered DFE as relevant and has not initiated any DFE activities.

B.   False start. The company has tried to involve product development in the environmental work, but the initiative failed.

C.   Picking low hanging fruits. They did make a start, but worked only with ad hoc decisions and short-term goals. The company involves product development in the environmental work. But the environmental aspect is only included on a detailed level in the design process. The improvements are fast and cheap.

D.   Integration. The company integrates DFE in its projects in such a way that improvements are realized and DFE is continued in all other product development projects too. The improvements are substantial and are achieved by changes in the total product life cycle system. Unfortunately, so far very few companies have been identified to belong to category D.

Statements

■   Companies' attitudes range from reactive (following legislation) to proactive but all focus on short-term financial benefits.

■   Proactive companies, perceiving DFE as combination of a lower environmental load with long-term new business perspectives, are hard to find.

■   Autonomous diffusion of DFE into the companies will be slow, until   motivational factors change.

Table 3.3 shows a number of obstacles in DFE. Only three of them will be addressed further below. They are:

1.   Management lacks vision and does not supply support;
2.   Companies lack basic structure for product development projects;
3.   It is not easy to balance the environmental efforts with others.

3.3.3 Management lacks vision and does not supply support

It is clear that the requirements for DFE and the potential changes that may occur in the product and its life phase systems may have large impact on both the consumption of organizational resources and the competitiveness of the products. This combination is perceived as being "risky" to work with and therefore attention and support from management is essential.

There is a risk for DFE-rituals or 'blind' improvement actions, like marking components with material codes, costing little and yielding less. This will happen if the management does not point out strategies to follow and does not deploy resources to secure that the improvements are not blind, but are in line with activities and cooperation with e.g. other companies dealing with the product after its use.

Dutch ECOdesign projects (1994) have shown that applying DFE can have various positive spin-offs, apart from reduction of environmental load and related (future) costs. DFE, when exploited to the most, leads to innovative thinking. This is because the   product design is approached from a new perspective.

DFE can result in new solutions on different levels of detailing (product range, concepts, structure and components). A spin-off with yet another character was shown during a Philips project on the development of a 'green' TV set. Working with DFE resulted in an increase of employees' motivation in general. Awareness about these positive spin-offs of DFE should be included in the managerial mindset.

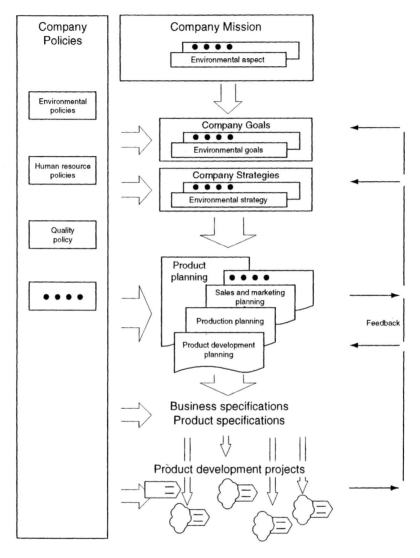

Figure 3.6 Relation between environmental decision-making on strategic and operational level (Keldmann, 1994).

Environmental Management Systems as a carrier for DFE

It is valuable to consider the Environmental Management Systems (EMS) to find an entrance to develop vision and support for DFE. DFE has never really been incorporated in any EMS.

EMS policy - An EMS provides a frame for stating a long term environmental policy in relation to the company's products and services. EMSs also serve as carriers for cascading environmental goals through the company, showing the directions for realizing the environmental strategy. An EMS should supply at least the organizational structure, norms and measurement and control systems.

It can be valuable to strive for integration of the DFE policy, expressed in goals, means, actors and prerequisites, in the determination of the general EMS policy. Figure 3.6 shows the linking between the long term environmental/business decisions (strategic level) and the specifications for products and projects in product development (operational level).

Actual EMS operationalisation - At this moment, operationalisation of EMSs is the responsibility of the company's environmental department. It boils down to end-of-pipe measurements, process improvements and so called 'good housekeeping'. Today most EMSs focus on the processes inside the company, strive for control and reduction of waste streams and consumption of materials and energy. EMSs in industry so far have a reactive character, responding to environmental legislation and regulation. Sometimes they end up in being paper tigers, since EMSs imply increase of bureaucracy.

The internal focus, bureaucracy, process and legislation orientation and reactive character of the EMSs don't correspond to the requirements for implementing DFE in product development. Therefore operationalisation of DFE should be the responsibility of the product developers. Implementation of DFE needs powerful and creative 'change agents', since DFE deals with the properties of the product itself, and therefore touch the heart of the company.

This leads to the conclusion that determining the company's DFE policy should be in line with and can be supported by the general long term environmental policy, stated in the EMS. But operating and managing DFE should not be the task of only the environmental department. They should share this responsibility with the product managers and developers so the EMS policy is operationalised and translated in 'product development language'.

Referring to DFX

The focus of a certain development project can be proposed by e.g. the R&D function, but the decision is in the hands of the company's management. If this focus is set, it depends on the level of ambition which tools will be selected to support the project.

This stresses the fact that it is absolutely necessary to ensure the commitment of the company's management, which can be a hard job since the arguments to apply DFE are mostly external and results are long-term and difficult to measure. Only a few companies are stimulated by moral aspects or see competitive advantages. We must however be aware that motivational factors will not remain as they are today, but will change over time.

In other DFXs, the stimuli arise inside the company and the results of DFX efforts can be expressed directly in increased market shares instead of the relatively 'vague' environmental improvements in DFE, so the comparison is limited.

It may be valuable though to have DFE stimulated and supported by structures and institutions which in the past only concentrated on quality improvement. In this way creating DFE visions is stimulated by the existing attention for increase of product quality, which recently seems to grow, at least The Netherlands.

Suggestions

☐ Demonstration projects, monitored by experienced institutes, (partly) financed by governmental funding and sector organizations, can develop competence in the area of DFE. When insight in the technical, financial, market and organizational consequences of strategic decisions is developed, insecurity decreases and management motivation to follow increases.

☐ Branches should coordinate their activities so they are able to influence the development of legislation, recycling infrastructure etc. When a couple of companies within a product sector make the first DFE step, others will get motivated and follow.

☐ Many companies are used to stating their long term environmental policy on end-of-pipe issues and process improvements in an Environmental Management System (EMS). DFE can profit from this policy generating structure, when DFE policy is developed in the framework of the EMS.

Product system level

Product system level

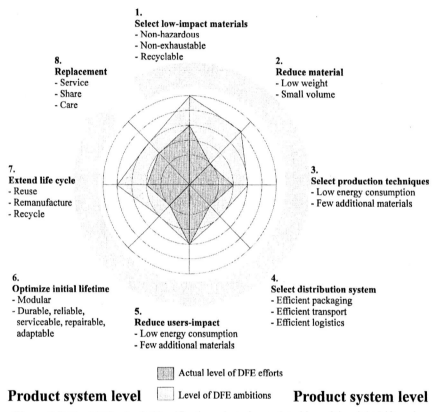

1.
Select low-impact materials
- Non-hazardous
- Non-exhaustable
- Recyclable

8.
Replacement
- Service
- Share
- Care

2.
Reduce material
- Low weight
- Small volume

7.
Extend life cycle
- Reuse
- Remanufacture
- Recycle

3.
Select production techniques
- Low energy consumption
- Few additional materials

6.
Optimize initial lifetime
- Modular
- Durable, reliable,
  serviceable, repairable,
  adaptable

5.
Reduce users-impact
- Low energy consumption
- Few additional materials

4.
Select distribution system
- Efficient packaging
- Efficient transport
- Efficient logistics

▨ Actual level of DFE efforts

▢ Level of DFE ambitions

Product system level

Product system level

Figure 3.7 The LiDS-wheel: Classification, clustering and ranking of the eight Lifecycle Design Strategies (Van Hemel, 1994).

Experience shows that companies are motivated to obtain certifications of the ISO 9000 series concerning quality management. These companies will probably head for certification, like the British Standard 7750, of their environmental management system too. If DFE policy is integrated in or derived from the company's EMS policy, DFE can benefit from the certification opportunities.

Many DFE tools for environmental analysis and optimization of disassembly procedures have been developed by now, but tool development focused at strategic decision support is somewhat neglected. If management is motivated to support DFE, it still has to make strategic decisions on which DFE paths to follow. The options for improvement show a great variety in levels of ambition and environmental and innovational potential.

The various paths, called 'Lifecycle Design Strategies', are classified by Van Hemel (1994) in the LiDS-wheel, Figure 3.7. Some improvement strategies ask little investment in terms of time and money and can be realized on short-term. In most cases these improvement directions will be preferred, because they ask little investment and are perceived as "low risk" options. Some other strategies however can yield much more environmental and innovational benefits, but ask more attention, involve more people and will be realizable on a longer term.

Clock-wise, the sequence follows roughly the product life cycle; in the other direction one can identify the phases of the product development process, starting at product system level, ending at product component level. Since in general changes at product system level need more attention than changes in product details, the strategies are roughly ranked according to the complexity of their realization. However, it will be clear that the necessary efforts for realization of the different strategies depend on the character of the product and the company, and therefore this ranking-to-complexity is not always valid.

The aim of the LiDS-wheel is to serve as a communication tool, supporting the process of strategic decision making. At the start of a project it gives an overview of the possible directions and serves a an idea-generating technique. Later, it visualizes and documents the company's environmental product policy.

Statements

- Radical DFE is perceived as 'risky'. It consumes organizational resources, alters the product and influences the way of doing business.
- DFE has various positive spin-offs like innovative power, resulting to new solutions since the product is perceived from another angle.
- Without direction and consciousness DFE-work ends up in DFE-rituals.
- Development of DFE vision can and should be supported by the experience and structures of Environmental Management Systems.
- The environmental policy should be traceable and visible in the specifications of product development projects.

3.3.4 Companies lack basic structure for product development projects

Implementation of DFE is easier when companies have a structured approach to product development in general. Each company will establish its specific structure, but a minimum of structured approach in projects is, in our opinion:

1) Specifications. For evaluating design activities (product specs/business specs);
2) Phases. For separating and defining clearly the main activities in the project;
3) Milestones. For systematic review and control of project results, goals and preconditions.

Without the structure, the DFE application will get the project "out of balance" since this issue will dominate. A first risk of this is that the environmental concerns are not well 'balanced' mutually, leading to sub-optimal environmental improvements. Solutions for one environmental problem should not increase another environmental problem.

What next can happen is that R&D invests much in developing environmental improvement proposals, which will never be commercialized since the decisions are not shared by the people who take care of the realization of the product.

This is illustrated in some of the EcoDesign cases (1994). In the realization phase some important environmental decisions were neglected. This seems to be partly due to discontinuity in the process; the people in charge of the product's production preparations, have not participated in the EcoDesign project. The other side of the story is that only in the production phase the investment decisions must be made. When at that moment the risks are perceived too high and the market opportunities too low, investment proposals are withdrawn. In cases like these, environmental priorities should have been set differently to be realizable.

Referring to DFX

In practice, development projects are less structured and well-defined as we may think.

However, we see DFX elements being implemented in product development practice. Facilitators for this are the availability of a shared knowledge base (located e.g. at specific institutes), clear procedures and training.

DFXs in general have each their specific target and only cover parts of the product development process. 'Blind' application of a specific DFX tool will affect the attention to the general properties of the product. It would be valuable, if the DFX tools could assist in providing the right context and conditions for their proper application.

Companies and design consultancies have each their specific way of (not) structuring their development process. Therefore they appreciate flexible DFX procedures which deliver knowledge and support the process. However, they should not force them to re-arrange the structure of the process. DFX tools, to support activities in specific phases of the process, have a higher chance to be applied correctly than tools for which a restructuring of the product development process is necessary.

Suggestions

- ☐ It should be clear to companies which tools and data-bases (materials, legislation, demonstration projects) are developed and what is the place and relevance of these tools according to their own product development structure.
- ☐ DFE procedures, like the Dutch PROMISE step-by-step approach for DFE (1994), can be more widely introduced via demonstration projects and training, to enable reactive companies to apply DFE principles in their own product development structure.
- ☐ Companies should have easy access to DFE databases and external support, organized by governmental agencies, sectorial organisations or academic institutions.
- ☐ Experiences with DFM show that designers need structures to group their suggestions for improvement. In DFE procedures, a structure must be supplied for this. A suggestion is to classify them in eight categories of the LiDS-wheel, Figure 3.7.

Statements

- ■ A structured approach in product development facilitates DFE handling, because integrating environmental issues includes consideration of all life phases and requires trade-offs with other virtues (cost, quality, flexibility etc.).
- ■ If a clear DFE procedure is introduced, the general product development process may become better structured.
- ■ DFE procedures are valuable for education and insight on the planning and bottlenecks in a DFE-oriented product development process. However, the procedures must be

constructed such that they easily fit in existing development practices, and don't force the product developers to re-arrange these.

3.3.5 How to balance the environmental efforts with others

The early phases in a product development project are especially important for creating a good balance between the efforts on environmental and other issues. Gatenby and Foo (1990) state" The team needs to understand how a design's attributes affect all Xs, so that one DFX concern is not inadvertently optimized at the expense of another. Similarly, the team must consciously evaluate trade-offs between DFX and performance or functionality considerations".

Two activities, which we call 'Setting the stage' and 'Concept modelling', play an important role in this.

Setting the stage:

This is an activity in relation to product planning and it addresses the issues 'around' the project, which the project group seldom has authority to address.

The main issues are:

1)   Defining the project's and the new product's role in the realization of the company's business concept. The team gets insight on what is critical in realization of the business concept. This secures that the project members have the right picture in mind when communicating solution alternatives, and that they know how to evaluate their fit with the business concept. From an environmental perspective it concerns the interpretation of the long term environmental decisions into the specification of the product and the project.

2)   Determining the innovational focus in the project and which parts of existing products to reuse. This includes communicating which competencies in the company are available and which could be of benefit to the particular project. This gives the team members a clear perception of the 'degree of freedom' they have in seeking new solutions. From an environmental perspective this is relevant, because the degree of freedom has impact on the level of ambitions in the environmental improvement work. Small product changes will seldom yield substantial environmental improvement.

3)   Determining the thematical focus in the project (e.g. environmental performance) and introducing tools for assisting the improvement work. The communication of a thematical focus and its priority in relation to other concerns is important to secure that the improvement activity gets the right level of attention. Else there is a risk that the focus activity gets too little attention. But too much attention is not good either; the project can get 'blind' to the project basics when the focused improvement activities are the most exciting to work on. Concerning the environmental issue it is important to consciously build up the environmental mindset of the project members and the role of the project work in relation to the company's general environmental work.

An important part of this stage setting is communicating the issues mentioned above, via project start-up seminars where the issues are presented and discussed with all project team members.

Concept modelling:

In practice, conceptual models of the product are used to facilitate the evaluation of the specific solution in relation to technical and commercial feasibility. Often this serves as the basis for authorization of the project. When incorporating environmental concerns along with other concerns, it is important for decision makers to evaluate the effects of the - environmentally motivated - product changes. One advantage of addressing the environmental issue at the conceptual design level is that the improvement effect is much larger than at the detailed level. A second is that conceptual considerations yield a larger degree of freedom in integrating environmental and commercial concerns. However, the use of quantitative LCA methods has, so far, made the designer focus on new environmental solutions, but only at a detailed design level. To make decisions on which environmental design strategies to follow, concept modelling is needed of the product and its surrounding systems, in paragraph 3.3.5 called the product lifecycle system. The environmental part of this modelling will provide the context for evaluating the improvement activities, facilitating making the right trade-offs between different product properties.

Another benefit of the modelling is that it improves communication between the product developers and the company management concerning technological and environmental solutions. Thereby it increases the chance that appropriate improvement options penetrate in the long term environmental product policy of the company.

Communication

Special attention in these phases is necessary because of the complexity of communication about environmental issues between the various company departments. The considerations and information on the improvement options perceived from the technical side, must be communicated to the marketing department and to the managers who make the final decisions. When deciding for one of the innovational environmental improvement strategies it is necessary to gather not only the product developer and his technical assistant around the blackboard, but also representatives from the marketing, management and production department, and even suppliers and main customers.

Referring to DFX

When looking at general DFX experiences for solving the trade-off problem by creating a structure to ensure the right balance between environmental and other issues, we should focus on DFXs of the DFX_{Virtue} type, like DFQuality and DFCost, instead of to $DFX_{Lifephase}$ type. As stated in 3.2.3, comparison with $DFX_{Lifephase}$ types is not suitable, since this type will lead to less dilemmas, focusing all attention on just one lifephase of the product.

DFE seems to have a lot in common with DFQ. They both need involvement of not only one company department (like production planning or actual production) but also other departments such as management (strategic decisions), marketing etc. Like DFQ, DFE will have impact on suppliers and customers. In both DFE and DFQ uncertainty exists on which product characteristics actually define (environmental) quality. Long-term results of DFE are difficult to predict and to communicate. Therefore, it may be difficult to convince people to start implementing DFE.

Suggestions

☐ A major cause of imbalance between environmental and other issues is the company's uncertainty about how legislation will develop and how the market will respond. Therefore companies could cooperate more to be able to set common goals and influence legislation.

☐   DFE can be exploited as marketing instrument to create unique selling points, if the company can prove its claim. Environmental product demands are naturally incorporated into business.

☐   Some approaches of the DFX$_{Virtue}$ types can be 're-used' in the field of DFE. In this sense the House of Quality can be converted to 'House of Environmental Quality' (Luiten, 1994) , to structure communication while balancing environmental with other demands.

Statements

■   Decision making in DFE is complex since communication, both company-wide and inter-company, is necessary.

■   Without providing decision makers with insights into DFE consequences of products and the companies' businesses, they will obstruct or stay passive. Executing demonstration projects can be a solution to this.

■   Facilitating DFE decision making in the early phases of projects is essential for making substantial environmental improvements. At strategic level the 'environmental stage' should be firmly set. Concept modelling, including the environmental aspects of the product lifecycle system, is necessary for evaluation and communication of the technical and commercial feasibility of the improvement solutions.

■   The use of quantitative LCA methods as modelling technique so far has made the designer focus on new environmental solutions, but only at a detailed design level.

3.4 SUMMARY

This chapter explores the relations between Design For Environment (DFE) and different DFX approaches, developed and implemented in the past to improve engineering solutions in product development processes. The X stands for Manufacturing, Assembly, Safety, Cost, Quality etc. The aim is to look back at the characteristics of and experiences with these approaches, to be able to define what aspects are relevant for the development and implementation of DFE.

First some general findings are presented on the various DFX tools and how they are applied in design practice. A first statement is that whenever tools for product designers are developed, their should be a clear picture of the characteristics of the potential users.

Then is discussed to which extent DFE can be designated as a new DFX approach, which will diffuse in practice like its predecessors. In general, companies are free to apply DFX elements if they are considered beneficial. Some companies are already convinced of the benefits of DFE, like cost reduction due to increased efficiency or better customer relations. However, we state that DFE has a combination of specific characteristics due to which it will not diffuse autonomously. The majority of the companies will need extra stimuli, such as low-cost demonstration projects, legislation and financial rewardings.

Some specific DFE characteristics are listed below.

☐   The effects of DFE activities on the environment, on the company's organization and business perspectives are difficult to understand, foresee and communicate. This is partly due to the uncertainty according to development of legislation and market response.

    Making it easier for the managers of product development projects to understand, estimate and visualize the implications of their decisions, by using e.g. modelling techniques, would be of great help for diffusing DFE in industry. Methods should be supplied for handling information, assisting the early decision making.

☐ Many companies consider the stimuli and environmental demands for DFE as external, not resulting in benefits for their own business. Other DFXs result in direct benefits for the organization.

DFE has innovational power, but it must get enough freedom and attention to be able to express and exploit this power. It should not be seen as a threat, but as a challenge, since DFE can create new market possibilities. It should not be compared to regulatory end-of-pipe or process improvements, which lead to bureaucracy and not to new product attributes.

☐ DFE has far reaching effects on the company, since it may influence the properties of the product. Therefore resistance is expected.

It is important that the company's policy on all environmental activities, including DFE, is coherent, long term based and in line with the existing sets of values in- and outside the company. When the company's environmental policy is coherently constructed it can be translated 'all the way down' to the initial decision on the DFE related focus and tasks in product development processes.

We argue that DFE should be approached from a broad perspective; it not only considers implementing environmental demands at the operational designers' level, but refers to the necessity of making decisions on product policy at a strategic level too. Because of the focus on details, too often now DFE is interpreted as Design for Recycling or Disassembly only, leading to sub-optimal solutions.

A couple of DFE tools are now being developed and introduced. These tools so far are meant for environmental analysis of products, and lead to improvement options which stay close to the reference product. To ensure more innovational, far reaching product development, tools or procedures are necessary which lead to strategic re-thinking of the product, support strategic decision making, and stimulate communication in- and outside the company.

For this, especially the relation between DFE and DFQ and their possible integration is worth studying. There seem to exist interesting parallels concerning their intangible character and impediments for implementation.

Finally, an overview of the main impediments for implementation of DFE is presented, followed by elaboration on three of the impediments and suggestions for their removal. One of the suggestions is to obtain a coherent overall environmental policy in a company by starting from the existing acquaintance with and structures for Environmental Management Systems (EMS). The EMSs can act as vehicles to diffuse DFE in industry. Therefore DFE should very early be introduced in EMSs as a means to realize the general environmental policy.

The overall conclusion of the chapter is that DFE seems to have a combination of characteristics which make it only to a certain extent comparable with its DFX family members. These distinctive elements of DFE are going to be the most challenging to get to grips with in the coming years.

REFERENCES

Andreassen, M.M., Hein, L. (1987) *Integrated Product Development*, IFS Publications Ltd., Springer-Verlag, London.

Crul, M. (1994) *Milieugerichte produktontwikkeling in de praktijk*, NOTA, SDU Den Haag.

Dutch Office of Technology Assessment NOTA (1994) *PROMISE Handleiding voor Milieugerichte Produktontwikkeling*, SDU Uitgeverij, Den Haag.

Gatenby, D.A., Foo, G. (1990) Design for X (DFX) - Key to competitive, profitable products, *AT & T Technical Journal*, May/June.

van Hemel, C.G. (1994) Lifecycle Design Strategies for Environmental Product Development, *Workshop Design-Konstruktion*, IPU, Technical University of Denmark, Denmark.

Keldmann, T. (1994) Challenges in Environmental Decision Making in Product Development, *WDK-Workshop Evaluation and Decision in Design*, Technical University of Denmark, Denmark.

Luiten, H. (1994) *Integratie kwaliteitsgericht en milieugericht produktontwerpen*, Faculty of Industrial Design Engineering, Technical University of Delft, The Netherlands.

Meerkamm, H. (1990) *Fertigungsgerecht Konstruieren mit CAD-Systemen*, Konstruktion 42.

Mørup, M. (1994) *Design For Quality*, Dissertation, Institute for Engineering Design, Technical University of Denmark, Denmark.

Olesen, J. (1992) *Concurrent Development in Manufacturing-based on dispositional mechanisms*, Dissertation, Institute for Engineering Design, Technical University of Denmark, Denmark.

Olesen, J. (1993) Introduction to Workshop DFX, *WDK-Workshop 'Design for X'*, TU Denmark, May.

Riele, H. te; Zweers, A. (1994) *ECO-design: Acht voorbeelden van milieugerichte produktontwikkeling*, NOTA, SDU Den Haag.

Rogers, E.M. (1983) *Diffusion of innovations*, Third Edition, The Free Press, New York

U.S. Congress, Office of Technology Assessment (1992) *Green Products by Design: Choices for a Cleaner Environment*, OTA-E-541, Washington DC.

U.S. Environmental Protection Agency (1993) *Lifecycle Design Guidance Manual*, EPA600/R-92/226, Cincinnati.

Walley, N., Whitehead, B. (1994) It's not easy being green, *Harvard Business Review*, May-June.

DESIGN FOR COMPETITION:
THE SWEDISH DFX EXPERIENCE

Margareta Norell; Sören Andersson

This chapter presents some results from a number of investigations concerning impact on concurrency and efficiency from the use of Design for X (DFX) tools in product development in Swedish industry. The project has been performed through an inter-disciplinary research programme with cooperation between Department of Machine Design at Royal Institute of Technology and Department of Psychology at Stockholm University. The studies were carried out in industrial sectors of mechanical and electro-mechanical products. Studied tools include DFA - Design for Assembly, FMEA - Failure Mode and Effects Analysis, QFD - Quality Function Deployment and EPS - Environmental Priority Strategies. The results have proven valid for product development in general. Important results are:

- No tool will create concurrent engineering unless the organization of work is adapted to a high degree of co-operation between functions, e.g. marketing, design and manufacturing, and competence domains.
- Use of the tools and/or information technology in product development could give excellent support to concurrent engineering if the implementation is made with regards to co-operative work.
- Design for competition demands a simultaneous focusing on both product development process and the persons in the process.

4.1 PROJECT BACKGROUND

A rapidly changing world puts the adaptation of the industrial organization in focus. The competition and the desire for survival of the business are contributing to the demands for a higher degree of efficiency in every part of industrial business. The product development process has been found to play an important role and has consequently gained an increased interest.

Product development includes complex combinations of technical, economical and marketing activities. A model considering that process is the Integrated Product Development model, which has its origin from works of Olsson (1976), Andreasen (1983) and further developed by Andreasen and Hein (1987). The model has been used with success as a guideline for product development in many Scandinavian industries (Mekanresultat, 1985).

Product development here refers to the whole process of product realization, including synthesis and analysis of new product concepts. Key factors in the process are time, quality and cost. The challenge is to find key factors by a convergent controlled and predictable product development process and to avoid a divergent "chaotic" process.

Surprisingly companies describe their process more as the divergent example. Lack of relevant input from market and customers is described as the main reason. Systematic knowledge transfer between market and engineering functions in the very start is of major importance. Multi-functional teams in project start, few limitations and successively more focused work is a strategy for more convergent product development processes.

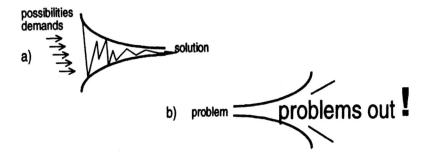

Figure 4.1 (a) Ideal, convergent product development process, (b) unwanted, divergent product development (Norell, 1992).

Short lead time is a competitive factor: parallel activities in the process is one way to reduce the calendar time. However, parallel activities cannot alone yield an efficient process. Many other aspects must be fulfilled before a competitive product development process is obtained. Communication and co-operation between different groups and functions are highly important factors. The necessity to consider all tasks in the process, also human aspects, has therefore gained increased interest.

Northern European industries have had reasons to rethink and intensify the work with competitiveness during the past years. Although systematic approaches to design have had considerable impacts on efficiency and competitiveness, further development is necessary.

The decisions in early phases of the development have been raised as important activities in the process. The decisions should include considerations concerning for example market needs, quality, manufacturing, life cycle. Japanese companies have shown to be highly competitive in that respect by "doing right from the start" (Womack *et al.*, 1991).

In order to improve the efficiency of different phases of the development process, several support tools have been introduced and used, some of them computer based. Most of these tools have considerably improved the development process, mainly by doing different tasks faster. However, it is not clear to what extent these tools have influenced the process efficiency. The process efficiency is rather a question if the tools have supported and/or improved the results of the different activities in the process.

The concept of Concurrent Engineering (CE) concerns itself with product development work carried out in parallel processes and with a high degree of co-operation between different domains. CE shows a lot of similarities with the Integrated Product Development process. The concept includes aspects of both the process and the individuals. This makes it interdisciplinary. Three major ingredients in Concurrent Engineering are:

- Organization and management supporting integrated methods of working.
- Use of efficient methods for support in product development.
- Use of relevant information transferring systems.

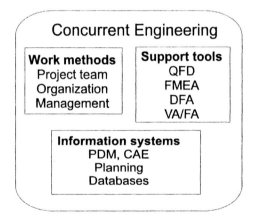

Figure 4.2 The concept of Concurrent Engineering includes methods of working/organization, support tools for special applications and information transferring systems (Norell, 1992).

Of the three areas in Concurrent Engineering the work method has a major importance. An integrated method of working requires an organization and its management involved in and motivating co-operation between different domains.

Different methods can be used to simplify the transition to a higher degree of co-operation. By creating new, unclaimed forums for discussion, prestige-related barriers can be diminished. Product development work in cross-functional teams has shown to be efficient. Skilled project managers should create the same objectives for everyone in the project.

4.2 PROJECT DESCRIPTIONS

An interdisciplinary research programme, MANDECO, was initiated in 1988 in co-operation between the Department of Machine Design at the Royal Institute of Technology and the Department of Psychology at Stockholm University.

4.2.1 Project Objective

The objective of the programme is to develop knowledge concerning efficiency in product development processes regarding both technical and organizational questions. The purpose of the studies presented here was to investigate and analyze practical use of support tools in industrial product development. Areas of questions were:

- What characterizes successful implementation and use?
- How do the support tools impact on product, project efficiency, concurrency and co-operation in the process?
- Are there any effects on learning and competence development observed to be dependent on work with support tools?

4.2.2 Scope Of Design Tools

The choice of studied support tools has been made with the demands that the tool should:

- Address a concrete problem in product development.
- Be used by several functions and persons.
- Have a potential to be a "bridge builder" - a forum for co-operation.

Based on these criteria the following tools are chosen and studied:

1. DFA - Design for Assembly, according to Boothroyd and Dewhurst (1989), is used to point out parts in the design or concept which need further attention for assembly cost reasons. When analyses with DFA are performed, every detail in a product is considered regarding handling, fitting and necessity. The method includes tables with similarities to time studies, MTM. (Boothroyd and Dewhurst, 1989)
2. FMEA - Failure Mode and Effects Analysis is a method which is used to find and judge potential sources of error in products or manufacturing processes. FMEA also includes a judgment of how serious the consequences of a presumed error would be and the possibility that the error is discovered (IEC Standard, 1985).
3. QFD - Quality Function Deployment is a method to translate the customers demands of the product to technical demands. QFD is used in a matrix, "the House of Quality", which is filled in with information of customer demands, aims, benchmarking, priorities etc. One objective with a first step QFD-evaluation is to accomplish a well rooted specification of demands (Sullivan, 1986).
4. EPS - Environmental Priority Strategies is a valuation system for executing quantitative data for Life Cycle Assessments (LCA). It is a new tool and still under development. With EPS the "total environmental load unit" (ELU) for a product or a system can be calculated and judged early in product development processes (Ryding and Steen, 1991).

4.2.3 Scope Of Industries

The industrial companies included in the interview studies were chosen with the following criteria:

- Companies with a clear formulated ambition to make the product development process more efficient.
- Companies with at least two years of experience of work with the methods (excluding EPS).
- Companies with at least one interviewee driving the process of implementation of support tools.

Eight Swedish companies, all with a large share of export sales, took part in the studies. All companies are developing and manufacturing their own products, mechanical and electromechanical, of very different types and complexity. Product examples include diagnosis and analysis instruments, chain saws, pumps, etc.

4.2.4 Methodology

The investigations referred to in this chapter have studied the implementation and use of the four support tools listed previously. A number of product development projects at the eight participating companies have been included in the studies and the data have been collected by the researchers mainly by interviews.

Before each study, the purpose of the study was carefully described for the people directly involved in the development project as well as for those who are not directly involved in the project but have other interests as experts or managers. The importance of including persons representing different functions in the interview study was particularly put forward at this stage.

For each tool about ten persons representing different functions were interviewed. Besides people representing design, about 20% represents production, 20% quality and 10% marketing and product planning respectively.

The interviews were semi-structured according to a pre-developed interview guide, which covered the following tasks:

- background and function of the interviewee,
- education and experience with the actual support tool,
- reasons for choosing the particular tool,
- the phase of the development process where the tool has been used,
- preparation and use of tool,
- demands and wishes regarding the use of the tool,
- advantages and disadvantages of the tool,
- effects of using the tool,
- influence on cooperation,
- plans for the future.

Each interview was tape recorded and afterwards transcribed and coded by at least two researchers independently. The coded protocols were then condensed and analyzed and provided the basis for the reported findings.

4.3 INVESTIGATION FINDINGS

4.3.1 Support Tools In Product Development

In the studies, QFD, FMEA and DFA have been applied in manual versions, even though computerized variants exist. A generally expressed desire is to use computerized tools for documentation and reuse. However, it should be noted that many persons interviewed have mentioned manual analysis is especially advantageous for team-building. The EPS-tool is, however, implemented in a PC version and no manual variant is used in the studied companies.

From a technical point of view, there is an opinion that the reasons for using different types of support tools in product development are to support and guide designers individually. If they do, the tools are beneficial for the efficiency of a particular designer but not necessarily for the whole project. It has been shown from the results that in order to support the efficiency of the project, the support tools should themselves constitute a platform for communication.

The studied tools are all perceived as very efficient and relevant in the domain they address. Furthermore the results from the studies show that they all can improve the interaction between people involved in a product development project. The most reported communication improvements are: between market and design - QFD, design and manufacturing - DFA, and design and quality function - FMEA (Norell, 1992). But a functional integration is not automatically obtained after implementing a method, the integrative effect is strongly dependent on the ambition of the implementation and further use.

T - Team-building in design work

P - Product design review

A - Analysis of product features

G - Guidelines for design work

Figure 4.3 The GAPT model (Norell, 1993; Hovmark and Norell, 1994).

4.3.2 The GAPT Model

It is shown in the studies that the tools can be implemented and used on different levels, from guideline level up to team-building. The different levels can be described by the GAPT model, as shown in Figure 4.3. According to that model, product development support tools can be used on four different levels: Guidelines; Analysis of product features; Product reviewing; and Team building level (Norell, 1993; Hovmark and Norell, 1994).

Level G - Guidelines. When a support tool is used on level G, no formal analysis is carried out during the product development project. The designer relies on his/her experience and knowledge about designing products and uses the tool just as a "checklist".

Level A - Analyzing features. One or a number of formal analyses with the tool are carried out during the product development project with the purpose to focus on the actual "main" problem.

Level P - Product reviewing. The evaluation of the analyses of a design with the tool is enlarged and the participants will consider the results and the product from several perspectives.

Level T - Team-building. The support tool may act as a catalyst for team-building. The use of a tool could introduce a new neutral language understood by all team members. This has shown to lead to a higher quality of professional communication.

The GAPT model is found to be a useful instrument for analyzing the conditions for use of a support tool and its effect on the efficiency of the product development process.

4.3.3 Effects On Process Convergence

Applications of DFA, FMEA and QFD, in general, shorten the total development time (Norell, 1992). Several companies report that DFA gives products built up of less components and therefore can be assembled in shorter time. At the interviewed companies, FMEA has been reported to diminish the number of late errors and failure effects in the products. Furthermore, in the studied companies, the number of changes in the specification drastically decreased in projects where QFD has been a basis for the specification.

EPS may have a similar positive effect on product development. It is only possible to give some indications now, since the experiences of using EPS are rather limited and the tool is still under development. However, user opinions show so far that the tool gives relevant advisory support concerning environmental effects of a concept. It is probable that EPS will become an important tool in the future when demands for more environmental respect. Several interviewees reported that EPS may encourage co-operation particularly with suppliers (Ritzén and Norell, 1995).

QFD, FMEA, DFA and EPS are all contributing to the adding of more knowledge to the product in an early stage of the product development process and consequently may decrease the number of errors and changes during the process. The general recommendation is to use the methods as early as possible in the product development process.

4.3.4 Impact On User - Learning, Teams And Efficiency

Problem solving functions are not included in the studied methods. The tools can be characterized as "problem pointing" and leaves the creative work to the user. The risk is therefore minimal that the methods will cause an impoverishing effect on the work in the product development process. All the interviewed persons in the study have the opinion that the usage increases knowledge and competence. All four support methods are contributing to the learning in both depth and width in the domain.

It is possible, theoretically as well as practically, to perform DFA, FMEA and QFD individually. At individual applications, technical effects could be satisfactorily accomplished within limited areas. Yet the study shows that the greatest value lies in the meeting between different competencies over a qualified support tool. Cross-functional teams and a good tool to use in a joint project, increases the possibilities of qualified co-operation to a great extent.

All of the interviewed persons reported the increased co-operation as one of the most important effects with the usage of DFA, FMEA as well as QFD. By adding a broad competence to the project in an early stage, a comprehensive view of the product could be created which favours both the project quality and the process cycle time. This has a great value also in future development projects.

4.3.5 Implementation Observations

Experiences have shown that a high degree of awareness is demanded to reach success in making the product development process more efficient. If the process of increasing the efficiency is being strengthened with the help of support tools, a conscious strategy is also needed if the tools are going to have the presumed long term effects.

A generalized picture of the implementation of support methods are given in Figure 4.4, the MI-model. The model is developed from interviews concerning implementation and usage of support methods (Norell, 1992). During the first period, step 1, the support method is tested by a smaller group in a limited project. The participants are often well motivated to go through the test and the results are usually good. After that, a period of reflection follows. The use is decreasing since the pilot project is finished. The usage in step 2 depends on a number of factors and how the company acts. Either the usage decreases and the tool is forgotten, (the lower curve in step 2), or a more systematic usage is started within different development projects, (the upper curve). It is in the area between step 1 and step 2 (area A) that development is determined. Without certain measures the most likelihood is that the lower curve will be followed.

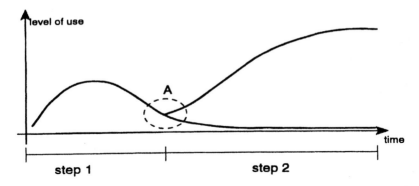

Figure 4.4 The Method Implementation (MI) model (Norell, 1992).

The decision of which one of the two curves that will be followed, should be taken at management level in the company. In the case of continuous usage, the decision should contain objective, application level and amount of resources. If the company is going to continue using the support method, an extra effort is needed in area A.

The extra effort can consist of various components in different organizations. A necessary but not sufficient condition of lasting usage, is that the method is well grounded in the responsible line manager and that a strategy for education and guidance is developed. The application should be adapted to the company's business, general rules can only partly be used. Routines should be developed for choices concerning if, when, and how, a method should be applied within the company.

4.4 INTEGRATION GUIDELINES

A concrete plan of action for starting the work toward integrated product development is described below. The plan refers to the methodology to use competent tools to accomplish co-operation, and how they can contribute to competitive integration. The guidelines are established from results and conclusions from collected data in interviews and observations. Following are eight steps towards integration:

1. *Choose the focus.* It is important that everybody knows what the problem is, what should be treated and what should be accomplished. It is better to focus narrowly than too wide at first. The focus should be expressed in general terms to be supportive to choices of method and project. To choose the focus is to critically examine the business!

2. *Formulate objectives and check points.* The overall objectives should be divided into sub-goals possible to measure or follow up. This gives possibilities to early findings of factors leading in the wrong direction or which cannot fulfill the desired aims. Objectives should be well documented and grounded in a group who is about to perform the work. It is very important that everyone participates and perceives a share in the objective. Occasions for the follow-up of objectives should be determined already in the stage of planning, since it is very common that the follow-up is defective. By clearly fixed occasions for follow-up, the time in between should be used with the largest freedom of action possible for those involved.

3. *Ground the decision in the group and the management.* It is important that the decision to start the usage of support methods or computer systems, which support co-operation in product development, is grounded at the highest management level in the company. The investment of resources is not likely to be refunded immediately but in the long term. Therefore it is important that the plan and objective are firmly accepted in the organization. In certain cases, a successful first step according to the MI-model has been performed in a purely operative level, without an out-spoken support from the management. However step 2 demands a clear sanction, where the experiences from step 1 could be a useful basis for decisions concerning step 2 (Figure 4.4).

4. *Choose method and pilot project.* To choose a suitable first method is not complicated if a thorough problem analysis is performed (point a). It is recommendable to start from the problem which is judged to be of greatest importance in the business. If changeable specifications of demands are a problem, QFD should be suitable. If frequent complaints are a problem, FMEA etc. Do not mix up the usage of several methods at the starting point.

5. *Select a responsible person and a group.* An important skill for the supervisor in the implementation is coaching capabilities. The role of the dedicated person should not be underestimated. He/she could with his/her enthusiasm perform very successful step 1-processes. It is very common that step 1 is started by an "informed dedicated person". However, to continue to step 2 with an established usage, the dedicated person's role probably must be more firmly grounded, and the introduction of the support method(s) become a main task in his/her work. The work group should consist of persons with different competence. Normally just one representative for every specialist domain should participate. The size of the work group should be limited and consist of 4-7 members. A good thing is if the method guidance is done by a person who is not

participating in the development work. It is important to have time for team-building. Resources for regular, personal meetings (especially in the beginning) increases the possibilities for a well functioning work group. Education in the support method should be performed within the project group if possible.

6. *Appoint external motor/catalyst if required.* In those cases where a suitable supervisor is not available in the organization, it can be advantageous to initially appoint an external force. All this for the reason of getting experiences from earlier cases into the project, until the experience is built up within the company. The advantage with an external supervisor is that he/she becomes more method and process focused and should/could not have a viewpoint on the operative development work.

7. *Measure and check with the objectives.* To be able to show the effects of the usage of support methods by checking with objectives, is crucial for lasting success. Easy measured results (for example total time, number of components, number of changes etc.) as well as judgments concerning learning, increased co-operation are important ingredients.

8. *Correct and increase the number of methods and projects.* By analyzing the results, the measurements and the judgments, a base for judging the proceedings is created. The activity, analysis - correction - proceed, is a condition for changeability and should always be made when standing at the starting point of a new project. At this point the number of methods could be increased if necessary.

4.5 SUMMARY

A very clear finding of the studies is that competitive design demands motivated persons in the process. Basic psychological needs for human beings are perceived autonomy, possibilities for a comprehensive view of the task and chance for development and learning in work.

There are different approaches to reach the desired result for a more integrated product development. However, there are a number of important, seemingly trivial, factors which can be expressed in the following way:

- Co-operation cannot be obtained without support in the organization and the management.
- Support methods and tools, not perceived as efficient, will not be used.
- Clear, common formulated objectives reduce functional and prestige-related barriers.
- People with authority to take responsibility are stimulated, motivated and report better results.

Those factors are basically about striving towards simultaneously focusing on the efficiency of the process as well as on the individual/group which is going to drive/participate in the process.

ACKNOWLEDGMENTS

The simultaneous focusing on both technical processes and individual demands need different kinds of knowledge to be mixed in research. This has been made possible by the double funding from Swedish National Board for Technical Development (Nutek) and Swedish Work Environment Fund. The authors acknowledge Dr Svante Hovmark at Stockholm University for his co-operative work during the studies and MSc Cecilia Hakelius for her valuable contributions.

REFERENCES

Andreasen, M.M. (1983) Integrated Product Development - a new framework for methodical design, In: *Proceedings of ICED 83*, Heurista, Zürich.

Andreasen, M.M.; Hein, L. (1987) *Integrated Product Development*, IFS Publications Ltd., Springer Verlag, UK.

Boothroyd, G.; Dewhurst, P. (1989) *Product Design for Assembly*, Boothroyd Dewhurst Inc., Wakefield, RI.

Hovmark, S.; Norell, M. (1994) The GAPT model: Four approaches to the application of design tools, *Journal of Engineering Design*.

IEC Standard (1985) *Analysis techniques for system reliability - Procedure for failure mode and effects analysis (FMEA)*, Bureau Central de la Commision Electrotechnique Internationale, Genève, Suisse.

Mekanresultat 85 02 08 (1985) *Integrerad produktutveckling - en arbetsmodell*, (in Swedish) Sveriges mekanförbund, Stockholm.

Norell, M. (1992) *Advisory Tools and Co-operation in Product Development* (Doctoral dissertationin, in Swedish), *TRITA-MAE-1992:7*, Department of Machine Elements, The Royal Institute of Technology, Stockholm.

Norell, M. (1993) The use of DFA, FMEA and QFD as tools for Concurrent Engineering in Product Development Processes, In: *Proceedings of ICED 93*, The Hague.

Olsson, F. (1976) *Systematisk konstruktion* (Doctoral dissertation, in Swedish), Department of Machine Design, Lunds Tekniska Högskola.

Ritzén, S.; Norell, M. (1995) LCA and EPS - experiences from use in industry, a pilot study, In: *Proceedings of ICED 95*, Praha.

Ryding, S.O.; Steen, B. (1991) *The EPS SYSTEM, a PC-based system for development and application of environmental priority strategies in product design - from cradle to grave*, Swedish Environmental Research Institute, IVL B 1002, Gothenburg.

Sullivan, L.P. (1986) Quality Function Deployment, *Quality Progress*, June.

Womack, J.P.; Jones, D.T.; Roos, D. (1991) *The machine that changed the world*, New York: Harper Perennial.

DEVELOPING DESIGN FOR X TOOLS

George Q. Huang

This chapter presents a generic Design for X (DFX) development framework, or DFX shell in short, which can be easily tailored or extended to develop a variety of DFX tools quickly and consistently. A set of formal but pragmatic "commonsense" constructs such as Bills of Materials and Process Charts are provided to convert the conceptual PARIX model, which has been outlined in the introductory chapter as a basic DFX pattern, into the DFX shell. Following are basic questions that must be addressed in this conversion:

1. How to represent decisions in designing products, processes and resources?
2. How to relate these decisions?
3. How to measure decisions and their interactions?
4. How to collect and display data necessary for above tasks?

Figure 5.1 shows a seven-steps procedure for developing a DFX tool using the DFX shell. The above question will be addressed at appropriate steps of this systematic procedure. Each step will be discussed separately in a section. Major issues are highlighted, approaches are explored, advantages are outlined so that they can be extracted for incorporation, and pitfalls and traps are flagged so that they can be avoided.

This chapter is prepared for those who are involved in developing DFX tools. Those who are involved in implementing DFX tools and those who generally want to know more about the subject may also find it highly relevant. One early warning is necessary that the DFX shell and the DFX/BPR shell to be discussed in the next chapter have not been fully prototyped, though intended, on computer systems. Sample screens are for illustrative purposes only.

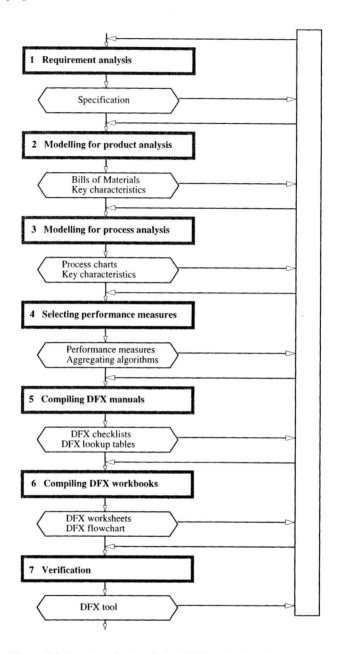

Figure 5.1 Procedure for developing DFX tools using the DFX shell.

STEP 1 - REQUIREMENT ANALYSIS

Like most product development projects, developing a DFX tool is customer-driven, following a cycle of continuous improvement. The cycle starts with the first step of investigating customer requirements and establishing DFX development specification.

Well-known DFX tools, such as Hitachi Assemblability Evaluation Method (AEM) (Shimada, Miyakawa and Ohashi, 1992), Boothroyd-Dewhurst Design for Assembly (DFA) (Boothroyd and Dewhurst, 1990), and Lucas Design for Assembly (DFA) (Swift, 1981; Miles, 1989), were developed by highly experienced practitioners with close collaboration between academics and industrialists. Requirement analysis was carried out, though not necessarily formally and explicitly, at the right beginning and new requirements were incorporated every time a new version was upgraded based on field experience.

Three major categories of key characteristics can be identified for developing DFX tools. They are functionality, operability and focus.

Functionality Requirements

A DFX tool must fulfil some or all of the following functions:

1. Gather and present facts.
2. Measure performance.
3. Evaluate whether or not a product / process design is good enough.
4. Compare design alternatives: which design is better?
5. Highlight strengths and weaknesses.
6. Diagnose why an area is strong or weak.
7. Provide redesign advice by pointing out directions how a design can be improved.
8. Predict "what-if" effects.
9. Carry out improvements.
10. Allow iteration to take place.

The DFX developer must be clear about which of the above functions should be included in the DFX tool under development. The first five functions are basic functions that should usually be provided by a DFX tool. The second five functions are more advanced features, available only in a few research DFX systems. Even well-known successful DFX tools do not perform these functions. Instead, they are left for the user to perform.

Operability Requirements

Functionality does not exist alone. It co-exists in pair with operability. By operability it is meant the ease of using the DFX tool to fulfil its functions effectively. Stoll (1988) proposes ten operability criteria for evaluating various DFX [DFM] approaches:

- ☐ *Pragmatism - Training and/or practice.* Concepts and constructs used should be already familiar to the user or easily learnt with little effort.
- ☐ *Systematic.* A systematic procedure ensures that all relevant issues are considered.
- ☐ *Data requirement and quantitative.* Product and process data must be easily collected and presented to the analyst or the analysis team to enable further actions.

☐ *Teaches good practice.* The use of DFX methodology teaches good DFX principle and formal reliance on the method may diminish with use.

☐ *Designer effort.* The designer or design team, a prime user, should be able to use the DFX tool effectively with little additional time and effort.

☐ *Management effort.* Management is not a prime user and therefore effective use of the DFX tool should not be totally dependent on management support or expectation.

☐ *Implementation cost and effort.* It should be distinguished between those changes and commitments that are required for implementing the DFX tool and those changes and commitments that are highlighted by the effective use of the DFX tool for necessary improvements.

☐ *Rapidly effective.* Effective use of the DFX tool should produce visible and measurable benefits.

☐ *Stimulates creativity.* Effective use of the DFX tool should encourage innovation and creativity, rather than impose restrictions.

The right balance between functionality and operability is pivotal to the success of developing a DFX tool. A sophisticated DFX tool with comprehensive functionality may be too difficult and time-consuming to operate. On the other hand, an over-simplistic DFX tool may be easy to use but fail to function effectively.

Focus Requirements

One of the distinctive strengths of implementing CE through DFX is the focus and the vision necessary for the analyst or the project team to make changes. Focus requirements play an essential role in achieving the right balance between functionality and operability. This focused approach tends to be widely preferred by industrialists and practitioners.

For a DFX tool to be practically functional, it should be applicable to a range of problems and its results must be reasonably accurate. That is, some degree of flexibility must be incorporated so that the DFX tool can be configured and customised to emphasise particular requirements under different circumstances.

Flexibility and focus are determined by the following factors:

☐ The target product sector must be determined, mechanical, electrical, electronic, etc. It would be beneficial to start with a narrow range of products and generalisation could be introduced once sufficient insights have been gained from tests and applications.

☐ In Design for X, variable X has two parts: $X = x + bility$. The suffix "*-bility*" corresponds to the performance metrics. Exact definition of the variable "*-bility*" is not given at this stage and will be discussed at Step 4 - Selecting Performance Measures. The x part represents one or more business process corresponding to one or more life cycle in product development. The x variable should be determined at this stage. For example, "x = total" and "*-bility* = quality" in "design for total quality"; "x = whole-life" and "*-bility* = cost" in "design for whole-life cost"; "x = assembly" and "*-bility* = cost" in "design for assembly cost" (or simply assemblability if other -bility measures such as assembly times are used); and so on.

☐ *Design* in Design for X is concerned with decision making activities, their outcomes - decisions, and their interrelationships in designing products, processes (activities), and systems (resources). Most successful DFX tools are based on interactions between products and processes (activities) with resources implicitly embedded in activities for consideration. This type of DFX tools are said to be capability-oriented or process-oriented. Alternatively, a DFX tool can be based on interactions between products and resources with activities implicitly embedded in resource centres. This type of DFX tools is said to be capacity-oriented or facility-oriented. With capacity-oriented DFX tools, product designers are able to explicitly or systematically incorporate the impact of new product introduction on the existing capacity and anticipated product mix of the manufacturing facility at the product design stage (Taylor, English and Graves, 1994).

☐ It must be determined at which stage of product design process the DFX tool is to be used. It has been widely acknowledged that the earlier the DFX principle is applied, the greater the benefits, and harder to apply it. This decision will have an effect on what data should be collected. If a DFX tool is to be used at the concept stage, then it should be based on major design decisions, not detailed decisions. If a DFX tool is used at detailed design stage, more information is to be collected, with the expectation of higher accuracy.

☐ It should be made clear how the DFX tool is to be used in design decision-making process. Very few research DFX tools are design systems which actually make design decisions. A few help and guide design decision making. This type of DFX tool is said to be on-line. Most existing DFX tools are used to evaluate design decisions after they are made. This type of DFX tool is said to be off-line. An on-line DFX tool checks its data/knowledge base to ensure that the design decision being considered will not violate the DFX rules. Tentative decisions which violate DFX rules are not included as final decisions. In contrast, an off-line DFX tool checks design decisions already made against its data/knowledge base to see if any DFX rules are violated. Those decision which violate DFX rules will be improved. Some efforts have been made to embed off-line DFX tools into design systems. Such tightly integrated design platforms would perform functionality similar to on-line DFX tools.

STEP 2 - MODELLING FOR PRODUCT ANALYSIS

Product modelling is primarily concerned with how to represent design decisions related to products, not how to make design decisions - decision-making activities. There are three general categories of product information:

☐ *Composition*. What constitutes a product?
☐ *Configuration*. How constituent components are related to each other?
☐ *Characteristics*. What describes constituent components and their relationships?

There is a wide selection of product models. Product information is available in a number of forms, such as technical illustrations, engineering drawings, and other associated documents. Although they are required in DFX analysis, they cannot be used as a base model in the DFX context to represent product design decisions concisely and incrementally.

The DFX shell exploits two concepts for product modelling: Bill of materials (BOM) and key characteristics. A bill of materials is a list of the items, ingredients, or materials needed to produce a parent item, end item, or product (Greene, 1987). The important role of a BOM in DFX tools lies in that it is the basis for data inputs and outputs. It is used for acquiring key characteristics of its components and their relationships.

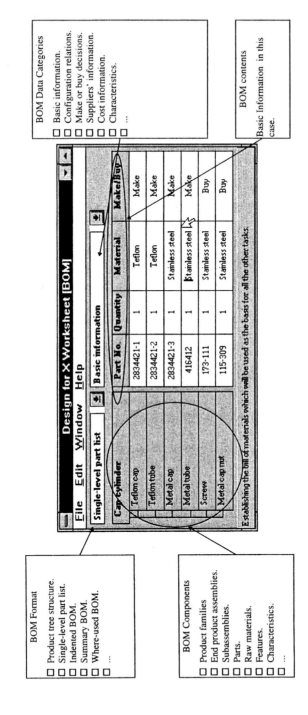

BOM Data Categories

- ☐ Basic information.
- ☐ Configuration relations.
- ☐ Make or buy decisions.
- ☐ Suppliers' information.
- ☐ Cost information.
- ☐ Characteristics.
- ☐ ...

BOM contents

Basic Information in this case.

Design for X Worksheet [BOM]

File   Edit   Window   Help

Single-level part list   Basic information

Cap Cylinder

	Part No.	Quantity	Material	Make/Buy
Teflon cap	2834421-1	1	Teflon	Make
Teflon tube	2834421-2	1	Teflon	Make
Metal cap	2834421-3	1	Stainless steel	Make
Metal tube	416412	1	Stainless steel	Make
Screw	173-111	1	Stainless steel	Buy
Metal cap nut	115-309	1	Stainless steel	Buy

Establishing the bill of materials which will be used as the basis for all the other tasks.

BOM Format

- ☐ Product tree structure.
- ☐ Single-level part list.
- ☐ Indented BOM.
- ☐ Summary BOM.
- ☐ Where-used BOM.
- ☐ ...

BOM Components

- ☐ Product families
- ☐ End product assemblies.
- ☐ Subassemblies.
- ☐ Parts.
- ☐ Raw materials.
- ☐ Features.
- ☐ Characteristics.
- ☐ ...

Figure 5.2 Product bills of materials.

Bills Of Materials

The concept of BOMs has been widely used by parties within manufacturing organisations. Its uses are to define the product and distinguish it from other products; to facilitate the forecasting of optional product features; to permit the master schedule to be stated in terms of the fewer possible end items; to allow easy order entry from customers; to provide the basis for product costing; to facilitate material procurement; to aid manufacturing planning and final assembly scheduling; and to permit efficient file storage and maintenance. A bill of materials is known as an engineering bill when used by the product design function; a planning bill by the process design function; a manufacturing bill by the production operation function.

Figure 5.2 is a typical BOM. It reflects product composition and to some extent product configuration. Configuration is generally modelled implicitly in the proposed DFX shell. Products are assorted into families, each of which consists of a number of similar products. A product is often a complex assembly of a number of low-level components: subassemblies and elementary single-piece parts. In product design and manufacturing planning, single-piece parts are usually further decomposed into features (Wierda, 1991). Usually, features are partial forms like holes, slots, pockets, notches, etc. Features play important role in human reasoning processes and in computer programs that try to do part of this reasoning. Designers will probably think of a design in terms of function-oriented features while process planners reason about manufacturing-oriented features. Depending on the level of reasoning, features are considered globally or in detail. Therefore, it is common to represent features in a hierarchy or taxonomy, just like the hierarchical product trees. Although purchased components are often themselves assemblies, they are usually treated as single-piece parts without further decomposition into elementary parts. Like single-piece parts, however, it is sometimes necessary to decompose purchased component assemblies into features. Raw materials are primitive components in BOM. A series of operations are performed to transform raw materials into finished components through different forms of intermediate components.

Because of the multi-level nature, there are several ways of presenting a product BOM. Single-level exploded bills of materials and related lists may be said to be views of the product structure looking "downward". A single-level bill of materials is simply the complete list of components going into one assembly, regardless of its level in the overall structure. A complete list of all parts for a product, from the completed item down to all purchased parts and raw materials, is simply a complete listing showing all assembly stage. The format is known as "indented explosion".

Information in a basic BOM includes: (1) a part number is a number that uniquely identifies a component; (2) a brief part description is a statement that identifies the part number; and (3) The part quantity per assembly is the quantity of that part number required to produce the assembly. A wide variety of other information can be associated with BOM items, for example, part material is specified by a number or description that uniquely identifies a raw material from which the part is made; "make or buy" decision is an indication if the part is made by the plant or purchased from outside, either partially or totally.

The format and content are largely determined by the intended use. In a computerised environment, these data can be easily retrieved and displayed in appropriate formats to satisfy different users in the organisation. The left-most Combo box in Figure 5.2 is used to specify/change the BOM format. The second Combo box can be used to specify the content of the BOM information content. For example, many DFX tools employ single-level exploded bills of materials - a flat part list.

Key Characteristics

Each component in a BOM, be an end product assembly, intermediate subassembly, a purchased item, an elementary part, or a low-level feature, is characterised by a set of attributes. A key characteristic is an attribute or parameter that significantly influences the aspects of the product: (1) properties such as strength, reliability, appearance, ergonomics, etc.; (2) life-cycle issues such as fabrication, assembly, operation, distribution, installation, service, retirement, etc.; and (3) competitiveness metrics such as quality, cost, delivery, productivity.

In general, key characteristics can be divided into several categories, for example geometry characteristics (shape, size, etc.), physical characteristics (weight, density, etc.), technological characteristics (tolerances, limits and fits, etc.), material properties (hardness, flexibility, etc.), and so on. Different DFX tools may require different sets of characteristics. For example, characteristics considered in Design for Assembly include product structure, component forms and shapes, limits and fits, component orientations, component symmetry, weight and size, component rigidity, etc.

Key characteristics can be associated with the product BOM in two ways. One is to treat key characteristics as a group of BOM content. This approach is particularly useful when BOM elements share similar characteristics. Alternatively, key characteristics can be associated with a product BOM as special tree branches (Liu and Fischer, 1994). This second approach is useful when BOM elements are described by different characteristics.

STEP 3 - MODELLING FOR PROCESS ANALYSIS

Process modelling in the DFX shell is concerned with (1) how to represent business process, (2) how to represent resources, (3) how to represent consumption of life-cycle activities by product elements, and (4) how to represent consumption of resources by activities. Composition, configuration, and characteristics of process activities and resources should be included in representation models. Clearly, process modelling is a key step in developing a DFX tool.

It would result in excessive work to require the DFX user to produce a process model which accomplishes all the above aspects. Therefore, some simplification is necessary in practice. As far as activities and resources are concerned, only one is explicitly represented as entities and the other is embedded as attributes or characteristics.

There are a number of process models: IDEF0 (Air Force, 1981; Harrington, 1984), GIM (Doumeingts, 1984; Chapter 7), and Process Charts (Gilbreth and Gilbreth, 1917; Carson, 1958). Both IDEF0 (*I*ntegrated computer aided manufacturing *Def*inition) and GIM (*G*RAI *I*ntegrated *M*ethodology) have originally been developed for modelling and designing computer integrated manufacturing (CIM) systems. They deliver clear indications of the flow of information between activities or workcentres.

Process charts have been in widespread use for modelling the flow of materials and process improvement, although IDEF0 and GIM can also be used for similar purposes. The DFX shell uses the concept of process charts as a base process model for their simplicity, clarity and ability of representing the flow of materials between activities and/or workcentres. In addition, fewer jargons are used in process charts. A process chart is a rudimentary process skeleton and details can be associated with it flexibly and incrementally. Two types of process chart are relevant: flow process charts (PC-F's), and operation process charts (PC-O's).

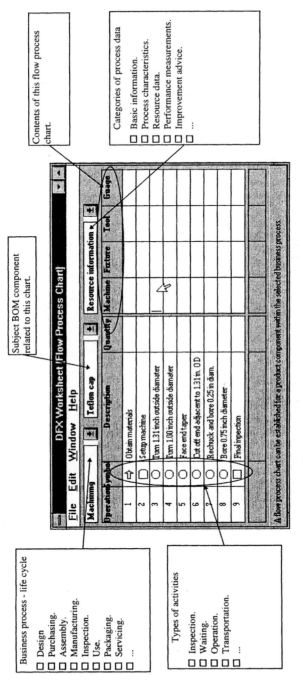

Business process - life cycle

☐ Design
☐ Purchasing.
☐ Assembly.
☐ Manufacturing.
☐ Inspection.
☐ Use.
☐ Packaging.
☐ Servicing.
☐ ...

Subject BOM component related to this chart.

Contents of this flow process chart.

Categories of process data

☐ Basic information.
☐ Process characteristics.
☐ Resource data.
☐ Performance measurements.
☐ Improvement advice.
☐ ...

Types of activities

☐ Inspection.
☐ Waiting.
☐ Operation.
☐ Transportation.
☐ ...

DFX Worksheet [Flow Process Chart]

File Edit Window Help

Machining ↑ Teflon cap ↨ Resource information ◄ ↨

Operation symbol		Description	Quantity	Machine	Fixture	Tool	Gauge
1	⇧	Obtain materials					
2	☐	Setup machine					
3	◯	Turn 1.31 inch outside diameter					
4	◯	Turn 1.00 inch outside diameter					
5	◯	Face end taper					
6	◯	Cut off end adjacent to 1.31 in. OD					
7	◯	Rechuck and bore 0.25 in diam.					
8	◯	Bore 0.75 inch diameter					
9	☐	Final inspection					

A flow process chart can be established for a product component within the selected business process

Figure 5.3 Sample screen of flow process charts.

Flow Process Charts

Basically, a flow process chart (PC-F) is a schematic model specifying the step-by-step sequence of activities during a process or procedure associated with an item. Only those data which are necessary for the DFX analysis are acquired to facilitate data collection and processing. Figure 5.3 shows a typical PC-F.

The content of a PC-F may vary widely. Information in a basic PC-F would cover the following items which describe the process and its activities: the names of activities, unique identification numbers of activities, and brief descriptions about activities. In addition, activity-specific information can be included in a PC-F, for example, feed rate, cutting speed, cutting depth, number of cutting, and length of feed can be associated with machining (metal cutting) activities. A variety of other information can be associated with a PC-F, mainly for outputting the results from the DFX tool.

There are two ways of associating resource information with PC-F's: One is to use workcentres, rather than activities, for charting. Resource data such as machines, jigs/fixtures, tools, gauges are then defined for each workcentre. Resulting process charts are often called route process charts (PC-R's). The other is to treat resources as one of the contents in PC-F's. For example, workcentres can be specified to indicate the locations where activities take place; machines, jigs/fixtures, tools, gauges used for activities can also be included.

Operation Process Charts

An operation process chart (PC-O) is a graphic representation of the points at which materials are introduced into the process, and of the sequence of activities such as inspections and operations. This type of process chart has also been widely used across an organisation for various purposes. One of them is to help planning a new product and coordinating the efforts involved in putting it into production. In the DFX context, PC-O's are particularly relevant and useful for modelling interactions between product elements and process activities in a straightforward fashion. That is, the consumption of activities by products from raw materials to finished goods is explicitly represented in PC-O's. In addition to product/process interactions, an extra merit of PC-O's is that they depict interactions between product BOM elements, i.e. when one is brought together with another.

Figure 5.4 shows a typical PC-O in relation to the product BOM as used in DFX. It can be seen from the figure that the standard format of PC-O's is modified to suit the needs specific to DFX:

- Standard PC-O's are usually based on only two types of activities, inspections and operations. In contrast, a DFX operation process chart would contain any activities which are relevant to the analysis, including those activities for material handling.
- A further extension is that workcentres can be charted instead of activities. In this case, the resulting PC-O's reflect the interactions between BOM components and resources.
- Standard PC-O's usually contain information such as work centres on the right-hand side and activity cycle times on the left-hand side. For clarity in DFX, however, little detail like this should be included except for necessary activity identities. Instead, details should be presented in corresponding flow process charts (PC-F's).
- Standard PC-O's must be rotated to the left by 90 degrees in order to suit the format of the product's bill of materials. By doing so, the interactions between BOM elements and process activities can be clearly indicated.

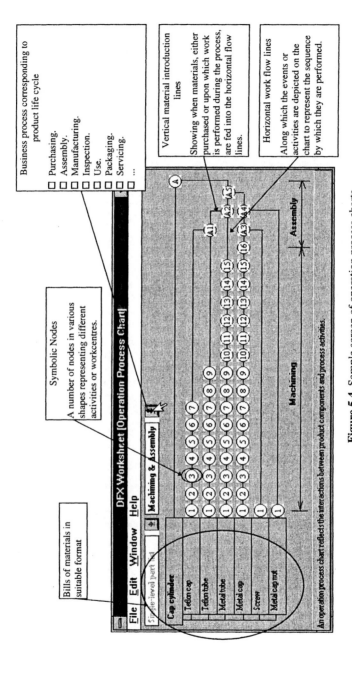

Figure 5.4 Sample screen of operation process charts.

STEP 4 - SELECTING PERFORMANCE MEASURES

Performance measures used by successful DFX tools differ widely from one another, even between those with the same x part in X ($X = x + bility$). For example, both Boothroyd-Dewhurst DFA and Hitachi AEM are tools for evaluating product assemblability. However, their definitions of assemblability, i.e. performance measures, are quite different.

This step - "Selecting appropriate performance measures", is concerned with the specific definition of the "-*bility*" part in X ($X = x + bility$). The following decisions need to be made:

☐   What affects the selection of performance measures and what is affected?
☐   What performance indicators should be used and how many are necessary?
☐   In what units should selected performance indicators be measured?
☐   How are the low-level performance measurements propagated to high-level measurements in relation to product components and/or process activities?
☐   How are performance standards established to assess if a design decision is good or bad.

There are number of factors that must be considered in selecting appropriate performance measures:

☐   Data availability.
☐   Desired functionality.
☐   Life-cycle focus of X.
☐   Process activities.
☐   DFX manuals.
☐   DFX worksheets.

The availability of information dominates the choice of some performance measures. Once selected, performance measures impose restrictions on the above factors as well. The choice of performance measures dominates the way that the DFX works and the collection of the data required in compiling DFX manuals.

What Performance Indicators and How Many to Use?

Maskell (1991) articulates the challenge of finding relevant performance measures for corporate businesses. Japanese companies tend to focus on performance measures at the level of workplace (Sugiyama, 1989). Olesen (1992) investigates performance metrics especially in the DFX context - the Universal Virtues. Close similarities can be observed between the elements of performance measurement systems mentioned by these people. They can be summarized by the following categories:

1.   Delivery performance and customer service
2.   Process time
3.   Production flexibility
4.   Quality performance
5.   Financial (cost) performance and risk
6.   Social issues such as environment, safety, etc.

Cost-related financial indicators seem to be universal. Almost all the other aspects can be measured using such indicators. For example, process cycle times and product quality can and are ultimately converted into financial measures. However, one recent argument is that non-financial performance measures - cost drivers, are often more relevant to particular decision-making activities although cost-based performance measures as provided by management accounting systems are of general use. This is particularly relevant in selecting proper performance measures for a DFX tool. Many successful DFX tools avoid directly using financially-based performance measures to define "$X = x + bility$". Financial appraisals or audits may be carried out before and/or after the DFX analysis. They would probably be part of project identification and effectiveness measurement (Steps 1 and 6 of the macro "Business Process Reengineering" procedure - see Chapter 6).

DFX tools use multiple performance measures. This may help viewing product development from different viewpoints. However, an examination of many successful DFX tools reveals that only 2-5 performance measures (such as activity time and special tool/equipment requirement) are used, plus a few overall measures (such as part count and number of processes).

How To Measure Performance? - Units of Measurement

Once appropriate performance indicators are selected, the next task is to decide upon the units by which each indicator is measured - the unit of measurement. There are wide variations among successful Design for Assembly systems in terms of the units of performance measurement. In general, they can be grouped into the following categories:

- *Absolute measurements.* Performance indicators can be measured by absolute units. For example, distance of movement can be measured in metres or feet; activity times in hours, minutes or seconds; costs in Stirling pounds, US dollars, or Japanese yens.
- *Relative measurements.* Performance indicators can be measured without any dimension. For example, dimensionless penalty scores or ratings can be used. They commonly use arbitrary 0-10 or 0-100 scales. These scales are often subjectively established and frequently contain personal opinions about what is good and what is bad, and the degree of each.

How To Aggregate Performance Measurements?

Both product and process structures are broken down into basic elements against which performances are measured. Once individual performance measurements are obtained, the next task is to aggregate them based on some algorithms to obtain overall performance measurements. There are two types of aggregation in the DFX shell:

- ☐ Horizontal aggregations of individual performance measurements of different activities associated with a product element. This type of aggregation is only possible when different types of activities are measured by the same performance indicators and units.
- ☐ Vertical aggregations of individual performance measurements of different product BOM elements consuming the same type of activity.

How To Establish Performance Benchmarks?

Once performance measurements are obtained, it is necessary to find out if a design is good or bad. This is done by comparing the performance measurements against the standards. It is important to note two distinctive types of performance standards. One is the aggregate standard for the aggregate performance of the product and process as a whole. The other is the individual standard for the individual performance of the product and process elements. There are a number of ways of establishing such a standard for a performance indicator:

☐ Best practice in the class.
☐ Competitor's performance.
☐ Historical internal records.

STEP 5 - COMPILING DFX MANUALS

DFX is data intensive. Experience indicates that collecting appropriate data is a bottleneck in carrying out a DFX analysis. To overcome the difficulty in data collection, successful DFX tools are equipped with DFX manuals. The contents in the manuals of a DFX tool determine the scope and effectiveness of its functionality. The formats affect the speed and efficiency of its use. Well-structured DFX manuals are easy to understand and follow.

It is a continuous effort to compile a manual for a DFX tool, whether it is paper-based or computer aided. Following questions should be addressed during the compilation:

☐ What data should be included in the manual?
☐ Where to collect the data?
☐ How should the data be represented?
☐ How should the data be used?

Where To Collect DFX Data?

Where to find the data and how to organise them into the desired format are the bottleneck in compiling a DFX manual. In some cases, data exist but need to be collected and processed before use. In other cases, data do not exist and must be generated and recorded. Following are just some means of collecting and processing data:

- *Textbooks.* Textbooks are usually the crystallisation of knowledge evolved over a long period of time. Data and knowledge are usually available in very general forms in textbooks (Matousek, 1957; Bralla, 1986).
- *Professional handbooks.* Many professional bodies produce handbooks containing invaluable domain-specific data in various forms (Ostwald, 1985). Some of them may be useful in developing DFX tools with some modifications. For example, MTM and MOST (Zandin, 1990) provides a basis for the Boothroyd-Dewhurst DFA and DFS manuals.
- *Field data.* Large volume of data can be collected from fields. However, these data should usually be carefully processed before they can be used. This is especially true when different organisations are involved.
- *Experiments.* Although usually expensive, data can be "manufactured" by experiments. Unlike field data, experimental data are relatively easy to process and ready to use. For example, Professor Dewhurst is leading such research to obtain time measurements for disassembly tasks (see Chapter 14).

DFX Checklists or Lookup Tables?

There are two general ways of collating DFX knowledge: checklists and look-up tables. In simplistic terms, a checklist is a grouped collection of rules and guidelines. Each checklist has a unique index number and a box to check off to indicate compliance or violation. The number of violations are noted. Well-known DFX tools such as Boothroyd-Dewhurst DFA, Lucas DFA and Hitachi AEM are said to adopt the second approach of using DFX lookup tables. One feature of lookup tables is the systematic grouping of logically-related DFX knowledge. Another feature of lookup tables is the introduction of qualifiers or quantification.

Checklists and lookup tables have been widely promoted as distinctive competing alternatives for compiling DFX manuals. However, a close examination into various successful DFX tools reveals that they are consistent. It can be illustrated that they are complementary and equivalent, if arranged appropriately. Patterning and quantification can be easily introduced in DFX guidelines. In fact, the compilation of most DFX lookup tables has been based on a collection of DFX rules. Most drawbacks such as difficult to use and qualitative ambiguity of the checklist approach disappear with the introduction of patterning and quantification.

The DFX shell provides a unified approach to compiling DFX manuals. Figure 5.5 shows a sample screens of a lookup table and a guideline. Following are important components of DFX lookup tables:

☐ *Table ID numbers.* Each DFX table has a unique identification number.
☐ *Table pattern.* Although a table can be uniquely identified by its ID number, patterns are often used to match the subject problem under consideration.
☐ *Row and Column ID numbers.* Table entries can be uniquely identified according to their row and column numbers.
☐ *Row and Column patterns.* Row and column ID numbers are established through their patterns. It is convenient, but not necessary, to use row patterns to match product characteristics and column patterns to match activity/resource characteristics.
☐ *Table entries.* The contents of DFX tables determine the scope and effectiveness of the DFX functionality. When the lookup table is set for measuring performance, the entries are numeric performance values or algebraic equations for deriving performance measurements. When the lookup table is set for troubleshooting, the entries are descriptions about potential problems. When the lookup table is set for advising on improvements, entries are description of possible actions.

In the DFX shell, guidelines and lookup tables are simply considered two different ways of displaying the same DFX knowledge retrieved from the data/knowledge base. A DFX lookup table with m rows and n columns can be transformed into $m \times n$ DFX guidelines. A DFX guideline has the following components:

☐ *Rule Identification.*
   Rule Identification = Table ID + Row ID + Column ID
☐ *Rule Pattern.*
   Rule Pattern = Table Pattern + Row Pattern + Column Patter
☐ *Rule Body.*
   Rule Body = Table Cell Entry

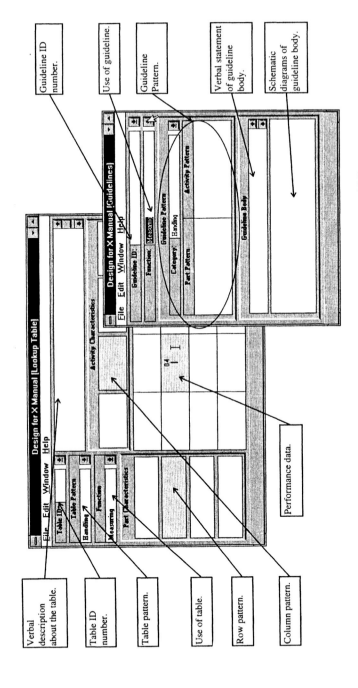

Figure 5.5 Sample screen of DFX manuals: lookup tables and guidelines.

Patterning

Entries in a DFX lookup table can be uniquely accessed using the table identification numbers, row identification numbers and column identification numbers. Similarly, a DFX guideline can be uniquely applied according to the guideline identification numbers. However, these identification numbers are not usually given ready for use. Instead, they are obtained through extensive patterning and pattern matching.

Patterning is concerned with the representation of conditions of DFX rules or look-up tables, and that of characteristics of products, processes, and resources. Pattern matching (See Chapter 6) is the process of comparing and contrasting the product, process and resource patterns against those DFX rules or look-up tables so that applicable DFX knowledge in the manual can be applied. There are three general approaches to patterning:

- *Geometric reasoning.* Patterning in many DFX tools are based on what can be classified as geometric reasoning. That is, geometric characteristics of products, processes and resources are described schematically.
- *GT coding.* In the field of mechanical engineering, sophisticated classification and coding systems exist with the development of Group Technology over the last several decades. Many countries have standard GT codes for some product sectors.
- *Parametric.* Key characteristics of products, processes, and resources are described symbolically by a set of (attribute, value).

All the patterning methods can be used for developing both paper-based and computer-based DFX manuals, in the form of DFX rules or look-up tables. The later two are more appropriate for automatic patterning and pattern matching. In practice, they are often used in combination, resulting in hybrid approaches to patterning and pattern matching.

STEP 6 - COMPILING DFX WORKBOOKS

Successful DFX tools are easy to use because they follow what can be called a workbook approach. Compiling DFX workbooks involves putting various DFX easy-to-use constructs devised in previous steps together in a way that the natural transition of attention from one area to another during process of DFX analysis is reflected.

Workbooks are not only documenting mechanisms but also guiding roadmaps. A typical workbook usually consists of comprehensive worksheets and systematic procedure(s). A systematic procedure is used so that the user can perform the DFX analysis in a logical order and without missing important aspects. The DFX procedure should reflect the logical flow of information in the DFX worksheet. This will be discussed in detail in the next chapter.

A comprehensive worksheet reflects and records the proceedings of the logical flow of the DFX activities. A DFX worksheet should provide main areas for inputs and outputs. However, it can be quite difficult to distinguish between inputs and outputs because DFX analysis is progressive. Everything asked by the DFX tool is used as input and also treated as output. For example, product models in the forms of bills of materials and key characteristics, and process models in the forms of process charts and key characteristics are both inputs to and outputs from the DFX tool. They have been discussed in Step 2 and Step 3 respectively.

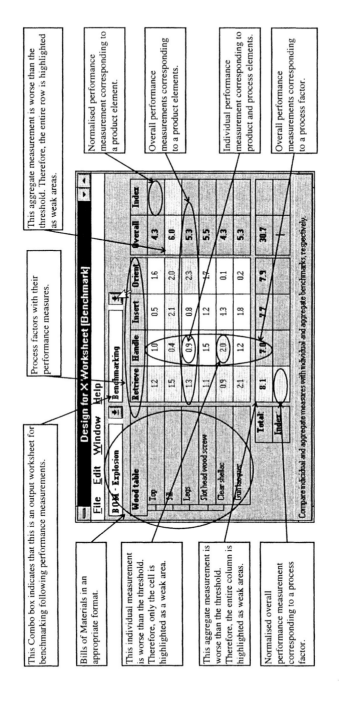

This Combo box indicates that this is an output worksheet for benchmarking following performance measurements.

Process factors with their performance measures.

This aggregate measurement is worse than the threshold. Therefore, the entire row is highlighted as weak areas.

Bills of Materials in an appropriate format.

This individual measurement is worse than the threshold. Therefore, only the cell is highlighted as a weak area.

This aggregate measurement is worse than the threshold. Therefore, the entire column is highlighted as weak areas.

Normalised overall performance measurement corresponding to a process factor.

Normalised performance measurement corresponding to a product element.

Overall performance measurements corresponding to a product elements.

Individual performance measurement corresponding to product and process elements.

Overall performance measurements corresponding to a process factor.

Design for X Worksheet [Benchmark]

File   Edit   Window   Help

BOM - Explosion   Benchmarking

Wood table	Retrieve	Handle	Insert	Orient	Overall	Index
Top	1.2	1.0	0.5	1.6	4.3	
Silk	1.5	0.4	2.1	2.0	6.0	
Legs	1.3	0.9	0.8	2.3	5.3	
Slot head wood screw	1.1	1.5	1.2	1.7	5.5	
Clear shellac	0.9	2.0	1.3	0.1	4.3	
Turpentine	2.1	1.2	1.8	0.2	5.3	
Total	8.1	7.0	7.7	7.9	30.7	
Index						

Compare individual and aggregate measures with individual and aggregate benchmarks, respectively.

Figure 5.6 Sample screen of DFX worksheet with areas highlighted after benchmarking.

The DFX shell provides a general output form. It is essentially a matrix. Product design decisions are represented to its left in the BOM form. Major process design decisions are represented to the top in the form of generalized (types of) activities or process aspects / factors. Performance indicators and units of measurements are also represented in relation to the corresponding process activities. A sample output screen is shown in Figure 5.6.

This output screen is capable of displaying a variety of outputs, ranging from performance measurements through causal diagnosis and effect predications to improvement recommendations. This can be easily done through the second Combo box. However, the labelling columns and rows remain the same. For example, when the second Combo box is set for presenting the performance measurements, entries to the relevant cells are the output data which measure the interacting (consumption) effects between product design decisions and process design decisions. After benchmarking, those cells whose values are worse than the individual thresholds and those columns (rows) whose aggregate values are worse than the corresponding aggregate thresholds are considered as weak areas and are highlighted. These shaded areas attract more attention and improvement actions should be recommended subsequently.

Finally, it is worth separating a DFX workbook from its DFX manual although both can be built in the same framework. The DFX manual usually contain proprietary knowledge which is not usually made available in the public domain. On the other hand, DFX workbooks are just means of identifying most appropriate knowledge and exploiting it and therefore should be made available in the public domain for the purpose of marketing and scrutiny. In a computer aided DFX environment, however, this separation is natural. The DFX workbook is simply the front-end user interface while the DFX manual is kept in the back-end protected database.

STEP 7 - VERIFICATION

Like any manufactured products which DFX tools are set to improve, a "right first time" DFX tool should be always aimed at. However, verification and testing is not a step which can be skipped. Continuous improvements should be made based on the experience from field and simulated tests. In fact, successful DFX tools have evolved considerably over the last decade. The objective of DFX verification is to identify the strengths and limitations of the DFX tool under development, to recognise opportunities and new requirements for further improvements and developments. The following questions should be addressed at this step:

- ☐ What should be verified and tested?
- ☐ What are the criteria for verification?
- ☐ How to conduct verification?
- ☐ How to improve the tool?

The entire DFX package, including the development specification, product and process models, performance measures, worksheets and procedures, and the DFX manual itself, should be subjected to tests. There are many factors that should be addressed during verification. A rule of thumb is to examine the DFX tool under verification according to the development specification established in Step 1 - Requirement Analysis. Some of the questions that must be addressed during verification are listed as follows:

- ☐ Does the DFX tool function as intended?
- ☐ Does it provide focus of attention?
- ☐ Is it general enough to cover the specified product/process range?
- ☐ Is it easy to find the data required by the DFX tool?
- ☐ Is the output adequately accurate and useful?
- ☐ Can the practitioners understand it?
- ☐ What is the level of time and effort for average practitioners?
- ☐ Does it serve as the media of communication and catalyst of co-ordination?

Tests should be carried out using a sufficient number and wide spectrum of test cases with full technical and managerial supports. The following are just some of the common means of verification.

- *Specialist/expert consultation.* Advice from experienced experts and specialists, developers or practitioners, is always valuable during the entire process of DFX development, if it is available. Every effort should be made to take advantages of this. It is an important part of verification. However, such advice cannot replace the whole verification activity.
- *Simulation.* Almost all DFX development projects exploit this technique to clarify what the tool is to achieve and illustrate how it is to achieve - the working principle behind it. Simulated cases may be invented solely for illustrative purposes. Just like the expert advice, this technique is more useful during the development process. It is not sufficient for final tests.
- *Benchmarking.* Early successful DFX tools such as Boothroyd-Dewhurst DFA, Hitachi AEM, and Lucas DFA have been widely used by developers to benchmark their own systems even for DFX tools outside the domain of assembly.
- *Retrospective case studies.* Past projects are selected and the DFX tool is applied to the projects as if they would not have been carried out. It is extremely useful to see if the DFX tool can highlight areas which have been encountered, and point out potential directions for improvement which have been performed. Suggestions from the DFX test analysis may or may not be considered. Chapter 2 presents such a case study.
- *Field improvement case studies.* The DFX tool is applied to actual on-going product improvement projects where actual (designs of) products and associated processes already exist. Verifiers should pay attention to possible contributions from the DFX verification analysis to the overall projects. Vast majority of existing DFX tools have been developed for product and process improvement.
- *Clean-sheet field case studies.* It would be desirable that DFX verification takes place in an environment where a new product is under development. Here the focus is on if the DFX tool helps generating better design decisions.

Outcomes from verification should be scrutinised when modifying the DFX tool. Care must be taken to maintain the right balance between functionality and operability. Such a balance can only be achieved through the following balances:

- *Balance between pragmatism and formality.* On the one hand, the use of a formal method does bring a high degree of clarity in revealing the interrelationships between the product design decisions and life-cycle activities. On the other hand, the formality usually requires training. In the DFX context, excessive effort of following formality strictly will detract the engineers away from the focus which the DFX is supposed to provide. Therefore, it is preferable to exploit methods which (1) may be already in use within an organisation; (2) is simple and easy to understand and use by the project team and; (3) is commonly familiar or easily becomes familiar to practitioners. Concepts such as Bills of Materials and process charts have been well taught in colleges and widely practised by almost all the personnel within organisations. After all, these common-sense concepts are themselves straightforward to understand, usually self-explanatory.

- *Balance between accuracy and data requirement.* For a DFX tool to be practically useful, its analytical results must be sufficiently accurate. There are several approaches to improve the accuracy. For example, it is generally believed that a sophisticated mathematical model considering numerous variables is likely to produce more accurate results in theory. In practice, this leads to increased complexity, and added difficulties for practitioners to understand and use it. Even if the complexity is avoided through an algorithm, practitioners tend to be sceptical towards the outcomes from the "black box" approach. In addition, sophisticated DFX models usually demand more data which may be extremely expensive to collect. It is important to note that DFX tools strike the balance between no quantification at all and complete quantification.
- *Balance between focus and flexibility.* DFX is both an analytical tool for evaluating design decisions and their relationships and a team tool for stimulating cooperation and communication. However, the detailed DFX shell should not be followed rigidly. Instead, the framework must be tailored and adapted to suit particular conditions.

SUMMARY

This chapter has further developed the basic DFX pattern outlined in the introductory chapter into a working DFX platform. A number of formal but pragmatic constructs have been used. Bills of materials are used to describe and analyse the overall product structure and product characteristics. Flow process charts are used to describe and analyse the overall process structure and process characteristics in relation to individual product elements. Standard operation process charts are modified to describe and analyse the overall process structure in relation to the product structure. Appropriate performance measures are used to evaluate the interactions between the elements of products, processes, and resources.

Seven steps are involved in using the DFX shell as a generic framework for developing a wide variety of DFX tools rapidly and consistently. Resulting DFX tools share a common understanding essential to integration and tradeoff analysis. An important feature is the overall balance between functionality and operability. This is achieved through the balances (1) between pragmatism and formality; (2) between focus and flexibility; and (3) between accuracy and data requirement.

The DFX shell not only facilitates the development of new DFX tools but also allows a common framework for their implementation. The next chapter describes a dynamic approach to implementing concurrent engineering by combining the focused application of DFX and extensive Business Process Reengineering (BPR).

REFERENCES

Air Force (1981) IDEF (Integrated computer aided manufacturing DEFinition) Architecture, Part II, Vol. IV - Function Modelling Manual (IDEF0), Air Force Materials Laboratory, Wright-Patterson AFB, Ohio 45433, AFWAL-TR-81-4023.

Boothroyd, G., Dewhurst, P. (1990) *Product Design for Assembly*, Boothroyd Dewhurst, Inc., Wakefield, RI, USA, (First Edition, 1983).

Boothroyd, G., Dewhurst, P., Knight, W. (1994) *Product design for manufacture and assembly*, Marcel Dekker Inc.

Bralla, J.G. (1986) *Handbook of Product Design for Manufacturing, A practical guide to low-cost production*, McGraw-Hill.

Carson, G.B. (1958) *Production Handbook*, Section 11: Process Charts, New York: The Ronald Press Company

Doumeingts, G. (1984) *Méthode GRAI, méthode de conception des systèmes en productique*, State Thesis, University Bordeaux 1, France.

Gilbreth, F.B., Gilbreth, L.E. (1917) *Applied motion study, A collection of papers on efficient method to industrial preparedness*, New York: The Macmillan Co., Inc.

Greene, J.H. (1987) *Production and Inventory Control Handbook*, Chapter 10: Product and Process Information, APICS (American Production and Inventory Control Society), McGraw-Hill Book Company, New York.

Harrington Jr., J. (1984) *Understanding the manufacturing process - Key to successful CAD/CAM implementation*, Marcel Dekker, Inc., New York.

Liu, T.H., Fischer, G.W. (1994) Assembly evaluation method for PDES/STEP-based mechanical systems, *Journal of Design and Manufacture*, **4**, 1-19.

Maskell, B.H. (1992) Performance Measurement for World-Class Manufacturing: A model for American companies, Productivity Press, Cambridge, Massachusetts, USA.

Matousek, R. (1957) *Engineering Design - A Systematic Approach*, The German edition by Springer-Verlag, Berlin; The English edition translated by A.H. Burton and Edited by D.C. Johnson, Published by Lackie & Son Ltd, London, UK.

Miles, B.L. (1989) Design for assembly - a key element within design for manufacture, *Proceedings of IMechE, Part D: Journal of Automobile Engineering*, **203**, 29-38.

Miyakawa, S, Ohashi, T., Iwata, M. (1990) The Hitachi New Assemblability Evaluation Method, *Transactions of the North American Manufacturing Research*, Institution (NAMRI) of the SME, the NAMR Conference XVIII, May 23-25, 1990, Pennsylvania State University, Dearborn, USA.

Olesen, J. (1992) *Concurrent development in manufacturing - based on dispositional mechanisms*, PhD Thesis, Technical University of Denmark.

Ostwald, P.F. (1985) *American Mechanist Cost Estimator, the comprehensive guide to manufacturing cost-estimating data and procedures*, McGraw-Hill.

Shimada, J., Miyakawa, S., Ohashi, T. (1992) Design for manufacture, tools and methods: - the Assemblability Evaluation Method (AEM), *FISITA '92 Congress*, London, 7-11 June, Paper C389/460, FISITA, SAE No. 925142, IMechE, 53-59.

Stoll, H.W. (1988) Design for manufacture, *Manufacturing Engineering*, January.

Sugiyama, T. (1989) *The Improvement Book - Creating the problem-free workplace*, Productivity Press, Cambridge, Massachusetts, USA.

Swift, K.G. (1981) *Design for Assembly Handbook*, Salford University Industrial Centre Ltd., UK.

Taylor, G.D., English, J.R., Graves, R.J. (1994) Designing new products: Compatibility with existing production facilities and anticipated product mix, *Integrated Manufacturing Systems*, **5** (5/6), 13-21.

Wierda, L.S. (1991) Linking design, process planning and cost information by feature-based modelling, *Journal of Engineering Design*, 2 (1), 3-19.

Zandin, K., (1990) *MOST*® *Work Measurement Systems*, Second Edition, Marcel Dekker, Inc., New York.

6

IMPLEMENTING DESIGN FOR X TOOLS

George Q. Huang

This chapter discusses a generic framework for implementing Concurrent Engineering (CE) by combining the focused application of Design for X (DFX) tools with extensive use of Business Process Reengineering (BPR). This framework is referred to the DFX/BPR shell. It provides a dynamic approach to transforming product development from a problem-prone sequential engineering environment to a problem-free concurrent engineering environment.

The DFX/BPR shell includes two 7-steps procedures. One is the micro DFX procedure for systematically applying a specific DFX tool. The other is the macro BPR procedure for tackling wider organisational issues. The micro DFX procedure is only one single (second) step in the macro BPR procedure. This embedding is necessary and advantageous. First, the micro DFX procedure provides the focus and vision necessary for the analyst or the team to build up momentum through tangible benefits such as improved quality, reduced cost, accelerated development, enhanced flexibility, and increased productivity. Second, the macro BPR procedure provides the mechanism for implementing radical changes and sustaining benefits - far reaching impacts on the efficiency and the way in which processes are operated. Third, the wide diversity of various DFX tools is compressed into a single step so that the macro procedure looks generic. Next, multiple DFX tools are applied in sequence from the micro viewpoint but simultaneously from the macro viewpoint. Finally, the DFX step in the macro BPR procedure prevents jumps to premature solutions without thorough analysis.

This chapter is prepared for practitioners who are involved in implementing Concurrent Engineering (CE) in general and Design for X (DFX) tools and Business Process Reengineering (BPR) in particular. DFX developers and those who generally want to know more about the subject will also find it highly relevant.

Table 6.1 Typical problems with "over the wall" sequential engineering

1.	Sequential activities resulting in protracted cycle times.
2.	Communication is inadequate, inefficient and/or ineffective.
3.	Focus on intermediate milestones.
4.	Extensive queues without priority control.
5.	Too many check points and wait too long to be checked.
6.	Priority given to crisis management.
7.	Scarce resources are wasted in fire-fighting, progress chasing, making changes, etc.
8.	"Inertia" is too high to be responsive.
9.	Design and production are insulated from customers and suppliers by other departments such as marketing and purchasing.
10.	Products are difficult to make, to service, to use, or to sell. Weak development of robust functionality. Weak design for producibility.
11.	Unnecessary technical complexity exists in products, processes and systems.
12.	Technology push leads to many great concepts but fails to meet important customer needs.
13.	Weak commitment to previous decisions, new and different, but not better.
14.	Specification is considered in isolation. Single feature optimization leads to sub-optimal solutions.
15.	Divergent interpretations of the specification.
16.	Lost and obsolescent information. Multiple, unsynchronized redundant databases are maintained by different functions. Lack of common modes of data management.
17.	Poorly-structured product development leads to poor coordination.
18.	Problems are discovered too late, resulting in panics and leading to "quick fix" solutions and compromises, and long and costly rework loops.
19.	Isolated automation of manufacturing processes such as CAD, CAM, CAPP, CAPM.
20.	Islands of expertise exist and human skills are narrow.
21.	Isolated management processes such as engineering change control and project management.
22.	Hierarchical structures lead to a situation where managers think and make decisions, and contributors work and enact decisions.
23.	Functional divisions work in a "black box" fashion, blocking the channels of communication.

6.1 "DESIGN FOR X"-DRIVEN CONCURRENT ENGINEERING

Shortcomings of sequential engineering and advantages of concurrent engineering in product development have become better understood. However, the transformation from a problem-prone sequential engineering environment to a problem-free concurrent engineering environment remains ever more challenging.

6.1.1 Transforming from Sequential Engineering to Concurrent Engineering

Product development is the heartland of manufacturing industries and battlefield of global competition. The product development process has not been the subject of much study until recently. There is much opportunity to improve it. Problems with the traditional product development process, or often referred to as "over-the-wall" sequential engineering are evident. Table 6.1 is compiled from a number of sources where problems typically plaguing product development are discussed (Clark and Fujimoto, 1991; Clausing, 1993):

An ideal product development environment free of these problems is Concurrent Engineering (CE). With CE, multi-disciplinary personnel works together to consider various competing issues in designing products, processes, and systems. The essence of concurrent engineering is both simple and subtle. Major characteristics of concurrent engineering compiled from a number of sources (Clark and Fujimoto, 1991; Clausing, 1993, Miles, 1989;

Youssef, 1994; Gatenby *et al.*, 1994) are listed in Table 6.2. Some of the factors in Table 6.2 are misinterpreted as pre-requisites for successful CE implementation, rather than the objectives that CE helps to achieve. If they are pre-requisites, then very few organizations can satisfy them just to start a CE project. For this simple reason, these factors are not considered as pre-requisites but objectives. If there is any pre-requisite for CE, it is the "good will" to improve.

6.1.2 Combining Design for X and Business Process Re-Engineering

The first point here is that DFX [DFM] is sometimes treated synonymous with CE. Youssef (1994) provides a good review of various definitions on. Unfortunately, it is difficult to appreciate much difference from their definitions and objectives without examining their working mechanisms. Gatenby *et al.* (1994) stress the difference between CE and DFX. They also tend to treat CE synonymous as BPR. There can be endless academic debate on these matters. One thing is clear that they share enormous similarity in terms of their objectives, implementation issues, difficulties, and even working mechanisms. In the discussion that follows, the term CE is used to describe an ideal environment especially for product development. BPR and DFX are two of the many ways of implementing CE. The focused application of DFX tools leads to the rationalization of *decisions* in designing products, processes and resources. The extensive use of BPR leads to the rationalization of *decision-making activities* in designing products, processes and resources. Better organized groups and streamlined activities are more likely to produce better decisions and prevent problems. DFX has been discussed in Introduction and Chapter 5. BPR is briefly introduced.

Table 6.2 Characteristics of Concurrent Engineering

1.	Use a full-time, co-located, core team with representation from different functions such as product, manufacturing, industrial, purchasing, suppliers, marketing, customers.
2.	Achieve effective and efficient teamworking based on individual skills.
3.	Develop trust among team-mates, strive for team consensus.
4.	Train personnel at all levels.
5.	Treat product development as a process subject to improvement.
6.	Structure product development, maximize transparency and concurrence.
7.	Define design process with marked milestone for review, hand-over, sign-off, project control and monitoring.
8.	Use pilot projects to gain insights.
9.	Obtain the support of the total organisation (management and employees).
10.	Obtain adequate resource.
11.	Have an open mind and the willingness to accommodate several different viewpoints.
12.	Focus on important aspects, solving real problems and removing their root causes.
13.	Document experience, and publicise the results, both benefits and lessons learnt.
14.	Start all tasks as early as possible.
15.	Utilize all relevant information as early as possible.
16.	Empower individuals and teams to participate in defining the objectives of their work.
17.	Achieve operational understanding for all relevant information.
18.	Adhere to decisions and utilize all previous work.
19.	Make decisions in a single tradeoff space.
20.	Make lasting decisions, overcoming a natural tendency to be quick and novel.
21.	Develop trust among team-mates.
22.	Strive for team consensus.
23.	Use a visible concurrent process.
24.	Follow up continuously to resolve open issues.

Table 6.3 Comparison between DFX, BPR, and DFX/BPR

Business Process Reengineering	Design For X	Business Process Reengineering/Design For X
Scope is wide and deep.	Scope is focused and deep.	Increasingly deeper and wider.
Solve big crisis problems resulted from merger, take-over, etc.	Solve problems at product development workplace.	Solve increasingly complex problems to prevent crises.
Potential benefits are substantial.	Potential benefits are dramatic.	Potential benefits are substantial.
Uncertainty and risk are high.	Uncertainty and risk are low.	Uncertainty and risk are manageable.
Confusion about what to do.	Clear vision for change.	Vision for change.
Project is difficult to control.	Project is easy to control.	Project is manageable.
Restructuring of decision making activities.	Focusing on rationalizing decisions of products, processes, and resources.	Better organization to make better decisions.
Lack of attention to details.	Details are attended to.	Details are attended to at reasonable levels.
Top down.	Bottom up.	Dynamic.
Organizational focus.	Content focus	Dynamic.
Lead time is long.	Lead time is short.	Lead time is reasonable.
Cost is high.	Cost is low.	Cost is reasonable.
Require intensive and extensive training.	Require intensive training.	Require intensive and less extensive training.
Concentrated on processes.	Concentrated on interactions between products, processes and resources.	Integrated consideration.

BPR is fundamental rethinking and radical restructuring of business processes to achieve dramatic improvements in critical, contemporary measures of performance, such as cost, quality, service and speed (Hammer and Champy, 1993; Johnsson, 1993). Main objectives are to make processes effective in procuring the desired results, make processes efficient in minimizing the resources used, and make processes adaptable in being able to adapt to changing customer business needs. Two extremes of BPR are radical and incremental. To the one end, radical BPR is to solve compelling problems or crises for survival. To the other end, incremental BPR is cautiously carried out through a series of BPR projects each of which follows a step-by-step procedure. In practice, companies operate with policies and strategies between the extremes.

The central theme of this chapter is to promote a dynamic approach towards concurrent engineering by combining DFX and BPR so that they complement with each other. Relative advantages and disadvantages are listed in Table 6.3. The main role of DFX in BPR is to provide the drive, focus, vision and concurrence necessary for BPR. On the other hand, the main role of BPR in DFX is to institutionalize good practice and make improvement permanent and continuous. In companies operating comprehensive BPR programmes, BPR seems to dwarf DFX. But it should be remembered that they have been enthusiastic DFX users with considerable DFX experience which helps BPR then and now.

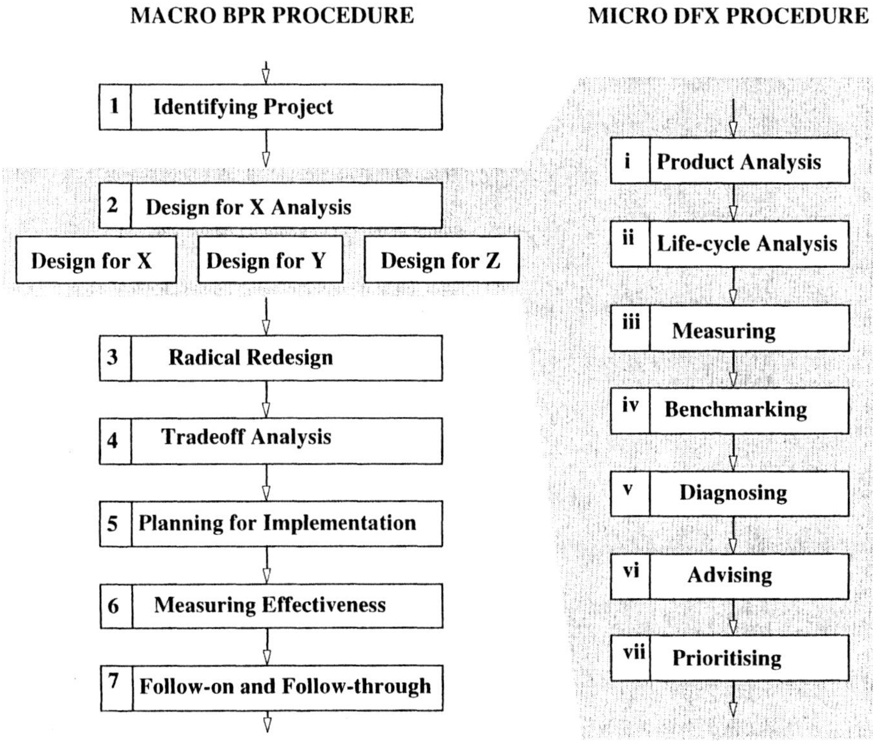

MACRO BPR PROCEDURE   **MICRO DFX PROCEDURE**

| 1 | **Identifying Project** |

| i | **Product Analysis** |

| 2 | **Design for X Analysis** |

| **Design for X** | **Design for Y** | **Design for Z** |

| ii | **Life-cycle Analysis** |

| iii | **Measuring** |

| 3 | **Radical Redesign** |

| iv | **Benchmarking** |

| 4 | **Tradeoff Analysis** |

| v | **Diagnosing** |

| 5 | **Planning for Implementation** |

| vi | **Advising** |

| 6 | **Measuring Effectiveness** |

| vii | **Prioritising** |

| 7 | **Follow-on and Follow-through** |

Figure 6.1 Macro BPR procedure and Micro DFX procedure.

The combination of BPR and DFX results in what is referred to as the DFX/BPR shell. It is a generic framework for transforming product development from sequential engineering to concurrent engineering. The DFX/BPR shell includes two seven-steps procedures: the micro DFX procedure and the macro BPR procedure, as shown in Figure 6.1. The micro DFX procedure, including seven steps, is mainly designed to apply a specific DFX tool systematically. On the other hand, the macro BPR procedure, also containing seven steps, deals with broader issues related to business processes.

6.2 MICRO DFX PROCEDURE

DFX tools are characterised by a systematic procedure which is easy to follow, a comprehensive worksheet which is logical to display data, and a proprietary data and knowledge base which is straightforward to look up. This section presents a 7-steps procedure for applying DFX tools developed using the DFX shell discussed in Chapter 5. A specific DFX tool does not necessarily include every step of the micro procedure or follow the order of steps presented. There are, as should be, variations in practice.

Step i - Product Analysis

Product analysis is the first step in the micro DFX procedure. The major object of this step is to collect and clarify information related to the subject product(s). A DFX tool usually specifies what product data it requires and how they are processed and reported. In the DFX shell, bills of materials (BOMs) are the common format of displaying product structure information. Other types of product data can be easily associated with the corresponding product BOM. Clarified product data are both inputs required by the subsequent steps and outputs from the DFX tool.

The process of product analysis is mainly that of collecting product data. More specific product data may be collected in later steps when they are needed. The following tasks may be involved in this step of Product Analysis:

1. Select a subject product with typical features in the target product family. It is helpful to obtain a product hardware to examine and understand its features.
2. Collect documents relevant to the product design, including assembly drawings, part drawings, service manuals, etc.
3. Identify all items in the product at appropriate levels of detail and make notes of item information including part number, part name, etc. Items should be identified by their names and/or numbers without omissions or duplications.
4. Establish the inter-relationships between system items at different levels.
5. Establish the inter-relationships between system items at specific levels.
6. Identify key characteristics for each system item from perspectives such as functional, physical, behavioural, etc. This can be deferred at later stages when the data are needed.
7. A number of straightforward analyses can be carried out at this stage, for example, counting total parts, counting parts of the same type, counting parts of different types, etc. Such simple analyses may reveal some problem areas in the product design already.

Step ii - Process Analysis

The second step in the micro DFX procedure is that of process analysis. It is mainly concerned with the collection, processing and reporting of process-specific and resource-specific data. A DFX tool usually specifies what process data it requires and how they are processed and reported. Just like product information, process information is also both inputs required by the subsequent steps and one of the outputs from the DFX tool.

The DFX shell suggests two types of process charts for process analysis: operation process charts and flow process charts. It would be advantageous to start with constructing an operation process chart to establish product-process interactions. Tasks below can be followed to construct an operation process chart:

1. A BOM of the product is established in an appropriate form. This can be directly imported from product analysis.
2. A decision is made regarding if process activities or workcentres are charted.
3. A business process corresponding to the chosen focus of the DFX tool is specified.
4. One of the parts making up the completed product is selected for charting first. The component on which the greatest number of activities is performed or the base part if the chart is to be used for laying out a progressive assembly line is usually chosen for this purpose. The BOM elements should then be properly ordered in relation to the first part chosen to start charting in order to produce a straightforward and clear PC-O. Otherwise, careless section and random ordering may result in complication and confusion because there may be too many intersections between the horizontal flow lines and vertical material lines.
5. A horizontal flow line is drawn next to the selected first element of the BOM from the left to the right. Related activities are added along the horizontal flow line until an additional component joins the first.
6. Draw a horizontal line corresponding to the next part to be fed into the first part. All associated activities before joining are added along this line until the next part is ready to join the first part.
7. Draw a vertical material line to show the point at which the second component enters the process.
8. Chart the activities which occur to the combined components along the horizontal flow line to the right until another part join it.
9. Repeat from Step (6) until all activities are charted for BOM elements.
10. Stop until all parts in BOM are brought in.
11. A number of straightforward analyses can be carried out at this stage, for example, counting total activities (operations), counting activities of the same type, counting activities of different types, etc. Such simple analyses may reveal some problem areas in the process design already.

If necessary, flow process charts can be used at this stage to acquire more specific process characteristics; otherwise, deferred to later steps when they are needed. The following tasks can be used to construct a flow process chart:

1. Decide if activities or workcentres are charted.
2. Select a subject BOM component.
3. Record the quantity of the selected BOM item handled.

4. Determine the business process which defines the starting and ending points of the chart.
5. Identify all the activities from the start to the end.
6. For each activity, record name, symbol, quantity, brief description, etc.
7. Repeat from Step (2) if other BOM components are to be charted.

Step iii - Measuring Performance,

Once the product and process information becomes available to the DFX analyst or team, their interactions can be measured in terms of the relevant performance indicators as specified by the DFX tool. This step may involve further activities in data collection and processing.

A DFX lookup table can be uniquely accessed according to its table number and the row and column numbers, as used in the Boothroyd-Dewhurst Design for Serviceability (Chapter 14). However, these numbers are not readily known to the analyst or the analysis team in practice. Instead, patterns are used to describe the entry conditions of the lookup tables and product and process characteristics (*see* Step 5 - Compiling DFX Manuals in Chapter 5).

Central to the performance measurement, and indeed diagnosing and advising, using DFX lookup tables is that of pattern matching, Figure 6.2. That is, patterns of lookup tables are compared against those of product and process characteristics. If a table pattern, and row and column patterns are matched by those of product and/or process characteristics, then the entry in the corresponding cell is the measurement data which will be entered into the appropriate cell in the DFX worksheet, or formulas for calculating performance measurements.

The following tasks are usually involved in this step of measuring performance:

1. Start with a BOM element at the lowest level.
2. Use a pre-set performance indicator.
3. Start with the first activity consumed by the BOM element.
4. Select an appropriate lookup table in the DFX manual by matching the table pattern against the problem description and performance measures.
5. Examine the row pattern against the characteristics of the product BOM element (assume that row patterns correspond to part characteristics)
6. Examine the column pattern against the characteristics of the process activity (assume that column patterns correspond to activity characteristics)
7. If matched, enter the measurement data from the lookup table in the DFX worksheet.
8. Repeat tasks 4-5 until all relevant activities are considered.
9. If there are more performance indicators, go to task 2.
10. Repeat tasks 2-7 until all the basic BOM elements are evaluated.
11. Aggregate overall performance measurements according to appropriate algorithms.
12. Report the performance measurements.

Step iv - Highlighting by Benchmarking

The fourth step in the micro DFX procedure is benchmarking and highlighting. The object is to address the question whether or not the subject product and process are good and what areas contribute to it. Benchmarking is mainly concerned with setting up standards and comparing the performance measurements against the set standard (Camp, 1989). Because performance is measured separately for individual consumption of an activity by a BOM element and total consumption of all activities of the same type by a BOM element, there are individual and aggregate benchmarks accordingly.

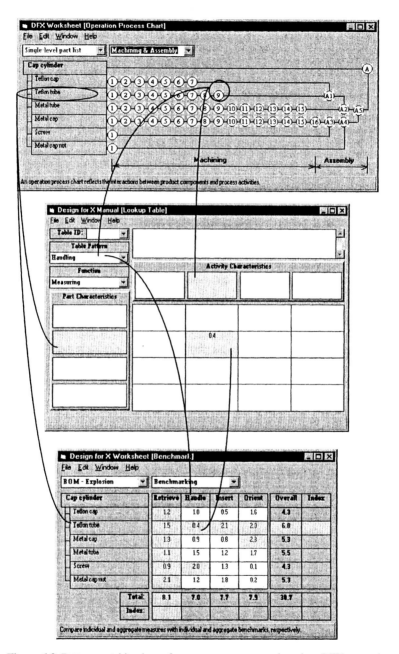

Figure 6.2 Pattern matching in performance measurement based on DFX manuals.

Once the performance standards and the measurements are available, the task of highlighting is straightforward. In general, areas where performance measurements are below standards are highlighted. Following are three ways of benchmarking and highlighting:

- If the individual performance measurement of a design decision is better than the individual benchmark, then this design decision is considered good enough in terms of the chosen indicator. Otherwise, it is highlighted as a problematic weak area.
- If the aggregate performance measurement of a product BOM element is better than the aggregate benchmark, then the design of this product BOM element is considered good enough in terms of all process aspects. Otherwise, all process aspects associated with this product BOM element are highlighted as problematic weak areas.
- If the aggregate performance measurement of a product design from a single process aspect is better than the aggregate benchmark, then the entire product design is considered good enough in terms of the chosen process aspect. Otherwise, all product elements are highlighted as problematic weak areas.

The following tasks are usually involved in benchmarking and highlighting:

1. Obtain individual and aggregate performance measurements as produced from Step iii.
2. Establish individual and aggregate benchmark thresholds.
3. Carry out individual benchmarking.
4. Carry out aggregate benchmarking according to process aspect or activity type.
5. Carry out aggregate benchmarking according to product BOM element.
6. Highlight problematic areas.

Step v - Diagnosing for Improvement

From performance measurement and benchmarking, it is known what is or is not good. To solve problems, it is necessary to know what causes the problem(s). This step is concerned with finding reasons why particular areas are weak (or strong). Very few DFX tools provide diagnosing facilities. In most cases, the human user is expected to accomplish this task. Cause-effect diagrams can be used to identify major causes for a problem.

It is assumed that the DFX manual provides knowledge for problem diagnosis. The following tasks are usually involved in this step of causal diagnosis:

1. Start with a BOM element at the lowest level.
2. Search through the corresponding row for highlighted areas.
3. For each highlighted problematic area, examine through the DFX manual by matching the patterns of product and process characteristics with those of the lookup entries.
4. If the cell is not empty, enter the cause from the lookup table in the DFX worksheet.
5. Repeat tasks 3-4 until all highlighted cells (problematic weak areas) are considered.
6. Repeat tasks 2-5 until all the BOM elements are evaluated.
7. Report on the diagnosing results.

Step vi - Advising on Change

This step is mainly concerned with exploring as many improvement alternatives as possible for each problem area. Not every DFX tool offers specific redesign advice. Many DFX tools leave the analyst or team in charge of redesign. Instead, they provide redesign objectives such as minimise the part number, etc. How exactly the subject product and process should be redesigned depends on specific circumstances. Table 6.4 lists some of the general techniques commonly used in redesigning products and processes (Osburn, 1963; Suzue and Kohdate, 1988). These techniques can be applied independently or in combination for best results.

Changes may take place to composition, configuration and characteristics at different levels of detail. For products, changes may be made across the entire product ranges, working principles / concepts, structures, subassemblies, components, parts, features, and/or parameters. For processes, changes can be made across product lines, business processes, procedures, steps, tasks, activities, and/or parameters. It is important to keep in mind that products and processes are closely interrelated to each other. A change in a product may well result in a series of changes in associated processes; and vice versa. This is the main reason for embedding DFX in BPR to maximize the benefits through considerate changes. One feature associated with advising on redesign is "what if" analysis, that is, to predict the potential effects of a proposed change on other areas of products, processes and resources.

The following tasks are usually involved in this step of advising on redesign:

1. Start with a BOM element at the lowest level.
2. Search through the corresponding row for highlighted areas.
3. For each highlighted problematic area, examine through the DFX manual by matching the patterns of product and process characteristics with those of the lookup entries.
4. If the cell is not empty, enter the causal description from the lookup table in the DFX worksheet. If the cell is empty (no advice is given), then the user has to think creatively about potential improvement actions.
5. Repeat tasks 3-4 until all highlighted cells are considered.
6. Repeat tasks 2-5 until all the BOM elements are evaluated.
7. Report on the diagnosing results.

Table 6.4 Techniques for redesigning products and processes

	Product	Process
Eliminate	Can any of the components be eliminated?	Can any of the activities be eliminated?
Integrate	Can one component be integrated with another component?	Can one activity be integrated with another activity?
Combine	Can the given components be combined in a better way?	Can a better sequence of activities be followed?
Simplify	Can components be simplified?	Can activities be simplified?
Standardise	Can components be standardised into one?	Can activities be standardised into one?
Substitute	Can any component be replaced?	Can any activity be replaced?
Revise	Can any component be revised?	Can any activity be revised?

Step vii - Prioritising

A DFX analysis may reveal a large number of problem areas in the subject product and process. There can be many causes and many alternative solutions for each problem. On the other hand, the resource available to the DFX analyst or team is always limited. The object of this step is to identify vital issues for further investigation and make improvements right away to many trivial aspects so that attention can be focused on important problems and promising solutions.

One of the most commonly used method for prioritization is Pareto analysis. Central to Pareto analysis are Pareto charts. Pareto charts are specialised bar graphs that can be used to show relative frequency of events such as products, processes, failures, defects, causes and effects, etc. A Pareto chart presents information is descending order, from the largest category to the smallest. Optionally, points are plotted for the cumulative total in each bar and connected with a line to create a graph that shows the relative incremental addition of each category to the total.

Prioritization should be based on some form of measurement data. Step IV - Tradeoff Analysis in the macro BPR procedure in the next section presents methods for thorough evaluation of the items in terms of chosen criteria.

The following procedure can be used to construct a Pareto chart:

1. Deciding which items to study and collecting data.
2. Tabulating data and calculating the cumulative number.
3. Drawing the vertical and horizontal axes.
4. Displaying the data as a bar graph.
5. Drawing a cumulative curve.
6. Creating a percentage scale on a vertical axis on the right side.
7. Labelling the diagram.
8. Examining the diagram.

6.3 MACRO BPR PROCEDURE

This section presents the macro BPR procedure. It follows a general process of problem solving (Bounds and Hewitt, 1995). The seven steps can be used to identify a problem, analyze the problem by identifying its causes and effects, generate potential solutions, select and plan a solution, implement the solution and evaluate the solution. This systematic approach is helpful in selecting the tool best suited to solve the problem and properly apply that tool.

"Design for X" analysis is only a single step in the macro BPR procedure, mainly for problem identification and analysis. The appropriate DFX tool should be selected and implemented properly to solve problems and make lasting improvements. There are overlaps between the micro steps and the macro steps, especially between (1) "Prioritising" in the DFX procedure and "Tradeoff Analysis" in the BPR procedure; and (2) "Advising on Redesign" in the micro procedure and "Radical Change" in the BPR procedure. Generally speaking, "Prioritising" and "Advising on Redesign" in the micro DFX procedure are more concerned with individual problem areas and their solutions. On the other hand, "Tradeoff Analysis" and "Radical Change" in the macro BPR procedure are more concerned with overall problems and solutions.

Step I - Project Management

It is most important and difficult to get started with a DFX project, especially when a company has no DFX, BPR or CE experience. If anything goes wrong in the beginning or preparation is done inadequately, the remaining six steps of the macro BPR procedure are likely to be less effective or at worst deliver incorrect solutions.

"Project management" is the first step in the macro BPR procedure and extends throughout the entire procedure. The following major tasks are involved:

☐ Project identification.
☐ Project definition and justification.
☐ Project organization.
☐ Project planning and scheduling.
☐ Project control and monitoring.

There are a number of good techniques for each of the above tasks. Many companies have their own ways of managing projects. Numerous textbooks on *Operations and Production Management* provide good coverage of these techniques. Therefore, they are not discussed here because of the space limitation. However, project identification is briefly discussed here.

Table 6.5 Quick audit sheet for problem identification

Brief descriptions of subject products / processes.		
Life-cycle business process	Decision	Comments
Design and development		
Piece-part fabrication		
Assembly	✓	
...		
Inspection and test		
Packaging and distribution		
...		
Recommending notes A Design for Assembly is recommended.		

Table 6.6 Checklists for problem identification

Performance metrics	Factors of production	3 Big problems
Is quality a problem? Is cost a problem? Is delivery (time) a problem? Is productivity (efficiency) a problem? Is safety a problem? Is morale a problem?	Is there a problem with materials? Is there a problem with machinery? Is there a problem with manpower? Is there a problem with the method?	Is there waste? Are there irregularities? Is the requirement unreasonable?

Identifying an appropriate project is itself a project. It usually starts with identifying what problems exist with products and their associated processes. It is not unusual for a consultant to receive a 50-pages long computer printout from the record - "This is our problem, do whatever you can". This is a totally inadequate attitude towards improvement. Rather, a simple audit is usually enough to identify a number of problem areas. Table 6.5 can be used as an audit sheet. For each life cycle business process, questions listed in Table 6.6 can be asked. Answers are recorded in the audit sheet. They will be used later as a guide for identifying key focus areas, setting targets, choosing DFX tool(s), and measuring achievements. There are many good ways of identifying and recording problems (Sugiyama, 1989).

Step II - Design for X Analysis

Design for X analysis is a major step of the macro BPR procedure. The 7-steps micro DFX procedure has been discussed previously. These steps guide the user from collecting relevant data through their proper processing to presentation. The DFX analyst or team can understand the elements of the problem and how they create the discrepancy that causes the problem. It provides a basis for formulating the potential solutions, and prevents jumping into the solution without rigorous analysis of the problem itself.

However, in addition to the seven steps in the micro DFX procedure, there are other issues that must be taken into account at this step of the macro BPR procedure:

1. Which DFX tool to use is a big decision which perhaps should be sorted out during the project identification. The selection of DFX tools is problem-driven and goal-directed, not solely determined by its availability. The introductory chapter provides a number of guidelines for selecting an appropriate DFX tool.
2. Multiple DFX tools should be used to analyze a problem from various aspects. In this case, the question is no longer which DFX tool to use but which DFX tool to start with and in what order other DFX tools are introduced. Those DFX tools which focus on product assortments and structures should be used before those DFX tools which deal with components, features and parameters. For example, Professor Boothroyd suggests that DFA should be used first and then DFM tools follow.
3. There are two basic variations of introducing multiple DFX tools. One is to apply one DFX tool within each cycle of the macro BPR procedure. A major advantage of this approach is the focus that the DFX tool provides for the analyst or team. A major

drawback is that design changes implemented according to one DFX tool may be contradictory with those suggested by subsequent DFX tools. Once a change is implemented, it is very difficult to change the change, causing embarrassment and confusion.

4.   The other variation is to introduce one DFX tool at a time until all of the selected tools are applied during the second step "Design for X analysis" of the macro BPR procedure. A main advantage is that conflicts among redesign suggestions can be resolved before they are implemented. Its disadvantages include the increasing scope of the project which may become too broad to manage.

Step III - Radical Reengineering

The aim of this step is to explore redesign alternatives or scenarios based on the results from the DFX analysis. It may be enough to reengineer either products or processes. It may be necessary to reengineer both products and processes. This depends on the nature and the degree of the problems highlighted by the DFX analysis. The outcome from this step is a set of change packages each of which consists of a sequence of snippets of change suggestions corresponding to problem areas.

Because of combinatorial explosion, the number of alternative change packages may be too big to manage. The method of morphological analysis (Zwicky, 1967; Norris, 1963) can be used to prevent this. Figure 6.3 shows a sample morphological chart (upper part of the figure) and the solution space (lower part of the figure). A morphological chart can be formulated for the subject product, its associated processes, resources or organizational structures. Problematic areas are listed in the first column of the morphological chart. All the conceivable solutions to these weak areas are listed in corresponding rows. The analyst or team can then select appropriate solutions to weak areas and combine them into potential change packages. Idea generation and selection requires creative thinking in morphological analysis. The chart systematically stimulates, encourages, and facilitate the creativity.

The following tasks are usually involved at this step based on morphological analysis:

1.   Generate as many conceivable solutions as possible for each problematic area. In some sense, this can be viewed as a continuation of Step 6 - Advising on Redesign of the micro DFX procedure. It is better to make an ideal change proposal rather than bound by present constraints.

2.   Select solutions most suitable for each problem area and exclude those obviously irrelevant to the problem. If necessary, the selection matrix method introduced in the next step - Tradeoff Analysis can be used.

3.   Suggested changes may well be related in some way. They may be revised first and then combined to form a better solution.

4.   Combine the selected solutions into overall solutions - change packages. Each of them is recorded and subjected to careful evaluation in the next step. In formulating a change package, it is necessary to investigate the interactions between the chosen change snippets for different weak areas.

   •   If two change suggestions are entirely independent of each other, then both of them should be included in a change package for further investigation.
   •   If one change suggestion is completely inclusive by another, then only the second needs to be included in a change package.

- If two change suggestions overlap, then both of them are investigated to form a new suggestion for inclusion in a change package.
- If two change suggestions are partially contradictory, a new change suggestion should be formulated to resolve the conflict while achieve their individual positive effects. The new change snippet is included in a change package.
- If two change suggestions are mutually exclusive (contradictory), a careful trade-off analysis is needed to resolve the conflict.

Step IV - Tradeoff analysis

The object of Tradeoff Analysis is to evaluate the costs and benefits of alternative change packages so that the most appropriate one can be selected for implementation. Alternative change packages can be evaluated on their own. This looks simpler because their constituent snippets are not analyzed explicitly or individually. However, the accuracy of evaluation may not be sufficient. To overcome this, individual constituent snippets of each change package can be included for thorough evaluation. This would result in higher accuracy but involves larger amount of analytical work. In practice, a balance needs to be sought.

One simple method for tradeoff evaluation is what is often referred to Pugh's selection matrix (Pugh, 1991), as shown in Figure 6.4. Across the top axis (first row) are all alternative change packages included for evaluation. On the vertical axis represented in the first column is a list of criteria against which evaluations are carried out. Criteria can be established in a number of ways, for example, to base on the criticality of the cause for the change; positive effect (benefits) of the change; negative effect (costs) of the change; or net gain of the change (positive effect minus negative effect). Assuming a team from different aspects are involved in evaluation, each member can ask one or more questions such as:

- Does this change has any effect on us?
- Is this change beneficial to us? If yes, to what extent it benefits us?
- Does this change have any adverse effect on us? If yes, to what extent it affects us?
- What do we have to do to implement this change?
- From the answers to the above questions, is this change worth implementing? Why?
- Identify how much resource can be expended in implementing any solution.
- How well will it be accepted by the customer?
- How long it will take to resolve the problem? Can we meet the schedule?

Quantitative answers to above questions are preferable. However, it is not always feasible in practice, because of the degree of uncertainty and the lack of information. Very often, qualitative analysis plays an effective role. Scores and ratings of various scales can be used. A datum change package is selected as the reference against which other alternatives are evaluated. If a change package is better than the datum in terms of a specific criterion, then a plus sign "+" is entered into the corresponding cell in the matrix. If a change package is worse than the datum in terms of a specific criterion, then a minus sign "-" is entered into the corresponding cell in the matrix. If a change package is equally good or bad as the datum in terms of a specific criterion, then a plus sign "=" is entered into the corresponding cell in the matrix. The total numbers of "+", "-" and "=" are counted for each alternative change package. A change package with the most "+" is considered the best and deemed for further investigation.

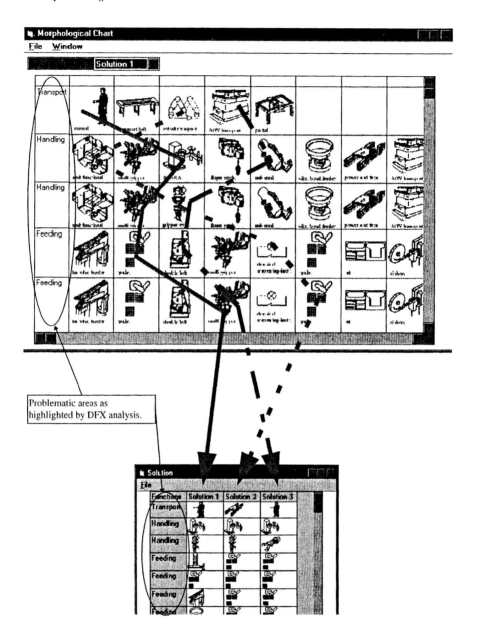

Figure 6.3 Sample screen of morphological chart for formulating change packages.

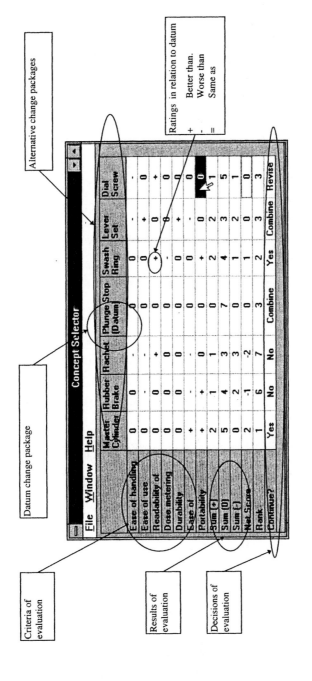

Figure 6.4 Sample screen of Pugh's selection chart for evaluating change packages.

This qualitative screening method is usually used for narrowing the number of change packages quickly and to improve them. If a satisfactory change package can be sought, then the DFX project can proceed to the next step of implementing the change package. If none change package is found satisfactory, then the project returns to the preceding step of radical reengineering. If potential is found in a change package but further modification is required, then it is revised, if necessary, re-evaluated for acceptance for implementation.

Step V - Planning for Implementation

This step is concerned with planning and implementation of the best change package resulted from the preceding step of Tradeoff Analysis. Having the best change package is one thing. Actually implementing it is quite another. Implementation of the solution is not easy. It will not just happen. Hard and conscientious work is required.

Implementing the change package can be considered as a project itself. Therefore, usual project management techniques apply. The following major tasks are involved:

1. *Preparing for implementation.* The main objective of preparing for implementation is to obtain support from the total organisation including both management and workforce. This is because everybody is likely to be affected by the changes from the DFX analysis. Without the support, the DFX project will not deliver the benefits promised by the DFX tool and expected by the DFX team.

2. *Planning for implementation.* The chosen best *change package* is turned into a *change programme* or a plan of actions through careful planning. Necessary resources are brought into the action to effect the optimum sequence of events: Who should do what, how, when and where.

3. *Executing implementation plan.* Upon approval, the change programme assumes the status of action. It may be worth considering breaking down the actions into smaller steps so that each step can be tested in a short trial. This applies especially when the problem and/or the solution is particularly complicated, or if the outcome of the solution has a high degree of uncertainty associated with it.

4. *Monitoring and controlling the implementation process.* The implementation must be executed according to the plan. It is not unusual to see unexpected problems popping up. This creates confusion and conflicts. It is essential to work quickly and diligently, with the help of people in other affected departments, to revise the plan and eliminate such conflicts in a way that works for all concerned.

Step VI - Measuring Effectiveness

Although each step of the macro BPR procedure and the micro DFX procedure is verified before closure, measurement of effectiveness of the overall DFX project is required to ensure that the desired results are achieved and sustained. Effectiveness can be defined as the improvement in performance measurement before and after DFX analysis and BPR programme. Such improvements can be observed by carrying out one or more of the following effectiveness audits:

☐ *Product / process audits.* The purpose of product / process audits is to measure the achievements of the DFX/BPR project by gathering performance data and comparing them against the targets set in the beginning. It is common to carry out another DFX analysis to the improved product and new processes as part of product and process audits. In fact, Hitachi AEM provides one performance measure especially devoted to effectiveness measurement (See Chapter 2). DFX results for the original design and the new design are compared for the difference in part count, activity count, etc.

☐ *Financial audits.* The purpose of a financial audit is to fully appreciate the tangible net savings of a DFX project. The net financial gain is the difference between the gross savings achieved by and actual costs for implementing DFX. Unfortunately, such financial audit can only take short-term factors into account such as reduced rework costs. It is difficult, if not impossible at all, to consider long-term financial gains for example achieved through increased sales and improved efficiency.

☐ *Organisational audits.* Organisational audits are concerned with long-term and intangible benefits of implementing DFX in product development in particular and BPR in general. Organizational audits are aimed at a much more efficient and effective organizations and operation. However, improvements such as communication and cooperation are extremely complex to measure. The scoreboard method presented by Carter and Baker (1992) may be useful in conducting organizational audits. One point worth making is that cooperation and communication do not simply mean sharing computer workstations and exploiting network facilities. DFA was once pushed by automation technology, it is now manual assembly where great savings are achieved. This applies to human communication and cooperation in contrast to computers.

Step VII - Follow-on and follow-through

Measurement of effectiveness assesses what has been achieved, and even more importantly, indicates what should be done next. It cycles back to the first step of the macro BPR procedure. Improvements must be made permanent, good practice must be institutionalised and new problems must be identified. There are several directions for further exploration:

☐ Repeat DFX analysis to review the approach to the way an improvement is implemented.
☐ Repeat DFX analysis to verify the results, whether positive or negative.
☐ Apply the same DFX tool to the same or similar product/process with new objectives.
☐ Apply the same DFX tool to different products/processes.
☐ Apply a different DFX tool to the same or similar product/process.
☐ Apply a different DFX tool to a different product/process.
☐ Inaugurate entirely new projects.

It is not unusual to note that a DFX project halts in the middle of implementation. One DFX tool is not fully implemented and another has already started. New made-up words *x-bility* become flavours of months. Real problems are often bypassed and no implementation is planned or eventually executed. In the end of the day, people are generally happy with every DFX project and problems remain as ever.

There are many reasons for such half-way implementation. Very often, barriers listed in Table 6.7 are mistakenly blamed as limitations of DFX tools. For example, some people complain that it is time consuming to conduct a DFX analysis; difficult to get people to participate actively in a DFX project; unclear where to collect data required by DFX analysis; incompatible with existing product development practice; too much paperwork; etc.

Table 6.7 Barriers for halfway DFX attempt

- Give in to problems associated with existing product development (See Table 6.1).
- Poor team selection and training.
- Absence of a team culture.
- Unclear project ownership.
- Lack of participant commitment and participation.
- Lack of management commitment and participation.
- No budget.
- Start off too big.
- Expect benefits too much, too soon.
- Lack of tangible (short-term) benefits.
- Lack of leadership.
- Resistance to change.
- Fear of consequences.
- Adverse or counter-productive policies, procedures, practices and structures.
- DFX is not understood.
- Resistance from design engineers.
- DFX tools hinder creativity.
- DFX tools do not take into account many manufacturing capacities.
- DFX is time-consuming to apply and tedious to follow.
- Lack of data or data are too difficult to collect in order to support DFX.
- Doubt about the accuracy of DFX results: enough to guide design decision making?
- Fear of considerable influence of manufacturing over product design.
- Fear of new responsibility in product design for the choice of manufacturing methods.

Quite to the contrary, these are exactly the problems DFX tools are designed to highlight. Creativity, reduction in paperwork, data availability and coordinated teamwork are more objectives than pre-requisites. DFX should be treated as barrier breakers: Any success in overcoming any of these barriers is flagging the success of implementing DFX. DFX is not about filling in forms with numbers, abbreviated phrases, and diagrams. It is the creative thinking that contributes to impressive improvements. The outcome from a DFX analysis is not acres of paperwork. But DFX dynamically transforms the product development process from a problem-prone sequential engineering environment to a problem-free concurrent engineering environment.

For example, most DFX tools have a modest requirement on input data. They are normally generated or required by activities other than DFX. For example, the part list of a product is generated at design and the bill of materials is used in production planning. They should be relatively easy to collect. If not, then this is exactly sort of problem highlighted by a successful DFX analysis.

Let us take teamwork as another example. DFX is a team tool and should be used as such. A team from different disciplines without DFX knowledge can hardly cooperate effectively to automatically create the effects that a DFX tool can deliver. The role of a team in the context of DFX is to contribute to what DFX tool demands and compensate for what the rest of the team is short of. DFX, in this case, cultivates the ground for better communications with a specific focus. DFX requires a team in place and helps the team to communicate and cooperate effectively and efficiently.

Professor Boothroyd (Chapter 1) and Professor Clausing (1993) have refuted many excuses which will undoubtedly continue to appear in practice. As Cohen (1995) indicates, those who become enthusiastic about DFX [QFD] are generally very creative in conceiving new applications. Those who dislike the DFX [QFD] formal structured approach are generally very creative in producing reasons why DFX [QFD] does not work.

6.4 SUMMARY - LEARNING BY DOING

This chapter has presented a DFX-focused framework, the DFX/BPR shell, for implementing concurrent engineering within manufacturing industries. The DFX/BPR shell consists of two systematic procedures. At the micro level, a DFX analysis is sequential in the sense that the subject product and process and their interactions are dealt with one by one, from performance measurement through root cause diagnosis to redesign. At the macro level, these DFX activities are regarded as concurrent in the sense that they are compressed into a single step. The micro DFX activity provides the necessary focus for the analyst or analysis team to concentrate on most important and relevant issues; concurrence for the product development team to cooperate; and vision for radical change in products, processes, systems and organizational structures.

Mixed feelings are natural. Where there is enthusiasm - "DFX works brilliantly for them to save ...!", there is scepticism - "Does it apply to us?". One of the most important hard lessons learnt by successful DFX users is to get DFX started first - learning by doing (Weber, 1994). The DFX process is cyclic and benefits are incremental. Initial rounds could be painful, time-consuming, and expensive. Once DFX practice becomes natural in day-to-day activities, there are fewer strangers.

Just as the DFX shell is a generic framework for developing a variety of DFX tools, the DFX/BPR shell is a generic framework for implementing concurrent engineering. Neither the DFX shell nor the DFX/BPR shell should be followed rigidly. Instead, they are flexible enough to be customized (tailored or extended) to best suit particular circumstances. Your DFX tools could work and be implemented in very different ways. But "Design for Modularity" (Chapter 17) also applies to DFX.

REFERENCES

Bounds, G., Hewitt, F. (1995) Xerox: Envisioning a corporate transformation, *Journal of Strategic Change*, 4 (3), 3-17.

Camp, R.C. (1989) *Benchmarking: The search for industry best practice that leads to superior performance*, American Society of Quality Control Press, Milwaukee, USA.

Cater, D.E., Baker, B.S. (1992) *Concurrent Engineering: The product development environment for the 1990s*, Addison-Wesley Publishing Company.

Clark, KB., Fujimoto, T. (1991) *Product Development Performance*, Harvard Business School Press, Boston, USA.

Clausing, D.P. (1991) *Total Quality Development: A step-by-step guide to world-class concurrent engineering*, ASME Press, New York, USA.

Cohen, L. (1995) Quality Function Deployment: How to make QFD work for you, Engineering Process Improvement Series, Addison-Wesley Publishing Company, Reading, Massachusetts, USA.

Gatenby, D.A., Foo, G. (1990) Design for X (DFX): Key to competitive, profitable products, *AT & T Technical Journal*, May/June, 2-13.

Gatenby, G.T., Lee, P.M., Howard, R.E., Hushyar, K., Layendecker, R., Wesner, J. (1994) Concurrent Engineering: An enabler for fast, high-quality product realization, AT & T Technical Journal, January/February, 34-47.

Hammer, M., Champy, J. (1993) *Re-engineering the corporation: A manifesto for business revolution*, Harper Business, New York, N.J.

Johnsson, H., McHugh, P., Penleburg, A., Wheeler, W. (1993) *Business Process Reengineering*, John-Wiley & Sons, Chichester, UK.

Miles, B.L. (1989) Design for assembly - a key element within design for manufacture, *Proceedings of IMechE, Part D: Journal of Automobile Engineering*, **203**, 29-38.

Norris, K.W. (1963) The morphological approach to engineering design, In: *Conference on Design Methods*, Edited by J.C. Jones, London: Pergamon Press, 115-141.

Osborn, A.F. (1963) *Applied Imagination: Principles and procedures of creative problem solving*, 3rd Edition, New York: Scribner.

Pugh, S. (1990) Total Design - Integrated methods for successful product engineering, Addison-Wesley Publishing Company, Workingham, England.

Sugiyama, T. (1989) *The Improvement Book*, Productivity Press, Cambridge, Massachusetts, USA.

Suzue, T., Kohdate, A. (1988) *Variety Reduction Program: A production strategy for product diversification*, Productivity Press, Cambridge, Massachusetts, USA.

Weber, N.O. (1994) Flying high: Aircraft design takes off with DFMA, *Assembly*, September.

Youssef, M.A. (1994) Design for Manufacturability and time to market - Part 1: Theoretical foundations, *International Journal of Operations and Production Management*, **14** (12), 6-21.

Zwicky, F. (1967) The morphological approach to discovery, invention, research and construction, In: *New Methods of Thought and Procedure*, Edited by F. Zwicky and A.G. Wilson, Symposium on Methodologies, Pasadena, Berlin: Springer, 316-317.

GIM: *GRAI* INTEGRATED METHODOLOGY
FOR PRODUCT DEVELOPMENT

Guy Doumeingts; Philippe Girard; Benoît Eynard

This chapter extends the conceptual GRAI model to include design activities of products, in addition to manufacturing systems. Section 1 reviews the importance of design management. Sections 2 highlights main features of GIM (GRAI Integrated Methodology). Its use in modelling products and product design processes is discussed in Section 3.

The GRAI/LAP - "Groupe de Recherche en Automatisation Intégrée/Laboratoire d'Automatique et de Productique" at "Université Bordeaux I" has been engaged in developing a methodology to design and specify Advanced Manufacturing Systems. The resulting methodology is referred as to GIM (GRAI Integrated Methodology). This methodology is based on the GRAI model which exploits systems theory and hierarchical theory (Le Moigne, 1977; Mesarovic, Macko and Takahara, 1970).

The GRAI model provides two different ways of looking at a manufacturing system. Firstly, a manufacturing system is divided into two parts: (1) the Control System and (2) the Controlled System. Secondly, a manufacturing system is divided into three subsystems: (1) Physical, (2) Decisional, and (3) Informational. The result of such multiple perspectives of system analysis is the clarified user and technical specifications of the manufacturing system.

GIM has been used in French industries such as Aérospatiale, SNECMA, GIAT, CLEMESSY, and other European industries such as British Aerospace, Pirelli, SAAB SCANIA. GIM was primarily developed for the modelling and design of manufacturing systems. However, it can also be used for modelling the product design process and product design activities. This chapter reports on this latest development.

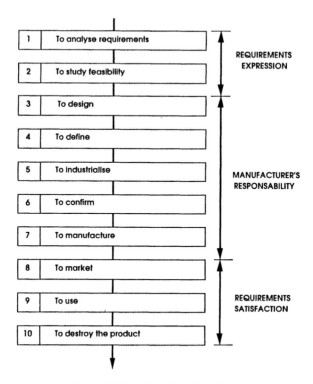

Figure 7.1 Traditional product life cycle.

7.1 PRODUCT DESIGN MANAGEMENT

Figure 7.1 show a typical life cycle of product development. It involves 10 general activities arranged in sequence. Activities 3-7 are usually within the responsibility of the product manufacturer. There are many problems associated with this sequential arrangement, for example restrained data, high costs, long lead times, and low quality. It is necessary to amalgamate the set of the activities because we can note that 75% of costs are committed during the product study when only 5% of costs are incurred (Petitdemange, 1991). The aim of simultaneous engineering is to synchronize these activities (O'Grady *et al.*, 1991). This permits us to decrease cost and time while maintaining the optimum quality. During the design activity it is possible to meet the constraints on the down-stream activities. The design step amalgamates other activities to design, to define and to industrialise.

Figure 7.2 is the definition of our product design process. It translates the customer requirements (specifications) into product definition and manufacturing process definition. The complexity of the design process increases and makes it more difficult to manage. Kusiak and Wang (1993) present a qualitative analysis of the design process. The purpose of concurrence analysis is to reorganise the design process and to determine potential groups of activities that can be performed in parallel.

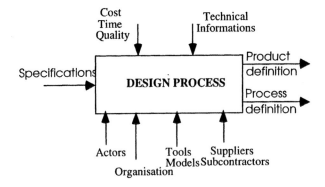

Figure 7.2 Design process.

Product design is influenced by many orgsniasational factors such as the flows of authority, materials, information and decisions within a company (Mintzberg, 1984). The design process allows us to elaborate all the information related to the product. This process is decomposed into three types of activities which must be organised and synchronised:

- **Design activities** are characterised by the necessity of an iterative process to obtain a solution. The solution objective increases the iterations to obtain several propositions satisfying the specifications.
- **Execution activities** are characterised by predetermined execution conditions. The solution objective increases simultaneously the technical performances and the economical performances.
- **Management activities** are characterised by the necessity to know precisely the decisions conditions in order to define the appropriated management measure. The solution objective is to obtain a management process to co-ordinate and to synchronise the execution activities.

The term *macro-process* is used here to reflect the hierarchical nature of the organizational structure. The macro-process can be decomposed into several design steps (Decreuse *et al.*, 1994). Each step is named *macro-activity* which represents a group of design activities. Figure 7.3 shows that each step of the design process is an iterative process represented like a knowledge spiral. Iteration is necessary when some product specifications are not satisfied.

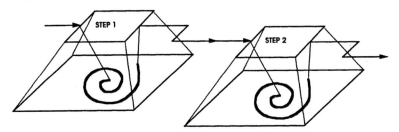

Figure 7.3 Design Macro-process.

7.2 GIM: GRAI INTEGRATED METHODOLOGY

The GRAI model is based on theories of complex systems, hierarchical systems, organization systems, and the theory of discrete activities. GIM has been developed for analysing, designing and specifying manufacturing systems in a context of integration. The GRAI approach is characterised by the following three elements (Doumeingts, 1984; Doumeingts, Vallespir, Darricau and Roboam, 1987; Doumeingts and Vallespir, 1992):

- Reference models
- Modelling formalisms
- Structured approaches

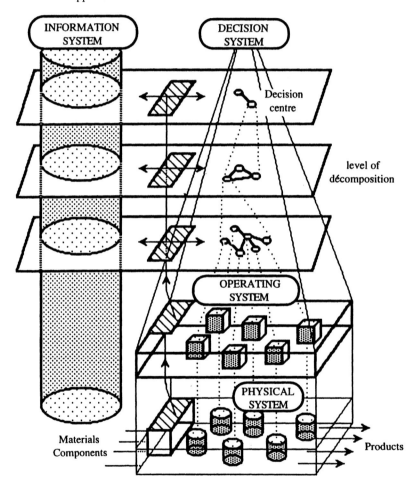

Figure 7.4 Reference model: structure of a manufacturing system.

7.2.1 Reference Models

A model is an abstract, simplified representation of reality. A manufacturing model can only represent a set of selected elements concerning the domain studied, and in agreement with defined objectives. A system can be represented by different kinds of models based upon different points of view. A good model should amplify the important characteristics and conceal the details which are not considered as important at a given abstraction level. In the domain of manufacturing, models are supported by mathematical formalisms, languages and/or graphical tools. A reference model is a conceptual and generic model of manufacturing system which describes the functionality, the structure, the components and the behaviour of the manufacturing system.

Figure 7.4 shows the GRAI conceptual reference model. A manufacturing system is usually very complex to analyze, to understand, to improve and to design without necessary coherent decomposition. The GRAI conceptual reference model helps defining basic concepts, relations between concepts, the structure and the design rules. It uses various criteria of decomposition derived from the systems theory (Simon, 1960; Le Moigne, 1977), hierarchical decomposition (Mesarovic *et al.*, 1970), organizations theory (Mintzberg, 1982) and from theory of discrete activities (Pun, 1984).

Let us first consider two sub-systems on a dynamic point of view: the controlled sub-system (called the Physical Sub-System) and the control system (called Production Control System). The controlled sub-system transforms raw materials into the "Products" sold to the customer. The transformation is accomplished through the Resources (machines, people, etc.). For the Design Function we must note that this "materials flow" is actually "information" (we will discuss this point in Section 7.3). This controlled sub-system could be decomposed again into departments, services, sections, cells, etc.

The control sub-system is first decomposed in two parts: decision and information. We will describe later the difference but now we have a simple equation:

decision = information + (objectives, decision variables, criteria, constraints).

The decision part is decomposed according two criteria: coordination (vertical decomposition), synchronisation (horizontal decomposition):

- The coordination criterion decomposes the decision part in decision making levels: strategic, tactical, operational. We make also the distinction between periodic driven decision and event driven decision (we call this last domain the *operating system*).
- The synchronisation criterion allows to synchronize the function "Management of Products" with "Management of Ressources". Usually the synchronisation is performed through the function "To Plan". Other functions could be synchronised with the previous ones: quality, maintenance, engineering.

The information part is structured according to decision part: it stores, processes, transfers all the informations needed by the manufacturing system. It is the link between the controlled and the control sub-systems and with the environment.

Figure 7.5 shows a reference model for a decision centre. This model aims at conceptualising the operations at a decision centre, in a steady and perturbed state as well. We find again a local decomposition into three sub-systems: physical, information, decision. We have therefore defined: the various activities of a decision centre, its decision frame (variables and decision limits), decisions made by the decision centre and information used by it.

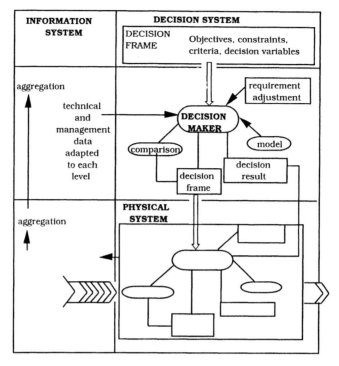

Figure 7.5 Reference model: structure of a decision centre.

7.2.2 GIM (GRAI Integrated Methodology) Formalisms

A modelling formalism is a concept to represent pieces of knowledge that have to be transmitted unambiguously. It allows to build models according to associated concepts. The theoretical basis for modelling formalisms can be found in the graph theory, the languages theory, logical structures, etc. These modelling formalisms that are used, are often associated with graphical tools and allow to describe the manufacturing system. A good diagram is often better than a long speech.

Composing the manufacturing systems is closely linked to and strongly dependant on the sub-systems presented Figure 7.4. If we modify one of them, we have to adjust the other. To take into account this aspect of co-ordination, it is necessary to build an integrated methodology which allows to analyse the global system. GIM covers the three domains of a manufacturing system shown in Figure 7.4. On the other hand, our methodology has to cover all the abstraction levels: the conceptual level (C) which defines *what* to do (functional and semantical analysis), the structural level (S) which defines *who*, *where* and *when* (taking into account the organizational options), and the realizational level (P) which defines *how* (technical options).

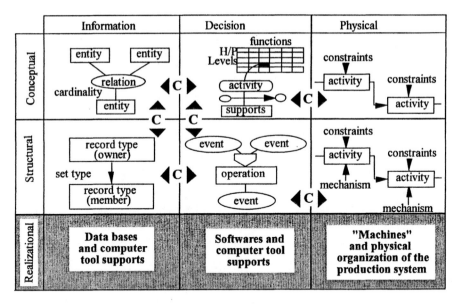

Figure 7.6 GIM formalisms.

Figure 7.6 shows formalisms used for each case:

- for data: the same formalisms than Merise: entity / relationship at conceptual level and specific formalisms (in relation to the type of database chosen) at structural level;
- for process: GRAI grid and GRAI nets at conceptual level and Merise data processing formalism at structural level;
- for physical system: IDEFØ, Stock resources

The letters "C" in Figure 7.6 between two domains on an horizontal line (1 and 2) point out the coherence procedures between these domains. Between two domains on vertical line (3 and 4), they point out the rules to translate a model from one abstraction level to another one.

The GRAI grid

The first axis (Figure 7.7) is related to a hierarchical representation of the whole structure of the production management system. The criterion of hierarchy is time. The second axis shows the intangible functions of a production management system. This framework makes appear the decision centres and the links between them (Doumeingts, 1984).

The GRAI nets

The GRAI nets (Figure 7.8) give the structure of the various activities of each decision centre highlighting decision activities and execution activities (Doumeingts, 1984).

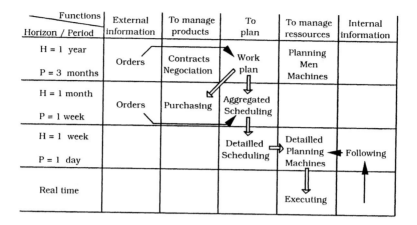

Functions Horizon / Period	External information	To manage products	To plan	To manage ressources	Internal information
H = 1 year P = 3 months	Orders	Contracts Negociation	Work plan	Planning Men Machines	
H = 1 month P = 1 week	Orders	Purchasing	Aggregated Scheduling		
H = 1 week P = 1 day			Detailled Scheduling	Detailled Planning Machines	Following
Real time				Executing	

Figure 7.7 GRAI grid.

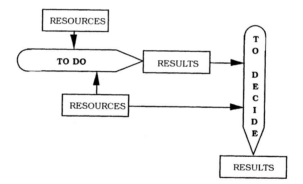

Figure 7.8 GRAI net.

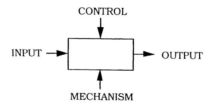

Figure 7.9 IDEF0 formalism.

<u>IDEFØ (Icam DEFinition)</u>

Figure 7.9 shows an IDEF0 formalism. It is made up of labelled boxes and arrows. Boxes represent the decomposition of the subject into parts, arrows connect boxes and represent interfaces or constraints between boxes. A control describes the conditions or circumstances that govern the transformation. A mechanism could be the person, machines, devices, information which carries out the activity (Mayer, 1990).

<u>Entity/relationship formalism</u>

The purpose of information modelling is to structure the memory of the company (Tardieu, Rochfeld and Colletti, 1983). Figure 7.10 shows the entity-relationship formalism. There exist different ways to model an information system, but the most elementary one is to identify information entity by its name, to describe this information by its attributes and then to establish the relationships between them.

ENTITY TYPE   RELATIONSHIP TYPE   ENTITY TYPE

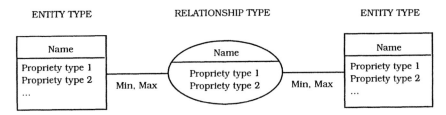

Figure 7.10 Entity-relationship formalism.

7.2.3 Structured Approach

Generally speaking, a structured procedure includes steps to be followed when applying a method to solve a problem. In manufacturing systems design, the structured approach should cover the whole life cycle of the manufacturing project: analysis, design, development, implementation, and operating. All these steps should be defined. The roles of particular actors must also be clearly defined.

  The application of GIM must be structured. Its use requires: a synthesis group composed of the main decidors of the considered doamin; an analyst-designer (or several if necessary); an expert on GIM; and the interviewed persons (the other users). The GIM application consists of two main phases:

- *the analysis phase:* to analyse the current system, to collect all data necessary for designing the new system and to improve the definition of objectives, to detect the inconsistency
- *the design phase:* to design the system from data collected during the previous phase, by analysing the inconsistencies between the current system and the reference model and taking into account the objectives and constraints of the future system. From the design phase, we elaborate the specifications of the future system.

  The first step of the methodology is the provision of a functional (IDEFØ) model which describes the functions of the system. The model will define the elements of the global system (the company) and the flows between them. Based on this functional model, we identify the scope of the manufacturing system. Then, we decompose this manufacturing system into a production control system and a physical system. The physical system is analysed and designed with the IDEFØ tools. The production control system analysed and designed with the GRAI and Entity-Relation tools.

After the global study, the design which details the physical system and the production control system and is undertaken. The coherence between the sub-systems, the checking and integration of them are achieved firstly at a global level (macro-integration) and secondly at a much more detailed level (micro-integration). The GIM procedure is summarised below:

INITIALISATION PHASE
.   first contact with the company management,
.   information and training,
.   definition of goals and study domain,
.   planning of the study.

ANALYSIS PHASE (existing situation)

Top down analysis
.   realisation of the functional model of the global system (IDEFØ),
.   realisation of the physical system model (IDEFØ),
.   realisation of the GRAI grid,
.   realisation of the Conceptual Information Model,
.   planning of the interviews.

Bottom up analysis
.   adjustment and validation of the physical system model (IDEFØ),
.   realisation of the interviews (GRAI net),
.   adjustment and validation of the GRAI grid,
.   realisation of Structural Information Model on a limited domain,
.   adjustment and validation of the Conceptual Information Model,

Check-up of the analysis
.   detection of inconsistencies on the GRAI grid and nets using formal rules,
.   detection of inconsistencies between models of data, process and physical system using coherence tools.

DESIGN PHASE (future situation)

Global design
.   identification of objectives and constraints,
.   inconsistencies analysis,
.   proposal for physical system model (IDEFØ), simulation if necessary,
.   proposal for new architecture (GRAI grid) = building up of Conceptual Decision Model,
.   proposal for a new Conceptual Information Model,
.   validation and adjustment of the different models using coherence tools,
.   realisation of GRAI nets,
.   elaboration of the Structural Decision Model,
.   elaboration of the Structural Information Model (using translation rules from Conceptual Information Model to Structural Information Model),

. simulation of the physical system using computer tools if necessary,
. adjustment and validation of the models.

<u>Detailed design</u>

. improvement of the Structural Physical Model,
. improvement of the Structural Decision Model,
. improvement of the Structural Information Model,
. validation of the models using coherence tools,
. detailed specification of the treatment,
. detailed specification of the data,
. technical choice adjustment,
. planning for development and implementation.

7.3 GIM FOR PRODUCT DESIGN

Similar to manufacturing systems, a product design system can be decomposed into three subsystems. The physical system transforms the product information in order to satisfy the requirements of market and production. The control system (or design management system) is split up into two systems: the decision system and the information system. The design management system with a decisional process synchronizes the set of design activities from customer requirements through product definition to manufacturing. It is necessary to include explicitly the decisional process for the product development and for the design activities.

7.3.1 *GRAI Structure* for Design Management

The physical subsystem of a manufacturing system is usually defined, e.g. humans, machine tools, raw materials, components, etc. In contrast, physical elements which make up the product are yet to be defined. Physical elements are part of the product information which is necessary to manage. An extension of GRAI grid is proposed.

Figure 7.11 shows the GRAI STRUCTURE. There are three axes: activity, product and time. The temporal axis is graduated in three levels of decision centres: strategic, tactical and operational. The activities are associated in macro-activities. The set of macro-activities represents the macro-process defined in 1.2. The product is described by three abstract levels: conceptual (C), structural (S) and realizational (R) defined in 7.3.2.

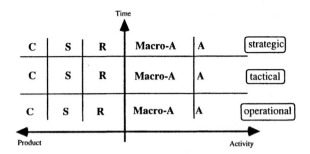

Figure 7.11 Overview of the GRAI STRUCTURE.

In the GRAI STRUCTURE we can identify the action plan (view of the action level) defined by activity and time axis. The action plan manages the design activities. The object plan (view of the object level) is defined by product and time axis. The object plan manages the product development. The design management system is defined by the simultaneous management in the action plan and the object plan (Girard and Doumeingts, 1994).

The tools and methods associated with GRAI grid permit the identification of decision centres. Each plan is a grid but they are linked to act simultaneously on design activities in the product development. To develop a product it is necessary to start a design activity. Its activity needs some resources which ought to be available. This depends on the other activities which work for the development of the set of products.

Figure 7.12 presents the different functions of the GRAI STRUCTURE. The object plan allows us to develop the product to meet the product specifications. Starting from a defined product state, we want to specify a new state with the knowledge of the previous state. The new state is known when the set of information which is sufficient and necessary to define it, is determined.

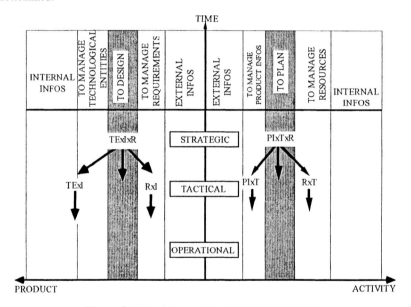

Figure 7.12 Functions of the GRAI STRUCTURE.

The function: to Design

This function determines the choice, the definition and the association of technological entities taking the requirements into account. It allows us to synchronise and coordinate the methods and the tools used in the product development.

The controlling criteria is not time, but the inadequacy of the information. The problem to solve is to choose the right strategy to suppress this inadequacy. It is necessary to choose an order in which to use methods or tools to satisfy a requirement in order to optimise time and obtain a good quality at lower cost. The vertical decomposition will be function of the degree of inadequacy. The higher level defines the smallest possible amount of information which can be determined, without limiting the remaining choices.

Its basic elements are Technological Entities (TE), Requirements (R) and Inadequacies (I).

The function: to Manage Requirements

A requirement exists when there is a demand. Demand exists for a new product or for an existing product performance not satisfying. A demand may appear at the time of the product development. These requirements are specific for each actor. It is necessary to express these requirements and to know which resources are available to solve it. Thus we can make a link with the action plan. The requirements are expressed for one activity and for one abstraction level.

Its basic elements are Requirements (R) and Inadequacy (I).

The function: to Manage Technological Entities

This function identifies, retrieves, modifies and creates technological entities. It must associate them to satisfy the product flows. The technological elements allow us to describe the product by their association. To exist, technological entities use various available resources.

Its basic elements are Technological Entities (TE) and Inadequacies (I).

The action plan allows us to manage the design activities, including using a design strategy.

The function: to Plan

This function determines the choice of a design action to define product information taking resources and knowledge into consideration. It synchronises and co-ordinates the design activities.

Its basic elements are Product Information (PI), Time (T) and Resources (R).

The function: to Manage Product Information

The product information is the product knowledge at one moment. It emphasises the problems to solve.

Its basic elements are Product Information (PI) and Time (T).

The function: to Manage Resources

The objective of this function is to optimise the use of the human abilities and material equipment. It is necessary to add the company know-how.

We can decompose this function into:

- to manage human resources
- to manage material resources
- to manage know-how

Its basic elements are Time (T) and Resources (R).

The GRAI STRUCTURE represents the action plan and the object plan. It allows to plan the design of the products in a company. This planing depends on the design requirements of each product but is realised simultaneously for all products.

7.3.2 The Product Model

Doumeingts *et al.* (1993) generalise of the Walrasian model for product modelling, resource modelling and for production control. As far as product modelling is concerned, the result of generalisation is a P_graph as shown in Figure 7.13. It is based on the relationships "goes-into" or "where-used" between entities. A node can represent a component / subassembly, or an activity to be performed. Likewise, the result of generalisation for resource modelling is a R_graph as shown in Figure 7.13. It reflects the organization of company resources in a hierarchical structure. Archs connecting P_graph nodes and R_graph nodes represent the "consumed by" relationships. The P_graph proposed here is not sufficient because the description of the products is only based on an architectural decomposition of product assuming that the product is known. During the product design phase is impossible to represent solution because they are unknown. The product specifications are known at the beginning. Therefore a functional product representation is better (Kusiak and Szczerbicki, 1992). A second remark is on the R_graph. The resources are humans and materials but in design activities the humans resources are the mains. Like the design process is iterative, based on mutual adjustment we suggest to group resources to solve each main problem.

A P_graph consists of two concepts: **technological entity** (T.E.) and **function** (Girard and Doumeingts, 1994). There are variety of technological entities, for example, geometry or assembly (Gama group, 1990). The function expresses the design objectives. A T.E. is defined by its interface with the outside. The connections and interactions are characteristics of functions. The technological entities and functions constitute a visualisation of the project progress. Figure 7.14 shows the evolution of a P_graph. Technological entities are represented in ellipses and functions in circles.

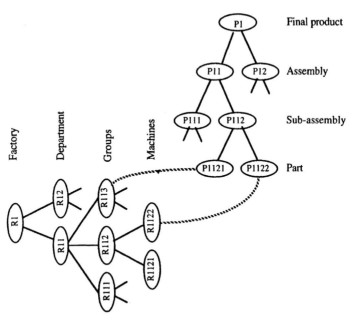

Figure 7.13 P_graph and R_graph.

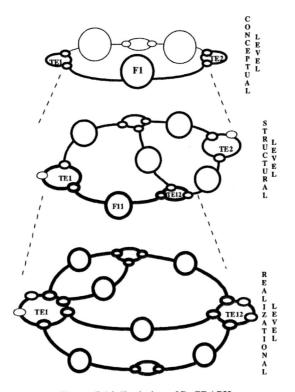

CONCEPTUAL LEVEL

STRUCTURAL LEVEL

REALIZATIONAL LEVEL

Figure 7.14 Evolution of P_GRAPH.

In the *conceptual* level the requirements and the final objective of design are defined. A functional analysis establishes and formalises the functions satisfying the product requirements. It is expressed by the proposition of potential solutions for each function. The *structural* level corresponds to the product architecture. A technical analysis gives a structure according to the functions defined at a conceptual level. It is a valuation of the solutions which finalises the definition of satisfying candidates to the requirements of considered solutions at conceptual level. The *realizational* level clarifies the answer of the functional requirements. Each structure is defined for giving a technological solution. We have to choose the solutions which propose a more precise product definition in accordance with the expected performance.

The third generalisation for production control adds another dimension: the time. Figure 7.15 shows a GRAI grid with three principal functions. The "to plan" function manipulates "Products" and "Resources" on the time (PxRxT), the "to manage Products" function manipulates "Products" on the time (PxT) and the "to manage Resources" function manipulates "Resources" on the time (RxT).

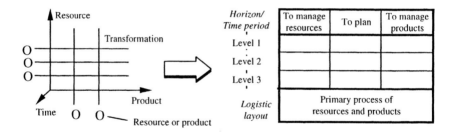

Figure 7.15 The Walras production model as production management system.

7.3.3 GIM Procedure for Product Design

The GIM product design procedure proposed is similar to the procedure for analysis and design of CIM systems discussed in Section 7.2. It is necessary to decompose the design system into two subsystems: the Design Physical System (DPS) and the Design Management System (DMS). DPS is the system which is piloted and DMS is the system which pilots. Marc Zanettin (1994) explains that it is possible to define these systems in two ways: (1) a design step oriented towards users and (2) a design step with technical orientation. During the design step it is necessary to take into account simultaneously the product, the processes and the organization. The proposed procedure tackles those three elements and their mutual influences. The product design procedure is not about which components to use but what functions must be satisfied by the product. The technological entities represent a part of the product at a specific level (conceptual, structural or realizational level). It is the knowledge of functions which define progressively the technological entities. The design procedure is progressive, from one state to another of the product (Kiriyama, Tomiyama and Yoshikawa, 1991).

Figure 7.16 shows the design procedure. Each step is subdivided into several types of activities: (1) design activities, (2) execution activities and (3) management activities. Decision centres can be identified in the action plan or the object plan.

TO SEARCH SOLUTIONS

The objective of this step is to generate solutions, it is recommended to use some creative methods. The choice of members of the group is very important because it conditions the possibilities to obtain solutions. They are chosen for their availability and for the type of the function to satisfy. This step is composed basically of design activities based on the mutual adjustment.

The different activities of the step: TO SEARCH SOLUTIONS

* to choose the work group
* to search product satisfying similar functions
* to imagine solutions to satisfy each function

On the Figure 7.16 circles represent functions. A big circle represents a product function at a specific state. A little circle represents a temporary sub-function, part of solution of a product function at a specific state.

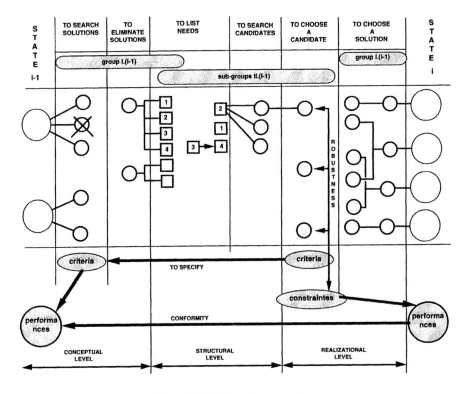

Figure 7.16 The design procedure.

TO ELIMINATE SOLUTIONS

Now we obtain possible solutions for each function. It is necessary to classify them to orient the design job. The classification of the solutions depends on technico-economicals criteria but also on managerial criteria. A multi-criteria methodology has been presented elsewhere (Doumeingts *et al.*, 1993; Ould Wane *et al.*, 1994). The valuation of the criteria is general and depends on the know-how of each member of the group. We eliminate temporarily some solutions to obtain a set of solutions that we develop.

The different activities of the step: TO ELIMINATE SOLUTIONS

*   To list criteria of comparison
*   to evaluate the criteria for each solution
*   to eliminate the solution which may not satisfy the performances.

These first two steps represent the conceptual level. Now the product is known with a set of solutions (solution function) in a position to satisfy requirements of the product in the state i-1. We search a product structure, solution of the conceptual level. It is the structural level.

TO LIST NEEDS

The objective is to define a design strategy to obtain in shorter time a good solution. To do that it is necessary to ordinate and to synchronise design. For each solution we identify the technological needs. In Figure 7.16 those needs are represented by a square. This step is basically composed by management activities.

   The different activities of the step: TO LIST NEEDS

- to identify the needs of all solutions
- to specify the action type to solve each problem
- to group the needs
- to ordinate and to synchronise the treatments of problems

TO SEARCH CANDIDATES

The needs are grouped by type and their treatments are planned. Now it is necessary to search a solution for each problem. It is a similarly step at the first step to search solution. We constitute specialised groups which ought to search candidates for a set of problems. The members of the groups are chosen according to their know-how for a type of problem.

   The different activities of the step: TO SEARCH CANDIDATES

- to choose the specialised work groups
- to search candidate satisfying similar problems
- to imagine candidates to satisfy each problem

   We prefer to use the term of "candidate" instead of "solution" to avoid ambiguity between steps which depend on different abstraction levels.

   We present now the final phase of the design procedure proposed. It is the realizational level. It permits to define a technological solution whose specifications specify the product at the state i-1. This solution is represented by a set of functions (big circles on the Figure 7.16) which define the product at the state i. This solution increases the definition of the technological entities which represent the product knowledge.

TO CHOOSE A CANDIDATE

This step is very similar to the step to eliminate solutions except the fact that we want to choose a solution candidate for a problem and not to restrict the research domain. First we evaluate the candidates and to class them. This allows to specify the criteria of the step to eliminate solutions and then to choose a solution. It is important to verify the robustness of the candidate considering the other candidates. The candidate can impose some constraints to its use. The different activities of the step: TO CHOOSE A CANDIDATE

- to list criteria
- to evaluate criteria
- to class candidates
- to verify the robustness of the candidate
- to identify the constraints of the candidate

TO CHOOSE A SOLUTION

With the set of candidates, the criteria of comparison and the initial performances, this step consists in proposing a more detailed solution conformed to the expected performances. It is the group formed during the first step which works. The different activities of the step - TO CHOOSE A SOLUTION include:

- to list the candidates chosen
- to aggregate the candidates in solution functions
- to verify the conformity
- to characterise the new functions defining the state i

7.4 SUMMARY

In this chapter, the GRAI approach to manufacturing system analysis and design has been reviewed and extended to adapt to new situations in product development. The GRAI STRUCTURE is an extension of the GRAI model. A product model has been proposed to support product development. A product can be decomposed into technological entities of varying abstraction levels from the chosen point of view. Design management takes place at two levels: the action level to manage the design activities and the object level to manage the product development. The series of design activities from the customer requirement analysis through the product definition to the manufacturing process are synchronised to optimise cost, time and quality.

REFERENCES

Bonami, M., De Henin, B., Boque, J.M., Legrand, J.J. (1993) *Management des systèmes complexes*, A reference book, Ed. De Boeck Université, Bruxelles.

Decreuse, C., Ferney, M., Feschotte, D., Jacobe, P., Zerhouni, S. (1994) Méthode d'identification de tâches de conception, In: *Proceedings of IFIP International Conference TC5-WG 5.3*, Valenciennes, France.

Doumeingts, G. (1984) *Méthode GRAI, méthode de conception des systèmes en productique*, State Thesis, University Bordeaux 1, France.

Doumeingts, G., Vallespir, B., Darricau, D., Roboam, M. (1987) Design methodology for advanced manufacturing system, In: *Working Conference at NBS: Information flow in automated manufacturing systems*, Gaithersburg, Maryland, USA, August 6-8 1987. Published in *Computers In Industry*, **9** (4), 271-293.

Doumeingts, G., Vallespir, B. (1992) CIM systems modelling techniques, In: *23rd CIRP Seminar on Manufacturing Systems*, WISU Verlag Aachen, Faculty Press International.

Doumeingts, G., Ould Wane, N., Girard, Ph., Marcotte F. (1993) Architecture and Methodology for Concurrent Engineering, In: *Proceedings of an International Conference: Towards World Class Manufacturing*, Phoenix (USA), 12-16.

Doumeingts, G., Chen, D., Vallespir, B., Fenier, P. (1993) GIM (GRAI Integrated Methodology) and its evolutions - A methodology to design and specify Advanced Manufacturing Systems, In: *JSPE-IFIP 5.3 Workshop on the design of information infrastructure systems for manufacturing*, Tokyo, Japan.

Dupinet, E. (1992) *Contribution à l'étude d'un système informatique d'aide à la conception de produits mécaniques par la prise en compte des relations fonctionnelles*, PhD Ecole Centrale de Paris.

The GAMA Group (1990) *La gamme Automatique,* A book reference, HERMES.

Girard, Ph, Doumeingts, G. (1994) Design management of mechanical products, In: *Proceedings of IFIP International Conference TC5-WG 5.3*, Valenciennes, France.

Kiriyama, T., Tomiyama, T., Yoshikawa, H. (1991) A model integration framework for cooperative design, Computer-aided cooperative product development, *Lecture notes in Computer Science 492*, Springer-Verlag, Berlin, 126-139.

Kusiak, A., Szczerbicki, E. (1992) A formal approach to specifications in conceptual design, *Journal of mechanical design*, **114**, 659-666.

Kusiak, A., Wang J. (1993) Qualitative analysis of the design process, In: *Proceedings of International Conference*, 1993 ASME Winter annual meeting, New Orleans.

Le Moigne, J.L. (1977) *La théorie du système général. Théorie de la modélisation*, Presses Universitaires de France, Paris.

Mayer, R.J. (1990) *IDEFØ function modeling*, Longmire Drive, Texas, Knowledge Based Systems, Inc.

Mesarovic, M.D., Macko, D., Takahara, Y. (1970) *Theory of Hierarchical, Multilevel, Systems*, Academic press, New York.

Mintzberg, H. (1982) *The structuring of organizations: a synthesis of the research*, Prenctice-Hall Inc.

Morin, E. (1977) La méthode, Tome 1: "La nature de la nature", Paris, Ed. Seuil.

Mony C. (1992) *Un modèle d'intégration des fonctions conception-fabrication dans l'ingénierie du produit*, PhD Ecole Centrale Paris.

O'Grady, P., Young, R.E., Greef, A., Smith, L. (1991) An advice system for concurrent engineering, *International Journal Computer Integrated Manufacturing*, **4** (2), 63-70.

Ould Wane, N., Bertin, R., Pun, L. (1994) Méthodologie multicritère d'aide au management de l'ingénierie simultanée, *IPMU'94 Conferences*, Paris France, 4-8.

Petitdemange, C. (1990) *La maîtrise de la valeur*, Edition AFNOR GESTION.

Pun, L. (1984) *Systemes industriels d'intelligence artificielle, outil de productique*, Editest, Paris.

Simon, H.A. (1960) *The new science of management decision*, Harper and Row publisher, New York et Evanston.

Tardieu, H., Rochfel, A., Colletti, R. (1983) *La méthode Merise, principes et outils*, Paris, Les Editions d'Organization.

Vallespir, B., Chen, D., Zanettin, M., Doumeingts, G. (1991) Definition of a CIM architecture within the ESPRIT 'IMPACS' project, In: *Proceedings of the fourth International IFIP TC5 CAPE '91*, 731-738.

Zanettin, M. (1994) *Contribution à une démarche de conception des systèmes de production*, Phd of University Bordeaux I, n° 1089, Bordeaux, France.

DESIGN FOR DIMENSIONAL CONTROL

Paul G. Leaney

This chapter presents a Design for dimensional control (DDC) technique. The idea of dimensional variability is first introduced. Some issues in the specification such as geometric dimensioning and tolerancing, interpretation, measurement and control of dimensional variation are outlined. Some endeavours in developing methodologies and tools to enable DDC and the control of variability in the broadest sense are indicated. Industrial efforts are briefly reviewed and advice on good practice is provided.

8.1 OUTLINING DDC

Design for dimensional control (DDC) refers to the total product dimensional control discipline which recognises and manages variation during design, manufacture and assembly. It aims to meet customer quality expectations for appearance and function without the need for finesse, by the shop floor operatives, in the manufacturing and assembly operations. It is part of a large and growing field of endeavour pertinent to the design and manufacture of products as diverse as, for example, cars, planes, printers, switches. In one form or another DDC is relevant to all manufactured goods. This burgeoning field is variously referred to as: DFV - design for variation; DM - dimensional management; DVA - dimensional variation analysis; etc. DDC embodies a range of tools and techniques and embodies an imperative for management to provide the appropriate organisation of engineering effort (affecting both the organisational structure as well as the process of developing products through design into manufacture using teams and goal directed project management) that is consistent with the tenets of concurrent engineering.

Major elements of production costs come from the failure to understand design for dimensional variation. This variation results in irreversible tooling and design decisions that forever plague manufacturing and product support. The aim of DDC is not to eliminate dimensional variation, as that is impossible, but rather to manage it. Managing variation (i.e. understanding and controlling variation) will lead to:

- Easier manufacture and assembly (e.g. less scrap and rework)
- Improved fit and finish
- Reduced need for shop floor finesse
- Improved flow-through (e.g. less work-in-progress)
- Reduced cycle time
- Reduced complexity (e.g. less design changes, simpler manufacturing operations)
- Increased consistency and reliability
- Better maintainability and repairability

8.1.1 Background

Often, in a manufacturing environment subject to pressing schedules, expedient measures are taken to get the product shipped. In such an environment the notion that many people will follow is 'if it is not broken then do not fix it'. What this notion of expediency often leads to is the inability to review the process by which things get done. This has left shop floor with a legacy of operator finesse that is largely overlooked. For example, drawings may be issued to the shop floor that contain dimensioning and tolerancing (D+T) information, but these drawings still need to be interpreted. Decisions need to be made on what measurements are to be taken and how. Production pressures will drive non-value added operations, such as inspection, to the minimum. If dimensional variation does not stop production, either in assembly or in product testing, then it is possible that no problems are perceived. However what this might hide is a high level of operator finesse. Operators build up expertise in dealing with variation and such expertise is often a source of pride, and deservedly so. What is lamentable is the product and process engineering that combines to demand such expertise from operators. This 'hidden factory' can, in fact, contribute to a variety of problems affecting operational efficiency and product quality. In addition it hides problems from the product designers. There is a need to open up the existence of this 'hidden factory' and put responsibility for its minimisation on the shoulders of both product and process engineers in equal measure. The aim of DDC is to provide the tools, techniques and management imperatives for doing just that. DDC should be seen as an engineering methodology combined with computer based tools used to improve quality and reduce cost through controlled variation and robust design.

The term robust design (Taguchi, Elsayed and Hsiang, 1989) relates to the design of a product or the design and operation of a process that results in functionally acceptable products within economic tolerances on economic equipment. This is usually done by using statistical analysis tools in conjunction with experimentation to allow empirical modelling of complex products or processes not easily modelled deterministically. The aim is to identify parameter values that make the product or process insensitive to natural variation encountered in the manufacturing environment or in the operation of the product. DDC is, for all intents and purposes, the application of robustness thinking to dimensional variation. The approach is to seek the best overall economic solution to achieving control of dimensional variation through appropriate product design in conjunction with process design and process operation such that the resulting variation does not give rise to any concerns or symptoms through manufacture and assembly (i.e. no finesse necessary), test/measurement or product operation (i.e. the product functions well and parts are interchangeable). Often this will be interpreted as meaning that product parts and assembly methods and fixtures should be designed to absorb areas not critical to product function thereby allowing larger tolerances without impairing product function. In reality DDC plays a key role in robust design and provides the cornerstone for linking related methods (e.g. DFA, SPC, Taguchi methods) together.

One particular area addressed by many commercially available DDC tools is the analysis or simulation of the variable dimensions in an assembled product. Some of the reasons are:

- To ensure interchangeability and assemblability of parts.
- To ensure appropriate clearances between adjacent parts not directly dimensioned or controlled during assembly.
- To determine the impact of variability upon aesthetics.
- To ensure that functional dimensions and tolerances are appropriate for manufacturing process and sequence. Assembly techniques such as selective assembly and shimming also require careful analysis so that they can be successfully implemented in manufacturing operations.

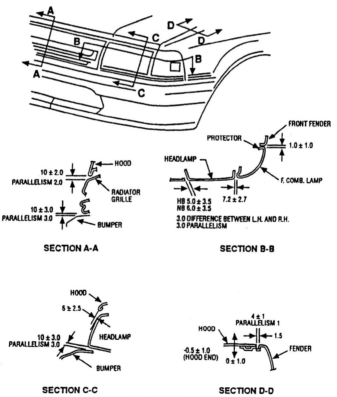

Figure 8.1 Example of dimension control requirements on an assembly.

Such tolerance analysis is based on information from design engineers about allowable variation for functional purposes and information from process and production engineers about manufacturing capability. An example of dimensional control requirements is illustrated in Figure 8.1. The necessary co-operation required between product and process engineers is facilitated, at least in part, by the use of a common language through dimensioning and tolerancing. The appropriate use of D+T has always been an important issue but now there is a growing demand to use D+T to understand how to define, verify and manufacture products through the discipline of DDC.

8.1.2 Dimensioning and Tolerancing - The Language of DDC

Dimensioning and tolerancing (D+T) standards are seen to provide the language and the practice for communicating allowable variation on drawings, with respect to the actual function or relationship of part features. Implicit in good D+T practice is the 'tolerance process independence dictum' which states that a designer should pay heed to 'define the result you want, not how to get it'. This principle of process independence is coming under pressure as it seems inconsistent with the doctrines of concurrent engineering (i.e. design the process with the product). This strain has come about because D+T annotation on drawings was often treated as a one way communication process, i.e. it came with the drawings issued over the wall from design into manufacturing. However, the discipline of DDC now provides the mechanism to use the language of D+T and to close the feedback loop from manufacturing back into design.

DDC is built upon the dimensioning and tolerancing (D+T) formalisms and syntax as communicated on engineering drawings according to various standards, for example ASME (1995a), BSI (1985), BSI (1990) and ISO (1983). Standards play an unusually important role in tolerancing and metrology because they have been based on evolving practice rather than scientific principles codified with rigorous mathematics. National and international standards thus provide a distillation of best practice. Lists of relevant standards appear in ASME (1995a) covering relevant ASME/ANSI standards and in BSI (1993) for British Standards and International (ISO) standards. Henzold (1995) also lists some German (DIN) standards and East European Standards with brief comparisons which highlight some differences, particularly between ISO and ASME/ANSI. However as time passes a process of commonisation occurs where there is agreement on best practice, for example Foster (1994) who outlines the metric application of GD+T techniques as based upon harmonisation of (US) national and international standard practices.

It is not the aim of this chapter to review D+T other than outlining the underlying purpose and principles. Two families of tolerancing schemes have been developed for industrial use, namely parametric and geometric. Parametric tolerancing is based on ordinary or size dimensions. There are three versions: worst case limit stacking, statistical tolerancing and vector tolerancing (Henzold, 1993). These schemes are called parametric because dimensions can be regarded as control parameters for an underlying mathematical representation. Geometric tolerancing was developed to ameliorate some intrinsic weaknesses in parametric tolerancing, particularly in relation to form. A tolerance of size, when specified alone, affects some degree of control of form but in many circumstances dimensions and tolerances of size, however well applied, would not impose the desired control. A different degree of control of form is required - this is covered by geometrical tolerances. A geometrical tolerance may be specified even if no special size tolerance is given e.g. flatness on a surface table. Geometrical tolerances should be specified for the requirements critical to functioning and interchangeability.

The aim of D+T, therefore, is to provide the language that means that geometries (i.e. form, size, orientation, location or position, waviness, roughness, edge deviations, surface continuity) of the geometrical elements of a part or assembly are completely defined and toleranced. It should leave no ambiguity as it assumes the person reading the drawing has no knowledge of part function. A typical engineering drawing with appropriate annotation is illustrated in Figure 8.2. DDC is concerned particularly with form, size, orientation and location as being the primary sources of dimensional variation.

EXAMPLE (FLANGE)

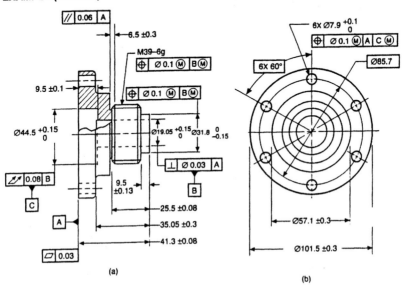

(a)

(b)

ASSEMBLY

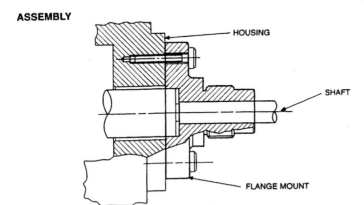

Figure 8.2 Example of a drawing using standardised notation for dimensions and tolerances
(Foster, GEO-METRICS III M, Copyright 1994 Addison-Wesley Publishing Company Inc. Reprinted by permission of Addison-Wesley).

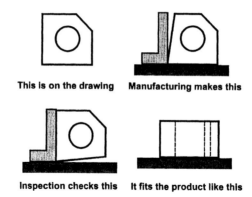

Figure 8.3 Need for communication: GD+T defines parts with respect to functional surface.

D+T and their measurement (metrology) are key technologies/tools in the quest for continuous quality improvement (Voelcker, 1993). However, the best use of tolerancing and metrology have been seen to suffer from an informality borne out of its evolution from shop floor practice (Krulikowski, 1991) . As such it is suggested (Srinivasan and Voelcker, 1993) that they require a level of human interpretation that is not compatible with modern computer aided design (CAD), computer assisted process planning (CAPP) and coordinate measuring machines (CMM) based inspection procedures.   This is manifest in two underlying phenomena:

- Methods divergence - the fact that measurements made by different methods (e.g. manual micrometer, functional gauge, CMM) or processed by different algorithms (as in CMMs) often yield different results.
- Specification ambiguity - that is ambiguity in the definition of what is to be measured.

It is certainly true that interpretation of GD+T by CMMs require a thought process quite different from that used in the traditional world of functional gauging. CMMs do not always interpret GD+T specifications correctly and some ambiguities in the standards became evident. In dealing with these problems, at least in part, a recent US standard has been adopted (ASME, 1995b). This ANSI/ASME standard provides the first efforts in providing the mathematical basis for the Y14.5 standard (ASME, 1995a). The aim of doing so is to provide unambiguous definitions. Measurement procedures consistent with the mathematical Y14.5 should be designed to assess conformance.

With regard to the standards a number of issues remain around the topic of defining assemblies and tolerancing for assembly. The Y14.5 standard (and other standards) focuses on part specification and virtually nothing is included about assemblies. Assembly drawings are almost always ambiguous unless covered with notes. Assembly drawings are often supplemented with an assembly process plan that specifies an assembly sequence and some feature-mating and feature-joining conditions. Some CAD systems do now provide the means to declare feature matings in a hierarchical graph structure (Allen, 1993). However defining feature-matings by themselves are not enough as matings often must be ordered in the same way that components of datum systems are ordered. On tolerancing for assembly it is possible to make use of the MMC (maximum material condition) criteria for assembling simple isolated features and the stack-up rules for one-dimensional chains but very little else is covered. Thus assembly analyses involving complex toleranced parts almost always default to Monte Carlo simulation (Turner, 1993) and practical, systematic tolerance synthesis procedures are being sought through research (Chase and Parkinson, 1991). Figure 8.3 simply illustrates the need for unambiguous communication between those involved in design, manufacture, inspection and assembly.

There is an increasing industrial demand, driven by the continued need for quality improvement and cost reduction, for variability control built upon the modelling, analysis and simulation of tolerance in product build. Tools now are being developed that provide the backbone of the discipline variously named as dimensional management, design for variation - otherwise referred to as design for dimensional control - DDC, and this is the subject of the next section.

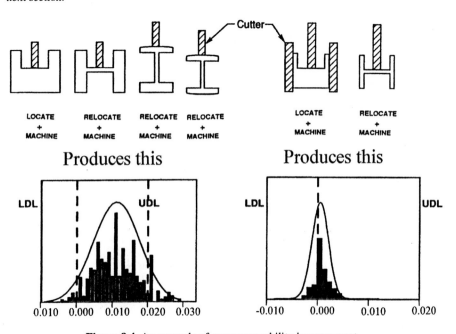

Figure 8.4 An example of process capability improvement.

8.2 TOOLS FOR THE CONTROL OF VARIATION

Robust design means that variability must be accommodated in design and controlled in production and mechanisms for doing this are woven through the entire production system. In the case of dimensional variation the primary control tool in design is D+T whereas metrology is the primary tool for assessing conformance to toleranced specifications and for gathering data for process control. However before outlining tools for the control of dimensional variation (e.g. tolerance analysis and simulation) it is instructive to outline a framework for DDC that dovetails with other related methods in promoting a robust design approach.

8.2.1 A Framework for DDC

One key to robust design is variability reduction. Here we are concerned with variability in size and form, of a product's design, rather than the make-up of the materials or their intrinsic characteristics like strength, fatigue performance or processibility. Complementary approaches to reducing dimensional variation include:

- Design for assembly. This focuses on assemblability, product simplification and part count reduction so that there are fewer parts to control, stock, plan, tool, produce and assemble.
- Cross functional product development teamwork. This is necessary to address questions of fabrication, assembly and tooling at the design stage. The team should also have responsibility to select and apply common datum's through design and manufacture.
- GD+T. This is a drawing language to communicate a part's functional requirement (of size and form), define common datums, controls tooling and assembly interfaces and provides uniform international interpretation.
- Process capability. The objective is to design parts with respect to known process capabilities (using SPC data for example) and to change design or process to achieve compatibility, for example see Figure 8.4.
- Tolerance analysis and simulation. Part geometry, assembly tolerance and assembly sequence/methods contribute to the statistical evaluation of dimensional variation for rapid evaluation of alternative designs (see Section 8.2.3).
- Design for manufacture. Incompatibility of design and process will be manifest in dimensional defects as measured against design tolerance requirements. Therefore DFM means identify part features, determine tolerance requirements, identify process for feature generation, select process capable of generating feature and, if necessary, change the design, improve the existing process or develop new process. For example, an increasing interest in high speed machining over conventional machining is based not only on the higher material removal rate but also on the reduced part distortion and better finishing; it also opens up the possibility of single machined parts replacing fabrications previously made from many parts and reducing dimensional problems.
- Key characteristics. A key characteristic is a measurable or observable attribute of a detail part, assembly, component or process for which it is desirable to minimise variation from a nominal value in areas that have a great influence on fit or performance. This approach can make full use of Taguchi and design of experiment (DOE) methods as well as SPC and statistical simulation tools. The aim is to guide design changes and process control efforts to areas of greatest impact.

Addressing the management of dimensional variation needs to be undertaken within a framework that embodies an holistic view of the product development and delivery process. Study of dimensional variation will draw in a wide variety of quality concerns that cover aesthetics, function, manufacture and assembly. For this reason all engineers in the value added chain, including suppliers and sub-contractors, should understand dimensional variation principles. This sometimes revolves around the use and interpretation of GD+T but the principles of DDC are much broader than that. DDC requires communication amongst people who share a collective, but clearly defined, responsibility to the whole product development and delivery process. This is sometimes manifest through the existence of a dimensional control team with wide ranging powers and a membership representing the broad interests. Success in DDC is built upon the three supporting elements of:

- **Tools** - e.g. GD+T, metrology and SPC, assembly tolerance analysis and simulation, CAD.
- **Organisation** - of the people involved in the engineering effort in design and manufacture.
- **Process** - e.g. establish a formal dimensional control process tied directly to the product development cycle

The actual balance between these supporting elements of DDC for a particular company will depend on the individual circumstance. It is of interest to note a difference of style between the US and Japan. For example, the 'big three' automotive producers centred in Detroit, Michigan (i.e. Ford, Chrysler and General Motors) have tended to lean heavily on the use of tools, particularly tolerance analysis and assembly simulation, which then drives a dimensional control process which in turn drives changes in the organisation of engineering effort, see Section 8.2.4 later. However the Japanese tend to centre their effort on the organisation and process. This is illustrated, for example, in Nissan's approach to the integration of design, manufacture and assembly (Imai, Shimizu and Araki, 1994). They recognise that the strict division of jobs according to function and duties of engineers in America and Europe has traditionally made it difficult to feed back information from the factory floor to the design process and to carry out tasks jointly as done in Japan. They go on to state that DFMA (design for manufacture and assembly) tools, by themselves, does not appear to be a method for aggressively incorporating manufacturing needs in the product design because product design engineers can only guess what the impact the product design might have on the manufacturing processes. In Japan the product design engineers are responsible for vigorously collecting from the production engineers all the information needed to execute designs for manufacture and assembly and for securing the latter's active involvement in the product design process. Production engineers for their part are responsible for presenting the production requirement in conjunction with conceptual studies of the manufacturing processes.

One area for such collective attention is dimensional variation and its control. They will collectively decide on the tolerances and reference points that are key factors for quality assurance through product build - as exercised, for example, on the installation points of the automotive body panels. These tolerances are not regarded as an issue that only pertains to the assembly processes. Rather, dimensional accuracy control is practised all the way back to the stamping process in order to raise the process capabilities for achieving the desired dimensional tolerances of the installation points.

The message from all this appears to be that Western style companies tend to use CAE and related tools to drive forward product and process integration in a way that may not be reflected in Japanese and other far eastern companies. This gives the DDC tools, and particularly assembly tolerance analysis and simulation, a particularly important role which will now be considered.

8.2.2 Assembly Tolerance Analysis

The objective of performing tolerance analysis is to determine if the design, manufacturing and assembly process optimally achieves final product build requirements. The accumulation (or stackup) of tolerances in an assembly come from the dimensional and geometrical variations of the constituent parts and from the variations that occur during, and due to, the assembly process and procedures. The term assembly tolerance analysis relates to the study of this variation. The tolerance analysis method may be classified by the dimensional space, and related degrees of freedom - dof, being considered:

- 1D   - 1 linear translation, dof = 1.
- $1^1/_2$ D - 2 translation, no rotation, dof = 2.
- 2D   - 2 translation, 1 rotation, dof = 3.
- 3D   - 3 translation, 3 rotation, dof = 6.

Tolerance stackup may be estimated by a worst case or statistical model. Consider the tolerance stackup of two unit sized blocks:

- Worst case, $1 +/- 0.01 + 1 +/- 0.01 = 2 +/- 0.02$
- RMS (root mean square), $2 +/- \sqrt{\{(0.01)^2 + (0.01)^2\}} = 2 +/- 0.014$

The RMS method has some statistical basis related to the fact that variation which fits a normal distribution can be 'stacked-up' by adding the variances where the variance is the square of the standard deviation (i.e. σ^2). For example, if the variation on the unit block follows a normal distribution then the resulting variation on the assembly of the two blocks would fit a normal distribution having a standard deviation that equals the square root of the sum of the squares of the standard deviations of the constituent blocks. It is not unusual to interpret the design requirements on the blocks of $+/- 0.01$ as representing $+/- 3\sigma$ on a normal distribution implying a reject rate of around 3 in a 1000 for a manufacturing process having a capability index, $C_P = 1$. Note that if there is any mean shift it is more appropriate to use the modified capability index C_{PK}. When quality programs talk of 6σ (six sigma) quality this means that the process capability (C_P) needs to equal 2 and the upper and lower design limits represent $+/- 6\sigma$ on a normal distribution implying a statistically insignificant reject rate akin to zero defects.

It is generally recognised that the worst case method for the estimation of tolerance accumulation overstates the observed variation on assemblies so that statistical methods are usually preferred. In estimation of tolerance accumulation both the magnitude and direction of tolerances need to be accounted for and this can be done using vector loop models or solid models. In vector models the dimension and tolerance vectors are arranged in loops or chains which relate to those dimensions which 'stack' together to determine the resulting assembly dimensions. Solid models are used, for example, to underpin the tolerance models used to simulate product build. Through such models it is usually possible to specify the particular measurements of interest on the assembly which will be given by a defined function and, with such specific measurements defined, there are several methods available for performing a statistical tolerance analysis. These include (Chase, Gao and Magleby, 1995):

- Linearisation method
- Method of system moments
- Quadrature
- Monte Carlo simulation
- Reliability index
- Taguchi methods

It is outside the scope of this chapter to review these methods other than to note that a number of alternative methods exist. However the Monte Carlo simulation method and the Direct Linearisation Method (DLM) of Chase et al. (1995) are worthy of further comment as these methods underlie some computer based DDC tools that are commercially available and receive broad acceptance in industrial practice.

The Monte Carlo simulation method is a widely adopted method that evaluates individual assemblies using a random number generator to select values for each manufactured dimension, based on the type of statistical distribution assigned by the designer or determined from the production data. By simulating a large number of assemblies the output can be represented as a statistical distribution either of the assembly variable (measurement) of interest or to represent the relative contribution to the final variation from the constituent parts making up the assembly. One distinguishing aspect of simulation is that it is not dependent on the algebraic manipulation of equations but rather a build up of samples. The samples are taken from a model based on defined Cartesian points. These points define the location and orientation (with respect to other features and datums) of the parts making up the assembly. The choice of which points to use is left to the person building the model but the work is eased if done directly from CAD data. In running the simulation these points have statistically based variation (due to the tolerance on parts and variation introduced during the assembly process) superimposed on their nominal values in building up overall assembly variation across hundreds or thousands of product builds.

The DLM method is quite different to simulation. It is, in effect, based on a kinematic model of the assembly. The kinematics present in a tolerance analysis model of an assembly is different from traditional mechanism kinematics. The input and output of the traditional mechanism are large displacements of the links. The links themselves retain constant dimensions. In contrast to this, the kinematic inputs of an assembly tolerance analysis model are small variations of the component dimensions around their nominal values and the outputs are the variations of assembly features, including clearance and fits critical to performance, as well as small kinematic adjustments between components. The small variations allow a very good linear approximation based on the first order Taylor series expansion of the assembly function - an equation representing the assembly dimension of interest.

The DLM method has some particular attributes. For example once the model is set up it is extremely quick to run on the computer as compared to a simulation run which may have to build hundreds or thousands of assemblies. This speed makes the DLM convenient to use for design iteration and optimisation. In addition the DLM method enables a degree of tolerance synthesis (or allocation) by making use of the inherent sensitivity analysis built into the mathematics of the model. However such sensitivity information can also be made available from a simulation model and generally what can be done with one tolerance analysis method can also be done with another. One advantage of the simulation method is that once a model has been built a number of measurements may be taken from one simulation run, although a sensitivity analysis would normally require a separate run of the model. The DLM method is focused towards one particular assembly function. A different function (or measurement) would require additional constraints to modify the model for a re-run, although the model would not need to be completely rebuilt. Arguably an assembled product that embodies some kinematic variation, such as a cam operated switch or a ball in a guide or race, would be more conveniently modelled via a kinematically based method such as DLM. The simulation method is accurate and has a non-linear capability not reflected in the linearisation method. However Gao et al. (1995) report that the linearisation method compares well to simulation apart from the non-linear capability.

8.2.3 Computer Based Tolerance Analysis Tools

The aim of this section is to preview a sample of tools to reflect their potential role and importance within a broader DDC framework. Clearly any tool used to evaluate tolerance requirements and effects would be most usefully used in the early stages of a product's development. To be useful through the concept design stages of product development it should include the following characteristics:

- Be able to bring manufacturing considerations into the design stage by predicting the effects of manufacturing variations on engineering requirements.
- Provide built-in statistical tools for predicting tolerance stackup, its relative contributory factors and percent rejects in assembly.
- Be capable of performing 2D and 3D tolerance stackup analyses.
- Be computationally efficient to enable convenient design iteration and optimisation.
- Use a generalised and comprehensive approach to be capable of modelling a variety of assembly applications and tolerance requirements.
- To embody a systematic modelling procedure that is accepted by engineering designers.
- Include ease of use as an equally important attribute, e.g. integrate with CAD and fully utilise a graphical user interface.

Mentioned here are three such tools that meet all the above criteria in various measure:

- VSA   - Variation Simulation Analysis (Texas)
- DCS   - Dimensional Control Systems (Chase, Gao and Magleby, 1995)
- TI/TOL - Texas Instruments / Tolerance Analysis (Srinivasan and Voelcker, 1993)

This list covers a number of software modules but the core tool for VSA and DCS is assembly tolerance analysis based on the Monte Carlo simulation method whereas TI/TOL is based on the linearisation method outlined in Section 8.2.2. VSA and DCS are Detroit based companies providing the 'big three' automotive manufacturers with extensive guidance in what they have originally identified by the term 'dimensional management' and some of this effort is reflected on later in Section 8.2.4. On the other hand TI/TOL grew out of efforts by Texas Instruments so that some of its market is focused on those with interests in mechanical aspects of electronically based products. However the market for these generic tolerance analysis tools is continually growing so that the functionality, development and application of such tools will also grow. It is useful to briefly outline the state of these three particular tools and the role that is promoted for their use. Focus is given to the 3D tool although generally 2D and/or 1D versions are also available.

VSA. VSA promote the idea that tolerance analysis without a well structured DDC program will have little value in the organisation. Tolerance analysis is just one tool used to support the DDC process. The following six point plan must also be established:

1. Well defined product dimensional objectives.
2. The ability to determine if the product and process optimally meets product build objectives. This includes knowledge of design and manufacturing alternatives and their impact on quality and cost.
3. Accurate product and process documentation
4. A measurement plan that supports and accurately reflects the product and process intent.

5. Manufacturing capability that achieves design and process intent.
6. A well defined production to design feedback loop.

VSA is a strong advocate of setting up a dimensional control group that should have program level or complete product cycle responsibility and be cross functionally funded from both design and manufacture (Craig, 1992). Amongst the software tools they provide are VSA-GDT and VSA-3D available on PC and UNIX platforms. VSA-GDT uses a math rule base derived from the appropriate GD+T standard to produce warning messages if a geometric feature on a part is not controlled or constrained correctly according to that standard. Among other things it aims to minimise ambiguity. VSA-GDT prepares appropriate information (features, tolerances, constraints, degree of freedom) for the VSA-3D tool. The VSA-3D tool provides the simulation which runs a math model of the assembly and is written in VSL (variation simulation language) which is a programming language developed specifically for that purpose. VSL has its roots with the VSM (variation simulation method) of General Motors from the late 1970's and early 1980's. A large portion of the VSL model is interactively created using a 3D graphics pre-processor. Although VSA-3D can be a stand-alone tool it is also available as an integrated part of certain CAD systems. For example the simulation software is fully integrated into the CATIA CAE system, courtesy of the Valisys tool of Technomatix. Valisys is used to define nominal features for the CAD database, relate GD+T call-outs to nominal features and check syntax for individual features. The VSA software is used to: check GD+T for consistency; create varied component geometry using tolerance rule base; define assembly sequence; define assembly method; define measurements; and perform analysis. The overall modelling and analysis flow is illustrated in Figure 8.5.

DCS. DCS promote the idea of a dimensional control procedure for any company wishing to develop their own dimensional control program. They also use the phrase 'dimensional management' to embody all that is necessary to enable the dimensional control procedure. Like VSA, the advice from DCS has been originally targeted at the automotive sector (for example, stamping and body construction) although their interests are broader than that. DCS define their dimensional control procedure in 10 steps:

1. Identify and document dimensional quality goals.
2. Team consensus and signatures.
3. Develop strategic plans to achieve all dimensional quality goals.
4. Determine global tolerance and major datums for major sub-assemblies.
5. Generate datums and tolerances for all parts and assemblies, statistical simulation, work towards buy-in from all team members - this is the key engineering phase.
6. Optimise design/process through 3D analysis.
7. Verify prototype tool and fixture designs - validate gauge and fixture capability.
8. Evaluate prototype results.
9. Verify production tool and fixture designs - validate gauge and fixture capability.
10. Support during pilot, launch and production.

Feedback from steps 8 and 10 lead back to step 6 so that in an iterative way the product and process can be redefined as necessary. These 10 steps show the general precedence of activities and is tightly managed against a defined program timing plan.

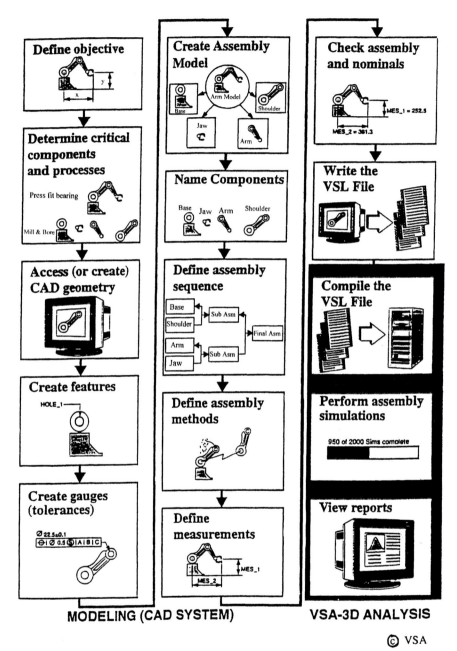

Figure 8.5 VSA Flowchart of modelling and analysis.

The statistical simulation tool has a major contribution to make from step 5 onwards although earlier steps embody some necessary decisions in preparation for that. DCS have a software product called 3D-DTS (dynamic tolerance simulation). This is a stand-alone PC based program that operates on graphic representations imported into the program via an IGES file. The software enables all the functionality for building and running models using menus. It is promoted on ease of use and good graphic capability. The graphic nature of input and output provide a focus for product and process engineers in communicating, analysing and resolving shared problems. It does not support its own programming or model building language as this is seen to be unnecessary.

TI/TOL. TI/TOL is a software product that has a full title of TI/TOL 3D+ Tolerance Management System. It was created using Pro/DEVELOP and is fully integrated with the CAD system Pro/ENGINEER and works directly off the solid model design database. Particular functions include 6-sigma optimisation tools to allow trade-off between product requirements and manufacturing process capabilities. It is promoted as a graphically based system (like DCS) that is intuitive and easy to use and does not require additional programming. Although addressed at the same type of use as that of 3D-DTS and VSA-3D the TI/TOL tool is not based on simulation but rather on the linearisation method as outlined in Section 8.2.2. TI/TOL is sold as a software product that would appeal to individual engineers or groups of engineers that need a fast and intuitive tool for the solution of technical tolerancing problems. Although not promoted on the back of a broader support service for dimensional management (as is DCS and VSA) the TI/TOL tool is aimed at providing any company with the basis of a disciplined and structured approach to tolerancing as the catalyst for further in-house developments in achieving DDC.

8.2.4 Some Industrial Efforts

The potential impact of DDC may become apparent by reflecting on some industrial endeavours. The aim of this section is to indicate the scale of effort by some automotive and aerospace manufacturers and provide some insight into their approach.

A large initial impetus came from the automotive industry based in Detroit. GM were involved in an in-house method called VSM in the 1970's and this method, in one form or another, continues today. However their initial effort flagged the need for dimensional control and identified simulation as an appropriate tool. This can be seen as having spawned the DCS and VSA companies. The Detroit based demand for DDC, or dimensional management, is so large that it also supports another company called Trikon but this company concentrates on providing dimensional control resource rather than marketing any tolerance analysis tools. Together these three companies provide dimensional control engineering resource to industry that literally amount to hundreds of people. Initially their biggest customer was automotive but aerospace is growing fast. Certainly a study of the companies developing their own DDC approach reflects the terminology of the consulting companies and, for example, there is repeated reference to: either a six point plan or a ten point plan; dimensional management; dimensional control procedure; dimension control team or group.

Ford has a whole Dimensional Control Department (D.C.Dept.) which is centred in the Body and Assembly Operations (BAO) at Dearborn. They regard the welded car body as the foundation of the car and all its assembly including powertrain, chassis and trim. A typical dimensional control issue might involve body construction and interior trim. For example if the instrument panel (dash) is attached to one sheet metal part and the consul attached to another then does the dash and consul match? These issues cut across functional divides and one key role for the D.C.Dept. in BAO is to provide dimensional control resource (i.e.

engineers and leaders) to co-ordinate effort of many diverse people. In addition to the BAO interests in DDC the powertrain organisation in Ford follows its own principles. For example they have required suppliers to demonstrate dimension control by undertaking an assembly tolerance simulation. The dimensional control engineers and leaders will hold the dimensional control team together. They have responsibility for producing drawings, for locator and datums and for what gets measured and when and how but they do not control the tolerances. That responsibility lies with engineering design and has to be led for functional reasons. Product engineers provide specific design information, customer driven vehicle requirements, design architecture and the specified tolerances. Manufacturing and assembly engineers provide process information such as location schemes and process capability data. Collectively they develop the product design and determine the locators, measuring points and tooling process. Ford is reorganising itself via 'Ford 2000' in aiming to become a truly global company. The opportunity of this reorganisation has promoted the role and influence of the D.C.Dept. as being one of the more process oriented (or systems approach) based efforts in the management of engineering effort, as opposed to the more traditional functionally based departments. In addition the D.C.Dept., which was originally part of Ford's North American Operations, is playing an increasing role in quality and in the engineering systems of that part of Ford that used to be referred to as 'Ford of Europe'. For example, significant dimensional control efforts have been focused at Ford's luxury car maker Jaguar, in England.

DDC, under the banner of dimensional management, was used successfully by Chrysler in the LH series and Neon and Cirrus automobiles and has now been used on many further Chrysler platforms. A key to their dimensional control procedure is their Dimensional Co-ordination Manual (DCM) and little 'Blue Book' of requirements both of which are produced for each car line. The DCM includes stated design intent like door gaps and flushness, headlamp fit and other customer / market requirements. This rolls down to the particular requirements on product build and tolerance assignment and control. Dimensional control teams make up the 'customer objectives' and all are asked to sign off the agreement up front on what is required. The agreed objectives are carried around by everyone in the little Blue Book lest anyone forget. The objectives form is signed by, for example: engineering, process engineering, quality and reliability, dimensional analysis, stamping and part fabrication, manufacturing feasibility, design studio and assembly plant. The DCM is structured in (i) objectives; (ii) build strategy; (iii) major subassemblies; (iv) parts. There is seen to be healthy tension between this top-down approach and the tendency to design cars from the parts upwards. The DCM provides an excellent mechanism for an overview of a car program development and it helps drive a systems engineering approach. There is a core dimensional group but their engineers are co-located in engineering. Typically a car-line development takes 3 years and the number of full time dimensional control engineers assigned might be 9 in the first year, up to 20 in the second year and down to 9 or so in the third year. At any one time there could be 10, or so, car-lines altogether being developed. Education in DDC principles and GD+T is a major issue. Chrysler has taken a strategic move to a CAD system, CATIA, and now much dimensional control documentation is done on their CAD system including the DCM.

General Motors (GM) has a four phase management process for their vehicle programs: preliminary feasibility; design / process development; prototype build and evaluation; and pilot. The basis for this management process comes from Hughes Aircraft who were seen by GM to have very good 'systems engineering' experience through their need to win and execute large government contracts. The GM approach to dimensional control is superimposed on this process and particular effort is focused up front where it is possible to look at the car as a whole system. GM has been traditionally focused, and good at, detail design and piece part manufacture. Now they see the need to focus on designing assembled products. Their approach to dimensional control starts at the highest system level by first addressing 'vehicle objectives' as derived from customer needs and then deriving a 'vehicle build strategy' which requires a balance of trade-offs before producing a 'vehicle requirements' document after the balancing act. These requirements then become sacrosanct so that after this stage it is essentially procedural. The requirements are contained in a small book that can be carried around by individuals in a manner that reflects the Chrysler approach. It contains the relationship between the customer's and functional requirements as a dimensional technical specification. For a total vehicle program GM will try to keep the full time dimensional control engineers down to around 12 people. At any one time there may be 14 vehicle programs.

In some respects the aerospace industry has quite different products to the automotive industry. Aerospace products tend to be complex products with many parts which are produced in relatively low volumes. However automotive products are less complex but have higher production volumes. The commercial pressures on the automotive sector has forced them into cutting costs by cutting overheads. The biggest overheads occur in production operations. Due to the high volumes the cost of engineering a car is a relatively small cost, as a proportion of product cost, as compared to, say, the engineering cost of a completely new aircraft. Commercial pressures have driven the aerospace sector to consider their overheads. They are keen to learn lessons from the automotive sector such as the JIT (just-in-time) concept and lean production, and are doing so through efforts like the Lean Aircraft Initiative. Lean production not only cuts overhead costs in production operations it also makes those operations more reactive to demand with a shorter lead time. The engineering cost and engineering expertise in the aerospace sector are both high. It is recognised that costs can be cut out of the engineering process as well as applying that expertise to engineering cost out of the product. One significant way to do this is through dimensional quality improvement and the concomitant production easement that follows. DDC shows promise in this regard and many aerospace companies are keen to explore the possibilities.

One such company is McDonnell-Douglas Aircraft. DDC is a part of a broader interest in variability reduction. Their initial impetus for investigating DDC came from two areas: (i) to seek improvement in assembly; and (ii) from other people who have experienced DDC elsewhere with apparent success. The improvements sought in assembly were to be brought about by the approach to design. This initiative promoted the importance of GD+T and training became an issue to be addressed. One big advantage of GD+T was to establish common datums between design and tooling. The emerging DDC approach was originally worked up from the six point plan previously outlined in Section 8.2.3. Using an assembly simulation model was important to demonstrate that the design approach can affect assembly problems. One general motive is to help quantify 'manufacturability'. This is important because when manufacturing engineers are in engineering teams then everything else has quantifiable performance measures like strength, fatigue, weight, aerodynamics. In this regard DDC complements the techniques like DFA listed in Section 8.2.1. Many other aerospace companies are now exploring the possibilities of DDC. Other motivations for seeking good dimensional control in aircraft, and other aerospace products, include interchangeability of parts, repairability and serviceability - all of which impinge on life cycle costs.

To focus on the DDC efforts of the large companies should not be viewed as presupposing that these are the only organisations that benefit. It is certainly true that large organisations have a particular problem in cutting across the functional chimneys and DDC helps do that.

However the increasing recognition that DDC has relevance to the aerospace sector shows that it is not necessary to justify a commitment to DDC on the basis of high volume production. Even for a very small company that faces particular issues of product functionality versus process capability their implementation of DDC might revolve around one or two engineers driving a tolerance analysis tool. At least to drive the tool would require some support with appropriate information gathering and dissemination procedures however informal they may be in the small company. There are many industrial DDC scenarios between the two extremes and many individual companies are developing their own practices.

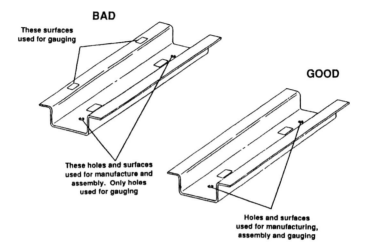

Figure 8.6 Guideline - use same locators.

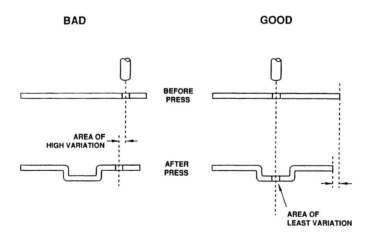

Figure 8.7 Guideline - position locator in area of least variation.

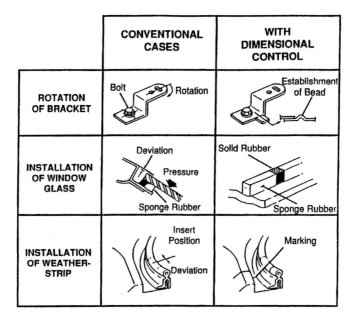

Figure 8.8 Guideline - other locator considerations.

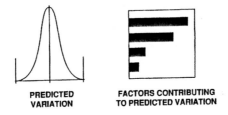

Figure 8.9 Assembly tolerance analysis output.

8.3. GUIDELINES OF DESIGN FOR DIMENSIONAL CONTROL

The aim of this section is to distil some generic good practice. This summary and brief list of guidelines should not be seen as either restrictive or comprehensive, but rather as illustrative. An individual company will need to develop its own approach to DDC depending on needs and circumstance.

DDC is a total product build dimensional control discipline which recognises and manages variation during design, manufacture and assembly to meet customer quality expectations for appearance without the need for operator fit and finesse. DDC should start during the product concept stages and continue until production stabilisation. It involves understanding of part function and manufacturing process capabilities, it involves selection of common locators and datums for manufacturing, gauging and assembly operations. It involves determining a consistent manner of measurement for parts and assemblies to understand and manage variation. Two major tools for DDC are GD+T and assembly tolerance analysis. DDC should be driven by teamwork - via a dimensional control team (DCT) which should be allowed to take responsibility for implementation. The following is a very brief list of good practice covering locators, measurement and use of DDC tools:

1. Identify significant features - these are the ones that have a major influence in the assembly and build process. To identify the significant features of any part, the team must first consider the whole product. When a concern is identified the process continues to the individual parts.

2. Locators are part features (holes, slots, studs, surfaces and edges) used to position the part during manufacturing and production. The same locator should be used for dimensioning, manufacturing, measuring (gauging), subassembly and assembly operations.

3. During the product design process the locators should serve as datums or origins from which part features are dimensioned. GD+T should be used to communicate the design intent to shopfloor operations. An unnecessarily tight blanket tolerance should not be used in the title block of engineering drawings.

4. Engineering should strive for zero defects on all drawings the way manufacturing does for all parts. The dimensional control team can help considerably in that.

5. Locators should be formed at the earliest stage of the manufacturing process (i.e. stamping or moulding) with other part features formed and dimensionally referenced to the locators. In stamping all the locator holes and slots should be established simultaneously in the first pierce operation and then used in subsequent stamping operations such as trimming and flanging. The same locators should also be used for subsequent assembly and gauging processes, see Figure 8.6. If, due to process constraints, common locators are not possible then transferred locators should be established relative to the principle (original) locators.

6. Locator type chosen should be holes/pins first, surfaces second and edges third and only if necessary.

7. Locator position should be placed in area of least variation, Figure 8.7.

8. Additional locator guidelines include:
    •   Establish the position of locators as early in the process as possible.
    •   Position locators in areas subject to the least amount of distortion during all operations including handling.
    •   Position in areas that are rigid and stable.
    •   Position in areas of simple contour (e.g. in faces perpendicular, and not oblique, to insertion direction).
    •   Position in areas suitable for establishing basic tools rests, holes, guides.
    •   Position in areas that allow parts to be processed in the same direction in all operations.
    •   Position to place variation where desired (e.g. using slip planes)
    •   Locators should be visible to the assembler.
    •   Hole/pin and slot/pin locators should be positioned as far apart as possible on the part or assembly to minimise rotation.

9. Other location methods where holes and surfaces do not apply then use a sensible approach to detail design. See Figure 8.8 for some examples.

10. Measuring points are selected after locators are established. Measuring points should be chosen to ensure the product meets design intent and to evaluate process capability. Measuring points should be related to the significant features identified and then used at tool buy-off, product acceptance, product prove out (during pilot) and process control (during production). They are the points on the product that are dimensionally important for fit, appearance and function concerns. They need to be determined for each part and subassembly as well as for the full product assembly. The measured data should also be used for resolving variation concerns.

11. If common locators for manufacturing and assembly / attachment are not feasible the part should be 'dual gauged'. This means that it will be set up and measured from both its manufacturing and assembly locators. This can be accomplished by either using one gauge with two set ups or by using two gauges.

12. To ensure meaningful use of data gathered in association with assembly operations, each part must be secured to the gauge or measuring fixture in the same manner that it will be secured during assembly operations.

13. Parts should be gauged independently as well as at the assembly level to isolate manufacturing and assembly issues. This is done because the root cause of assembly issues can only be understood if data on the parts, that make up the assembly, is available. Good assemblies cannot be made with bad parts.

14. Use assembly tolerance analysis to model product build and predict variation occurring in the assembly or subassembly. The two major outputs from the tolerance analysis is the predicted variation distribution and a Pareto chart providing a breakdown of the relative contributions to that variation from the various sources, see Figure 8.9. If variation is outside allowable product variation then the dimensional control team members can determine corrective action.

15. Assembly tolerance analysis should be carried out before the engineering release of drawings and the build of prototype tools and products so that the greatest savings in cost and time can be realised. When addressed early the dimensional control team can affect the product build strategy by changing the product (e.g. interface sections, slip planes, panel / part breaks, part reduction, materials) or changing the process (e.g. fixturing versus net build, stress free and distortion free, locators, assembly sequence, assembly methods and fasteners).

16. Document best practice for promulgation to other product programs.

8.4 SUMMARY

- Dimensional variation often emerges for the first time in assembly operations and those people in assembly operations build up an expertise in fit and finesse which constitutes a 'hidden factory'. The quality, cost and lead time implications of the 'hidden factory' are large and go beyond any simple measure of concessions or rework.

- The discipline of dimensional control should start at the beginning of any development program and carry right down to process control during manufacture and assembly operations. DDC needs to be incorporated within a structured product development and delivery process. It helps focus effort up front and provides a mechanism for communication.

- DDC provides the tools and the process to communicate design intent down through manufacturing planning and onto shop floor operation (i.e. roll down).

- DDC provides the context, incentive and means for characterising manufacturing and assembly capability that can be fed upstream in a development program (i.e. roll up).
- It needs commitment from product and process/production engineers as well as management. Training is important.
- Key elements of DDC include the use of GD+T, common locators, use of assembly tolerance analysis, documenting the design and manufacturing requirements up front and team based implementation from product concept stages through until production stabilisation.
- DDC links together efforts in, for example, robust design, SPC and process capability, DOE (design of experiments) and Taguchi methods, DFM, DFA, and key characteristics.
- An emerging strategy is based on DDC tools being supported within CAD packages.
- DDC minimises the need for shop floor finesse and improves product quality resulting in better customer satisfaction and less warranty claims.
- DDC saves money by shortening development lead times, by assembly easement and reducing rework necessary in producing tooling/dies.

REFERENCES

Allen, G. (1993) Tolerances and assemblies in CAD/CAM systems, *Manufacturing Review*, **6** (4), 320-327.

ASME (1995a) *ASME Y14.5M-1994 (revision of ANSI Y14.5M-1982), Dimensioning and tolerancing*, An American National Standard, Pub. ASME.

ASME (1995b) *ASME Y14.5.1M-1994, Mathematical definition of dimensioning and tolerancing principles*, An American National Standard, Pub. ASME.

BSI (1993) *BS 308 Engineering drawing practice: Part 1(1993) Recommendations for general principles.*

BSI (1985) *BS 308 Engineering drawing practice: Part 2 (1985) Recommendations for dimensioning and tolerancing of size.*

BSI (1990) *BS 308 Engineering drawing practice: Part 3 (1990) Recommendations for geometrical tolerancing.*

Chase, K.C., Parkinson, A.R. (1991) A survey of research in the application of tolerance analysis to the design of mechanical assemblies, *Research in Engineering Design*, **3,** 23-37.

Chase, K.W., Gao, J., Magleby, S.P. (1995) General 2D tolerance analysis of mechanical assemblies with small kinematic adjustments, *International Journal of Design and Manufacture*, **5** (4), Chapman-Hall.

Craig, M. (1992) Controlling the variation, *Manufacturing Breakthrough*, **1** (6), 343-348.

Dimensional Control Systems Inc., 3128 Walton Blvd., Suite 176, Rochester Hills, MI.48309, US.

Foster, L.W. (1994) *Geo-metrics IIIm: The metric application of geometric dimensioning and tolerancing techniques*, Addison-Wesley Publishing Co.

Gao, J., Chase, K.W., Magleby, S.P. (1995) Comparison of assembly tolerance analysis by direct linearisation and modified Monte Carlo simulation methods, *ASME Design Engineering Technical Conference*, Boston, MA. September 17-20.

Henzold, G. (1993) Comparison of vectorial tolerancing and conventional tolerancing, In: *Proceedings of the 1993 International Forum on Dimensional Tolerancing and Metrology,*

(Edited by V. Srinivasan and H. Voelcker), Dearborn, MI, June 17-19, 1993; Report CRTD-27, ASME, New York.

Henzold, G. (1995) *Handbook of geometric tolerancing - design, manufacturing and inspection*, John Wiley.

Imai, E., Shimizu, K., Araki, A. (1994) Productivity improvement through systematic approaches to design for manufacturability and assembly, *XXV FISITA Congress*, 17-21 October, Beijing, International Academic Publishers, SAE Paper No. 945204, 182-191.

ISO 1101 (1983) - Technical drawings: Geometrical tolerancing.

Krulikowski, A. (1991) *Fundamentals of geometric dimensioning and tolerancing*, Pub. Delmar.

Liggett, J.V. (1993) *Dimensional variation management handbook*, Prentice-Hall Inc.

Srinivasan, V., Voelcker, H. (1993) Introduction to special section on dimensional tolerancing and metrology, *ASME Manufacturing Review*, **6** (4), 255-257.

Taguchi, G., Elsayed, E.A., Hsiang, T. (1989) *Quality engineering in production systems*, McGraw-Hill International.

Texas Instruments, Concurrent Engineering Products, Defense Systems and Electronics Group, 6500 Chase Oaks Blvd, Plano, Texas 75023, US.

Turner, J.U. (1993) Current tolerancing packages, In: *Proceedings of the 1993 International Forum on Dimensional Tolerancing and Metrology*, (Edited by V. Srinivasan and H. Voelcker), Dearborn, MI, June 17-19, 1993, ASME Report CRTD-27, New York.

Variation Simulation Analysis Inc., Corporate Headquarters, 300 Maple Park Boulevard, St.Clair Shores, MI. 48081-3771, USA.

Voelcker, H. (1993) A current perspective on tolerancing and metrology, *ASME Manufacturing Review*, **6** (4), 258-268.

DESIGN FOR ASSEMBLY COSTS OF
PRINTED CIRCUIT BOARDS

Donald S. Remer; Frederick S. Ziegler; Michael Bak; Patrick M. Doneen

This chapter presents a software tool developed to help electronics designers predicting the costs of manufacturing circuit card assemblies and to enable them to make more cost-effective design decisions.

This tool can be used in a concurrent engineering environment to provide board density information, scrap and rework cost estimates and a breakdown of setup, labor and material costs for each step of the printed circuit board (PCB) assembly process. What-if analyses can be performed to compare the costs of using different component types, such as through-hole versus surface-mount components, or different manufacturing process alternatives, such as manual assembly versus automatic assembly. Within the software tool are activity-based models of the 56 processes that constitute the circuit card assembly manufacturing system. In addition to predicting the level of activity at each of the processes, the model also identifies and considers design decisions that lead to an increase in processing costs. Specifically, board density, solder characteristics, production batch sizing and the variety of components used in a design can lead to higher rework rates as well as increased setup and processing for certain operations. Verification was done by comparing predictions made by the model to actual historical cost data for 12 randomly selected boards currently in production. The average prediction error made by the tool is 6.68% for labor costs, 4.76% for component costs and 3.1% for density estimations.

9.1 BACKGROUND

In the manufacture of electronics products, there exists a critical link between design and production. In this environment, where 75 - 85% of a product's manufacturing costs are determined in the design stage (Boothroyd and Dewhurst, 1989), proper design decisions can cut costs dramatically. Increasing the manufacturability of designs can create a competitive advantage by decreasing downstream manufacturing costs, quality costs and time-to-market.

To make more cost-effective design decisions, engineers need more information about the impact that various design alternatives have on production costs. A joint project team from ELDEC Corporation and Harvey Mudd College Engineering Clinic has investigated this issue and developed a decision-support software tool for use during the design and development stage. This tool models the various costs associated with the production of circuit card assemblies and provides feedback on the time and cost needed to build the proposed circuit card. The program runs on any 80286-based machine or higher, requires at least 1MB of RAM, and runs under Microsoft Excel 2.1 or higher. The tool is a CustomExcel application, complete with its own menu bars and utilities.

The underlying algorithms are based on predicting the level of activity driven by the array of decisions made about the design and construction of the circuit card. Foster and Gupta (1990) discuss the implementation of an activity-based costing system in an electronics manufacturing environment. They identify 14 'activity areas' and the primary cost driver for each of these areas. Building from this work, we identify 56 processes available to assemble a circuit card, and identify multiple cost drivers for each of these operations. The algorithms used predict the level of activity at each of the operations by quantifying the cost drivers effecting each of the steps in the production system. With this information, the flow of the cards through the factory can be stimulated, resulting in a prediction of the time and cost to build the card.

9.2 THE PCB MANUFACTURING PROCESS

The circuit card assembly operation under study was broken down into 56 fundamental steps. This number was assessed by studying the flow of product through the factory and identifying the essential processes seen by a broad product mix. These processing steps include preparing the boards and components, placing the components on the board, soldering them, testing the boards, inspecting them, cleaning them and, if necessary, scrapping or reworking them. Preparing the components consists mainly of cutting and bending leads and adding sleeves or standoffs where needed. The components may be placed on the board in one of seven different steps. Surface-mount components may be placed on the topside or the backside of the PCB. Top-side surface-mounted components are reflow soldered, whereas backside surface-mount components are wave-soldered (since for the facility under study, all backside surface-mount components were on boards in conjunction with through-hole components). Most component are placed automatically, semi-automatically, or manually prior to wave soldering. Components which cannot be wave soldered are added by hand soldering after the wave-soldering step. Certain mechanical components are added in the final mechanical assembly step. Since the finished products are used in demanding aerospace applications, bonding and conformal coating are often needed.

Table 9.1 The PCB assembly manufacturing process under study

The following is a step-by step description of ELDEC's 56 manufacturing process operations for assembling PCBs. Note that the process begins with a bare Printed Circuit Board ready to be assembled.

Preparation

1. Manual Prep - Manual preparation of components, such as cutting and bending leads.
2. Machine Prep - Automated preparation of components.

Topside surface mount

3. Screen Print - Print design on board to place SMCs (surface mount components).
4. Place SMCs - Placement of surface mount components on the PCB with glue only.
5. Dry/Bake - Allow glue to dry, or bake, depending on glue used.
6. Preheat - Preheat board and components to accept solder.
7. Reflow Solder - Soldering the previously placed components on the board.
8. Clean - Remove and clean the excess solder and flux off the board.
9. Inspect - Inspection of the soldered components
10. Rework - Rework or scrap any improperly soldered components.

Through-hole component placement

11. Auto Insert - Automated insertion of through-hole components.
12. Semi-Auto Insert - Semi-automatic insertion of through-hole components.
13. Manual Insert - Manual insertion of through-hole components.
14. Manual Cut and Clinch - Manually cutting and clinching leads of components already placed on board.
15. Inspect - Inspect for defects from insertion
16. Rework - Rework or scrap any defects found in inspection.

Backside surface-mount component placement

17. Apply Adhesive - Apply adhesive for small SMCs.
18. Place SMCs - Place the small SMCs on the adhesive on board.
19. Cure Adhesive - Allow the adhesive to cure.
20. Inspect - Inspect the gluing of small SMCs.
21. Rework - Rework or scrap any defects during inspection.

Wave soldering

22. Mask (for add-ons) - Mask add-on parts from flow wave solder.
23. Flow Wave Solder - Wave solder all previously place components.
24. Remove Mask (for add-ons) - Remove the previously placed masks.
25. Clean - Remove and clean of solder and flux.
26. Secondary Lead Trim - Trim any protruding leads.
27. Clean - Remove and clean of solder and flux.

Table 9.1 The PCB assembly manufacturing process under study *(Continued)*

28. Inspect - Inspect the wave solder operation.
29. Rework - Rework or scrap any defects found in inspection.

Add-on component placement

30. Add-on - Place and solder add-on components on board.
31. Clean - Remove and clean of solder and flux.
32. Inspect - Inspect the solder operation.
33. Rework - Rework or scrap any defects found in inspection.

In-circuit testing

34. In-Circuit Test - Test board.
35. Failure Analysis - Testing under adverse condition.
36. Rework - Rework or scrap any defects found in inspection.

Bonding and marking

37. Bonding - Gluing of extra large components.
38. Part Mark - Labeling of parts on board.
39. Cure - Curing of glue
40. Clean - Clean off excess glue.

Environmental protection

41. Mask - Mask components not to be sealed.
42. Cure - Cure applied mask.
43. Clean (Aqueous) - Clean board with purified water for priming.
44. Priming (include bake) - Prime board and components to accept sealants.
45. Paralene - Coat board with paralene sealant.
46. Humiseal - Coat board with humiseal sealant.
47. Cure - Cure for sealants.

Final mechanical assembly

48. Demask - Remove previously place mask for sealants.
49. Final Mech. Assembly - Final assembly of manually placed mechanical components.

Final testing and inspection

50. Environmental burn-in - Simulate 90 days of use by heating.
51. Failure Analysis - Testing under adverse conditions.
52. Repair - Repair or scrap any defective boards.
53. Acceptance Test - Decide whether the repaired board is acceptable.
54. Failure Analysis - Testing under adverse conditions.
55. Repair - Repair or scrap any defective boards.
56. Final Inspection - Final test of completed PCB.

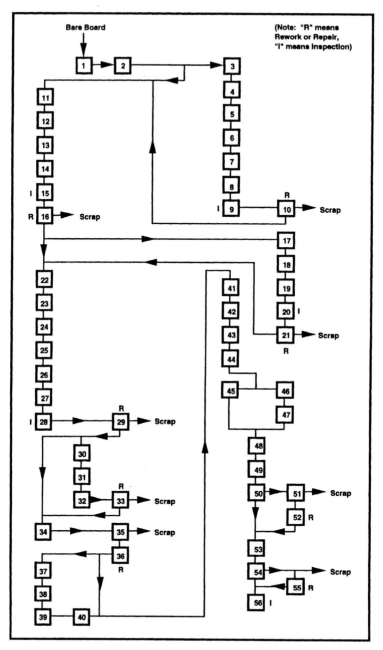

Figure 9.1 The PCB manufacturing process flow diagram.

Table 9.1 lists the 56 processes by work centers. Not all 56 steps are used by each design. For example, a circuit card without backside surface-mount components would not use steps 17-21. A general flow diagram of the operation, showing the 56 processes by number, is presented in Figure 9.1.

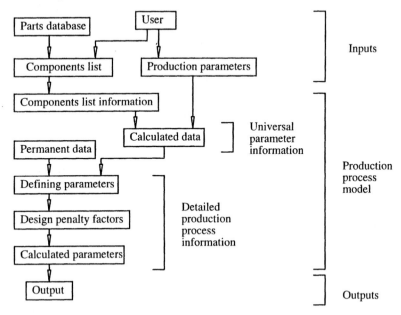

Figure 9.2 The process model flow of information.

9.3 THE PROCESS MODEL

The manufacturing model is driven by three groups of information. The first group, the *components list information*, consists of information derived from the list of parts to be assembled on the card. This component list is built by the user either from internal databases or by importing a parts list from an external file. The second group consists of the *production parameters*, which are variables not contained in the parts list, such as the number of boards to be produced and production batch size. This information must also be entered by the user. Finally, information about the costs of the production process steps, contained in the *production process model*, must be known. This model is split into universal parameter information, which applies to many process steps, and detailed production process information, which is particular to a single process step. Some of the universal parameter information costs are invariant (or permanent), and are retrieved from a database, and others are calculated from the information input by the user. The detailed production process information is made up of three sources. First are defining parameters, which determine the cost of each process step. Second are penalty factors. These are factors, such as component density, which have an adverse affect on certain processes owing to the introduction of added processing difficulty. Third are calculated parameters, which are derived from the defining parameters and penalty factors. The process model flow of information is shown in Figure 9.2.

9.3.1 Components List Information

The user must first input the components list, which reflects the parts to be mounted on the card. The components list includes part numbers, quantity of parts, component cost, component type and geometry of the component once mounted on the card. The component list can be created from internal parts databases, or imported from a compatible file. The components list information is calculated automatically from information retrieved from the components list. This information is shown in Table 9.2.

Table 9.2 The components list information

1.	Total number of components
2.	Total component area on the board
3.	Total volume of the components
4.	Average component area
5.	Total cost of the components
6.	Total number of leads
7.	Number of large surface-mount components
8.	Number of small surface-mount components
9.	Number of through-hole components
10.	Total required manual handling time
11.	Number of resistors
12.	Number of capacitors
13.	Number of diodes
14.	Number of microcircuits
15.	Number of transistors

9.3.2 Production Parameters

The second type of input needed by the model is the production parameters. The user inputs these by answering 26 questions about the production of the circuit card. This information is used as an input to the internal algorithms. There are five major types of production parameters.

(1) *Batch information*: this provides information on lot size, the size of the production run and accounting overhead rates.
(2) *Component information*: this is information on the preparation of components, where they are placed on the boards, and the manufacturing steps in which they are placed.
(3) *Environmental protection*: this covers the options associated with conformal coating.
(4) *Component options*: this consists of information about additional options to be added to the components, such as insulators, standoffs, sleeves and transistor pads.
(5) *Mechanical parts information*: mechanical parts are those added to the board which are not contained in the component databases, such as heat sinks, brackets and mechanical fasteners.

Table 9.3 is a complete listing of the production parameters that must be entered by the user.

Table 9.3 Production parameters

Batch information

Number of boards to be manufactured
Number of boards mounted on a panel
Number of batches
Percentage overhead applied to labor costs

Component information

Number of surface-mount components on the backside of the board
Number of machine-prepared components
Number of add-on components
Number of automatically inserted components
Number of semi-automatically inserted components
Number of manually inserted components
Number of bonded components
Number of final mechanical assembly components
Number of clinched leads
Number of fasteners for add-on components
Number of jumper wires for add-on components

Type of environmental protection

Number of layers of Paralene protection
Number of layers of Humiseal protection

Component options

Number of sleeves (for leads)
Number of transistor pads
Number of insulators (for leads)
Number of standoffs

Mechanical parts information

Number of brackets
Number of bracket fasteners
Number of heat sinks
Number of heat-sink fasteners
Cost of brackets, heat sinks, and fasteners

Table 9.4 Universal parameter information

Permanent data
Exponential factor for batches
Exponential factor for boards
Exponential factor for components
Exponential factor for leads
Time required to bend a lead
Time required to secure a fastener
Time required to handle and affix a sleeve
Time required to handle and affix a transistor pad
Time required to handle and affix an insulator
Time required to handle and affix a standoff
Time required to insert an electronic component
Time required to cut a lead
Time required to hand solder a solder joint
Time required to handle, insert, and secure a final mechanical assembly component
Cost of adhesive for one SMC
Cost of bonding material for one component
Cost of one Paralene layer
Cost of one Humiseal layer
Cost of a component mask
Cost of one solder joint
Calculated data
Number of components per board
Number of leads per board
Number of small surface-mount components
Number of large surface-mount components
Number of manually prepared components
Number of manually clinched leads
Number of surface-mount components on the top of the board
Total cost of components
Total handling time for components
Average cost per component
Average handling time per component
Time to handle and insert transistor pads and standoffs
Time to handle and insert sleeves and insulators
Number of automatically, semi-automatically, and manually inserted components and add-on components
Number of transistor pads and standoffs
Number of sleeves and insulators

Table 9.5 Detailed production process information

Original data	
Number of components added in process	
Setup required for process*	(min/board)
Capacity of process*	(min/board)
Labor required for process*	(min/board)
Down-time of process*	(% of setup and labor)
Labor rate for process*	($/min)
Throughput	
Throughput	(min/board)
Value Added Time	
Value Added Time	(min/board)
Setup	
Time	(min)
Time per board	(min/board)
Cost	($)
Cost per board	($/board)
Processing labor	
Time	(min)
Time per board	(min/board)
Cost	($)
Cost per board	($/board)
Material	
Cost	($)
Cost per board*	($/board)
Operation totals	
Time	(min)
Time per board	(min/board)
Cost	($)
Cost per board	($/board)
Number of boards which survive to undergo the current process	
Cumulative totals	
Time	(min)
Time per board	(min/board)
Cost	($)
Cost per board	($/board)
Attrition (due to scrap)	
Attrition	(% of boards)
Cumulative attrition	
Cumulative attrition	(% of boards)
Rework	
Rework*	(Number of leads and components/board)

* Indicates a parameter which helps define the process step cost.

9.3.3 The Production Process

The production process model consists of two groups of information. The first is the *universal parameter information,* which is information used by many process steps. Some of these parameters are permanent data; these are invariant and are retrieved from a database. The rest of the parameters are calculated data; these are computed from the components list information and production parameters.

An example of permanent data is the time required to insert an electronic component. This figure does not change from design to design. There are a number of manufacturing steps in which an electronic component may be added to the PCB. When an electronic component is added, the model accesses this number and adds it to the labor time for that particular step.

An example of calculated data is the number of surface-mount components on the topside of the circuit board. The model accesses this number whenever surface-mount components are placed on the top of the board.

The first four parameters of the permanent data are the exponential factors for learning-curve effects. One of these parameters is for operations performed on batches, one is for boards, one is for components, and one is for leads. The learning curve takes into account the fact that production cost declines as cumulated output increases (Ghemawat, 1985). Thus, the more times a worker performs an operation, the more efficient it becomes (up to certain limits). The time required to perform an action can be approximated as an exponentially decaying function of the number of times the action has been performed, written exp $(-aN)$, where N is the number of times the operation has been performed, and a is the exponential factor listed in Table 9.4. Data specific to the operation under study were used to determine the value of a.

The permanent data in Table 9.4 are derived from data specific to the ELDEC facility under study, from Boothroyd and Shinohara (1986), and from Boothroyd and Dewhurst (1989). The universal parameter information is listed in Table 9.4.

The model for the production process is defined by the *detailed production process information,* and applies to each of the 56 steps initially described in Table 9.1. This is permanently stored information derived from analysis at the manufacturing facility.

Each process has 10 defining parameters. These parameters include information about that particular process step. For example, the cost of material per board and the number of boards being processed in the step are defining parameters because the cost of a process step cannot be fully calculated without them.

Twenty calculated parameters useful to the user are derived from these 10 defining parameters. For example, the cost of material for all boards in a step is a calculated parameter because it can be calculated from the cost of material per board and the number of boards being processed (defining parameters). All 30 defining and calculated parameters are listed in Table 9.5; the defining parameters are italicized.

Design penalty factors make up a third category of detailed production process information. These factors were derived empirically from time motion studies and operations data at the ELDEC manufacturing facility studied. These factors represent important links between design and manufacturing (though certainly not the only links modeled in the tool). Current research is focusing on identifying and quantifying other critical design penalty factors. The four penalty factors quantified in the tool are summarized as follows.

(1) *Density penalty factor:* boards that have a high component density are found to cause a number of processes to be more time-consuming, and a number of defects more

prevalent. For example, manual insertion and hand soldering become more time-consuming and incorrect insertions occur more frequently as densities increase.

(2) *Batch size penalty factor*: consistent with the teachings of just-in-time manufacturing, larger batch sizes result in defects taking longer to detect. As a result, scrap and rework costs tend to increase as batch size increases. This effect is countered to some extent by the fact that setup costs decrease as batch size increases. One must also consider that inventory costs increase as batch size increases. By properly quantifying these three effects, the tool helps determine the batch size in which to build a particular design in the facility being modeled, so as to minimize cost.

(3) *Unique parts/total parts penalty factor*: operations research at the ELDEC facility found that when there was a large component mix (especially among components of similar type and geometry), there is an increase in insertion errors. This leads to higher rework rates and higher scrap costs (sometimes the wrong component inserted could result in a catastrophic failure). In addition, a high portion of unique part numbers drive up setup costs for 'prepping' and insertion operations.

(4) *Exposed solder penalty factor*: Higher soldering defect rates are found on designs with exposed solder traces as compared to boards that use a solder mask or a 'pads only' design. This is particularly a problem when backside surface-mounted components are used in conjunction with bare solder traces. Since component spacing can have a significant effect on this factor, company design standards are followed.

9.4 OUTPUTS

An important consideration when designing the tool was that the user must have easy access to a variety of information. It was decided that, in order for the tool to be an effective decision-support device, it had to provide immediate and quantified feedback to the user. There are five primary categories of output with which the user can interact.

(1) The first is intended to give the user a summary of the time and cost required to build the design being analyzed. The information is shown on both a per-board basis and for an entire production run, and is as follows: setup cost, labor cost, component cost, consumable materials cost, total cost, total manufacturing time, and value-added manufacturing time.

(2) The second type of output is a series of automatically generated graphs which the user can select to visualize the time or cost for manufacturing the design. Ten graphs summarize cost information and 10 graphs are generated to display the time required at the various processes to build this board. This time information can be particularly useful in helping predict resource shortages or production bottlenecks.

(3) A third level of output is a summary of the component list information (as shown in Table 9.2). This is mostly used for density analysis, and to help look at material usage by type of component.

(4) The fourth level of output is a display of the production parameters (see Table 9.3). This is just a summary of the user's answers to the 26 questions displayed in the table. This is essentially a list of the manufacturing assumptions underlying the cost estimate.

(5) A fifth output available to the user is a printout of the parts list of components assembled on the card. This is a 'bill of materials' for the design, and it changes as the design evolves through the development stages.

9.5 SAMPLE BOARD ANALYSIS

The following example board, BOARD B, was used to demonstrate some of the inputs and outputs to the program CCA_DFM (Circuit Card Assembly - Design For Manufacture). The following output was generated by the computer program. Here are the parts for BOARD B:

```
PARTLIST

CCA_DFM
Part list
All units are in dollars, inches, and minutes

Part Number                 Quantity
CMOS-069UBA                 1
CMOS-028                    1
CMOS-175BA                  4
CMOS-503BA                  6
MA17-04                     16
HY52-04                     16
AR23-14                     1
M38510/10101BPB             2
CCR05                       16
M39014/01                   137
M39014/02                   16
CSR13B2                     6
1N4148-1                    19
1N75?A-1                    2
1N361?                      2
1N4572A                     2
1N4625                      1
1N495?                      1
1N496?                      2
1N497?                      2
RLR07                       38
RNC55                       21
RNC60                       16
M83401/08(109)              5
2N3700                      6
```

The program can then use this parts list to run a board density/classifications analysis. In this analysis, the designer can look at the board density for this design, as well as a breakdown of the parts classification (number of resistors, capacitors, through-hole components, etc.).

Here is the density/classifications analysis performed by CCA_DFM for BOARD B:

```
Classification

Total Number of parts                   359       parts
Total Area for board                    40.799    in²
Total Volume for board                  6.279     in³
Board density                           0.114     in²/part

Total Component Cost for board          $859.85
Total Number of leads                   1834      leads
Total Number of large SMCs              0         parts
Total Number of small SMCs              0         parts
Total Number of Thru-Hole Components    359       parts
Total B/D handling time                 10.175    minutes

Total Number of Resistors (R)           75        parts
Total Number of Capacitors (C)          195       parts
Total Number of Diodes (CR)             28        parts
Total Number of MicroCircuits (U)       52        parts
Total Number of Transistors (Q)         6         parts
```

When a full producibility analysis is desired, the user must input some additional producibility data. This data includes information that cannot be captured in a static database, such as the number of parts inserted semi-automatically.

Here are the inputs used for the producibility analysis of BOARD B:

```
Producibility Data (used in latest calculation)

Batch Information
Number of Boards:                                        300
Number of Boards per Panel:                              6
Number of Batches:                                       60
% Overhead on Labor Costs:                               115

Component Information
Number of SMCs on Bottom:                                0
Number of Machine Prepped Components:                    157
Number of Add-On Components:                             4
Number of Auto-Inserted Components:                      0
Number of Semi-Auto Inserted Components:                     126
Number of Manually Inserted Components:                  229
Number of Bonded Components:                             3
Number of Final Mechanical Assembly Components:          0
Number of Clinched Leads:                                0
Number of Fasteners for Add-&On Components:              0
Number of &Jumper Wires for Add-On Components:           0

Environmental Protection
Number of Paralene Layers:                               0
Number of Humiseal Layers:                               1

Components Options
Number of Sleeves:                                       2
Number of Pads:                                          9
Number of Insulators:                                    1
Number of Standoffs:                                     0

Non-Standard Parts
Number of Brackets:                                      0
Number of Bracket Fasteners:                             0
Number of Heat Sinks:                                    1
Number of Heat Sink Fasterners:                          1
Cost of Brackets, Heat Sinks, and Fasteners             0.17
```

This producibility data, along with the previously calculated density/classification analysis, is used to perform the actual producibility analysis. These producibility calculations reveal the effects of a boards design - the breakdown of times and costs for a board.

Here is the producibility summary generated by CCA_DFM for BOARD B:

```
Summary:

5847.85    : Total Cost of Setup [$]
35079.06   : Total Cost of Labor [$] (w/ Overhead & Down Time)
1164.7     : Throughput of Production System    [minute/board]
24350.8    : Total Value Added Time [minutes]
275413.3   : Total Material Cost [$]
84189.7    : Total Time [minutes]
310492.3   : Total Cost [$]
294        : Number of Boards Successfully Completed
19.91      : Total Cost of SetUp per Board [$]
119.42     : Total Cost of Labor per Board [$] (Overhead & Down Time)
937.58     : Total Material Cost per Board [$]
286.6      : Total Time per Board [minutes]
1057       : Total Cost per Board [$]
```

The results can also be viewed step-by-step. This allows the designer to get a better look at where the costs/delays are being incurred.

Here is a partial printout of a producibility breakdown generated by CCA_DFM for BOARD B (additional columns breakdown the cost and time even further):

9	Operation	Time	Time/board	$	$/board
1	Manual Prep	658.9226	2.196 409	274.5511	0.91517
2	Machine Prep	1072.086	3.57362	446.7025	1.489008
3	Screen Print	1072.086	3.57362	446.7025	1.489008
4	Place SMCs	1072.086	3.57362	446.7025	1.489008
5	Dry/Bake	1072.086	3.57362	446.7025	1.489008
6	Preheat	1072.086	3.57362	446.7025	1.489008
7	Reflow Solder	1072.086	3.57362	446.7025	1.489008
8	Clean	1072.086	3.57362	446.7025	1.489008
9	Inspect	1072.086	3.57362	446.7025	1.489008
10	Rework (Touch-up	1072.086	3.57362	446.7025	1.489008
11	Auto Insert	1072.086	3.57362	446.7025	1.489008
12	Semi-Auto Insert	2296.409	7.675206	91057.18	305.0417
13	Manual Insert	3175.162	10.6191	255177.1	854.8573
14	Manual Cut & Clinch	3493.449	11.68539	255309.8	855.3016
15	Inspect	3840.671	12.84861	255454.4	855.7862
16	Rework	4524.8	15.1428	255870.3	857.181
17	Apply Adhesive	4524.8	15.1428	255870.3	857.181
18	Place SMCs	4524.8	15.1428	255870.3	857.181
19	Cure Adhesive	4524.8	15.1428	255870.3	857.181
20	Inspect	4524.8	15.1428	255870.3	857.181
21	Rework (Touch-up)	4524.8	15.1428	255870.3	857.181
21b	Mask (for Add-ons)	4525.067	15.1437	256354.3	858.8161
22	Flow (wave) Solder	4928.817	16.50788	261890	877.5202
22a	Rem Mask for add-on	5406.129	18.12061	262088.9	878.1922
23	Clean	5822.521	19.52751	262262.4	878.7784
23b	Secondary Lead trim	6111.873	20.50516	262383	879.1857
23c	Clean	8289.149	21.10414	262456.8	879.4353
24	Inspect	6404.889	21.4952	262505.1	879.5982
25	Rework	7272.945	24.44291	263958.1	884.5324
26	Add-on	8573.732	28.86006	267372	896.1252
27	Clean	9572.667	32.2522	267788.2	897.5386
28	Inspect	9739.156	32.81756	267857.6	897.7742
29	Rework	9739.156	32.81756	267857.6	897.7742
30	In Circuit Test	12619.36	42.60784	269057.7	901.8535
31	Failure Analysis	29543.71	100.1363	276109.5	925.8237
32	Rework	32364.43	109.7244	277284.9	929.8191
33	Bonding	33923.1	115.0208	278375.7	933.5267
34	Part Mark	38154.19	129.4047	280138.6	939.5192
35	Cure	38868	131.8311	280436	940.5302
36	Cleaning	39284.39	133.2464	280609.5	941.1199
37	Mask	39284.39	133.2464	280609.5	941.1199
38	Cure	39998.21	135.6728	280907	942.1309
39	Clean (aqueous)	40712.02	138.0992	281204.4	943.1419
40	Priming (w/bake)	42122.39	142.8932	281792	945.1394
41	Paralene (C.C.)	42122.39	142.8932	281792	945.1394
42	Humiseal (C.C.)	47291.79	160.4648	284534.3	954.4609
43	Cure	47886.64	162.4868	284782.2	955.3034
44	Demask	47886.64	162.4868	284782.2	955.3034
45	Final Mech Assembly	47886.64	162.4868	284782.2	955.3034
46	Env. Burn-in	58593.86	198.8823	289243.5	970.4682
47	Failure Analysis	61411.88	208.4708	290417.7	974.4634
48	Repair	67707.23	229.8911	303366.6	1018.523
49	Acceptance Test	71981.12	244.4332	305147.4	1024.582
50	Failure Analysis	74799.14	254.0217	306321.5	1028.577
51	Repair	81936.41	278.3188	309553.4	1039.579
52	Final Inspection	84189.74	285.9898	310492.3	1042.776
Totals		84189.74	286.6038	310492.3	1056.997

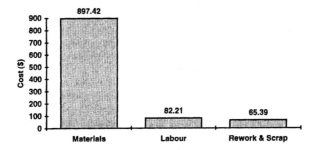

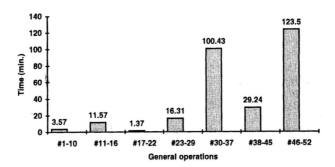

Figure 9.3 Graphs generated by computer program.

Obviously there are more significant figures than warranted. Graphs of the results are also available. CCA_DFM provides 18 graphs that the designer may view to look at the results in greater detail without having to look through pages of data. Two example graphs generated by CCA_DFM are shown in Figure 9.3. Finally, a producibility report may also be printed out, with a summary of the density/classification analysis, inputs to the producibility calculations, and the summary of the producibility calculations itself. The user is prompted for the board's name and designer's name before document is printed out.

Here is an example of the producibility report generated by CCA_DFM for BOARD B:

```
        Circuit Card Assembly - Design for Manufacture
              Density/Producibility Report

     Board Name:              BOARD B
     Report generated by:        TESTER
     Date:                    5/13/91

     Density/Classification

Total Volume for board            6.279        in³
Board density                     0.114        in²/part
```

```
Total Component Cost for board              $ 859.85
Total Number of leads                                   1834    leads
Total Number of large SMCs                  0           parts
Total Number of small SMCs                  0           parts
Total Number of Thru-Hole Components        359         parts
Total B/D handling time                     10.18       minutes

Total Number of Resistors (R)               75          parts
Total Number of Capacitors (C)              195         parts
Total Number of Diodes (CR)                 28          parts
Total Number of MicroCircuits (U)           52          parts
Total Number of Transistors (Q)             6           parts

Batch Information
Number of Boards:                           300         boards
Number of Boards per Panel:                 6           boards/panel
Number of Batches:                          60          batches
% Overhead on Labor Costs:                  115         percent

Component Information
Number of SMCs on Bottom:                   0           parts
Number of Machine Prepped Components:       157         parts
Number of Add-On Components:                4           parts
Number of Auto-Inserted Components:         0           parts
Number of Semi-Auto Inserted Components:    126         parts

Number of Manually Inserted Components:     229parts
Number of Bonded Components:                3           parts
Number of Final Mechanical Assembly Components:    0        parts
Number of Clinched Leads                    0           leads
Number of Fasteners for Add-%On Comp.       0           fasteners
Number of &Jumper Wires for Ad-On Comp.     0           leads

Environmental Protection
Number of Paralene Layers:                  0           layers
Number of Humiseal Layers:                  1           layers

Component Options
Number of Sleeves:                          2           sleeves
Number of Pads:                             9           pads
Number of Insulators:                       1           insulators
Number of Standoffs:                        0           standoffs

Non-Standard Parts
Number of Brackets:                         0           brackets
Number of Bracket Fasteners:                0           fasteners
Number of Heat Sinks:                       1           heat sinks
Number of Heat Sink Fasteners:              1           fasteners
"Cost of Brackets, Heat Sinks,& Fasteners": $0.17
Total Cost of SetUp                         $5847.85
Total Cost of Labor                         $35,079.06
Throughput of Production System             1164.7min/board
Total Value Added Time                      24350.8     minutes
Total Material Cost                         $275,413.30
Total Time                                  84189.7
Total Cost                                  $310,492.30
Number of Boards Successfully Completed      294         boards

Total Cost of SetUp per Board               $19.91
Total Cost of Labor per Board               $119.42
Total Material Cost per Board               $937.58
Total Time per Board                        286.6       minutes
Total Cost per Board                        $1,057.00
```

Table 9.6 A comparison of the model with historical results

(a) Labor Cost

Board	Historical (mean) ($)	Historical (std. dev.) ($)	Model prediction ($)	Difference (abs. value) ($)	Difference (std. dev.)	Difference (percentage)
1	129.80	11.54	110.50	19.30	1.67	14.87
2	81.57	4.67	79.49	2.08	0.445	2.55
3	112.05	12.32	89.57	22.48	1.82	20.06
4	91.06	5.94	87.01	3.96	0.667	4.35
5	102.52	6.61	103.07	1.18	0.19	1.15
6	168.42	15.09	158.20	10.22	0.677	6.07
7	67.01	4.18	69.96	2.95	0.706	4.40
8	99.50	7.24	84.95	14.55	1.76	14.62
9	45.17	3.48	45.73	0.56	0.161	1.24
10	135.21	11.86	142.78	7.57	0.638	5.60
11	98.32	6.22	94.81	3.51	0.564	3.57
12	127.18	8.64	125.00	2.18	0.25	1.71
			Average difference:		0.796	6.68

(b) Component Cost

Board	Historical (mean) ($)	Historical (std. dev.) ($)	Model prediction ($)	Difference (abs. value) ($)	Difference (std. dev.)	Difference (percentage)
1	388.12	19.32	361.66	26.46	1.37	6.82
2	262.52	10.77	268.58	6.06	0.563	2.31
3	276.38	12.82	259.64	16.64	1.29	6.02
4	343.16	18.38	320.64	22.52	1.22	6.56
5	513.68	27.41	498.55	15.13	0.55	2.95
6	1400.73	58.79	1444.81	44.08	0.75	3.15
7	174.55	10.56	161.35	13.20	1.25	7.56
8	189.95	10.12	189.21	0.74	0.073	0.39
9	101.44	9.39	110.92	9.48	1.01	9.35
10	803.51	41.84	778.32	25.19	0.602	3.13
11	384.29	19.87	378.65	7.59	0.38	1.47
12	295.33	15.09	273.58	21.75	1.44	7.36
			Average difference:		0.875	4.76

9.6 TOOL ACCURACY

One of the most important factors related to the usefulness of such a tool is whether or not its results are accurate and believable. To determine its accuracy, 12 boards currently in production (and ones that have been in production for at least six months) were randomly chosen for analysis. These designs were analyzed by the tool and the predictions were compared against actual historical operations and accounting data. Data comparing predictions against actual cost data for the 12 boards analyzed are presented in Table 9.6. The actual costs are determined by calculating the mean and standard deviation of the distribution of costs for each board over the previous six-month period.

As can be seen from the table, the average error in predicting labor costs is 6.68%. On average, this is 0.796 standard deviations from the mean of the distribution of historical labor costs for the 12 boards studied. The component cost prediction error is found to be an average of 4.76% and within 0.875 standard deviations from the mean of the actual historical distribution of the costs. The labor costs were found to have a larger error measured as percentage difference from the actual costs, but smaller as measured in standard deviation of this distribution. This reveals the fact that the facility under study was characterized by relatively significant variance in manufacturing processing costs for a given product over time. This suggest that a potential source of improvement could be found in better process control and improved utilization of processing standards. Where the material cost variance for a given product was smaller, this suggested a more controlled purchasing activity.

In addition to the data in Table 9.6, the average error in predicting setup costs is 11.58%, and 3.1% in predicting density. Since setup costs are generally small compared to total manufacturing costs (e.g. usually less than 10% of the manufacturing cost, excluding component costs), this prediction error is usually less than a couple of dollars. The density-predicting capabilities are of sufficient accuracy, and can be further improved by refining data for standard component footprints in the database. As a result of the verification, engineers and managers at the facility under study feel comfortable using the tool to predict costs during proposal developments. In addition, concurrent engineering teams find the tool to be a valuable means of facilitating discussions and weighing design and manufacturing tradeoffs that are based on data rather than opinions.

9.7 SUMMARY

A joint industry/academia project was undertaken to develop a decision-support tool to aid engineers in quantifying design tradeoffs and predicting the time and cost to manufacture a circuit card design. This tool utilizes algorithms that predict the level of activity at each step in the manufacturing system, as driven by the decisions about the design and production of the proposed circuit card. This tool allows engineers to try design alternatives, with immediate feedback given, to help lead them to the most cost effective designs. The accuracy of the tool's predictions compared against actual data for 12 randomly selected boards is found to be within 6.68% for labor costs, 4.76% for component costs and 3.1% for component density predictions. The tool is currently in use for cost estimating purposes and as an aid to cross-functional concurrent engineering teams. A more detailed report on the material in this paper is presented by Remer *et al.* (1991).

ACKNOWLEDGMENTS

The Harvey Mudd College Engineering Clinic Program would like to thank the ELDEC Corporation for sponsoring this project at Harvey Mudd College of Engineering and Science.

The authors would like to thank the following Harvey Mudd College students for their work on this project. Mark Harada was the second semester team leader and helped develop the model, Sung Lee was the first semester team leader and wrote most of the macro code, Sanford Leong created the output graphs, and David Mercer helped develop the model.

REFERENCES

Boothroyd, G., Dewhurst, P. (1989) *Product Design for Assembly*, Boothroyd-Dewhurst Inc., Wakefield, Rhode Island.

Boothroyd, G., Shinohara, T. (1986) Component insertion times for electronic assembly, *International Journal of Advanced Manufacturing Technology*, **1**(5), 3-18.

Center for Computer-aided Life Cycle Engineering (1991) *A Computerized Approach to PCB/PWB Design for Cost Effectiveness, Producibility, and Assembly*, Department of Mechanical Engineering, University of Maryland.

Foster, G., Gupta, M. (1990) Activity accounting: an electronics industry implementation, in *Measures For Manufacturing Excellence*, (Edited by R.S. Kaplan), Harvard Business Press, 225-268.

Funk, J.L. (1989) Design for assembly of electrical products, *Manufacturing Review*, **2**, 53-59.

Ghemawat, P. (1985) Building strategy on the experience curve, *Harvard Business Review*, March-April, 143-149.

Remer, D.S., Harada, M., Sung, L., Sanford, L., Mercer, D., Ziegler, F.S. (1991) *Circuit Card Assembly - Design for Manufacture: a cost estimation model for printed circuit board manufacturing*, Harvey Mudd College Engineering Clinic Report to the ELDEC Corporation. Harvey Mudd College, Department of Engineering, Claremont, CA.

DESIGN FOR INSPECTABILITY

Colin G. Drury

This chapter develops the ideas underlying Design for Inspectability, by considering in more detail the manufacturing climate, by introducing the human operator as a third component alongside product and process, and by defining a Design for Inspectability procedure. We show, with an example of this procedure, how product, process and person can be designed together to improve inspectability.

10.1 CONCURRENT DESIGN: MANUFACTURABILITY AND INSPECTABILITY

Industry's present moves towards concurrent design have emphasized the mutual dependence between product and process design (Niebel and Liu, 1992). The process has traditionally been designed to match the product, at least for chemical and flow manufacturing plants, but these are only a fraction of the total manufacturing capacity of a company or a nation. Most of our manufacturing facilities comprise small-batch manufacturing. Here, the process required to manufacture a new product must be assembled from largely pre-existing pieces of equipment. The equipment itself ranges from general purpose (machining centers, drill presses, sewing machines) to the product-specific (custom workstations for assembly or product test). Our goal in designing the process has been to achieve the savings associated with product-specific equipment without bearing its concomitant costs of customized design and rapid obsolescence. Design for manufacturability (Helander and Nagamachi, 1992) supplements this process design focus with a parallel focus on designing the product to make the most appropriate use of the production processes.

This concurrent design philosophy has been most closely associated with product-change steps such as machining, heat treating and assembly (Nevins and Whitney, 1989), but the steps involving decision are equally amenable to concurrent design. Decision steps are ones where the product itself does not change (except for relatively rare destructive tests, e.g., crash tests on automobiles or ultimate failure tests on aircraft wings) but instead the decision attributes pertaining to the product change. Examples of decision steps are:

1. Measuring the exact diameter of balls in a ball bearing, so as to be able to assemble a better bearing than random ball selection would produce.
2. Functionality test on an integrated circuit chip to determine whether it is fit for use.
3. Measurement of thickness of rolled steel so as to adjust the pressure in each rolling stage correctly.

With its emphasis on defect-free manufacturing (Luggen, 1991), industry is moving rapidly towards decisions about process (number 3 above) rather than decisions about individual items of product (numbers 1 and 2 above). Tighter control over processes requires a deeper knowledge of the process technology and a rapid, accurate way to determine the state of the output (product) so that process changes can be implemented well before any non-conforming items have been produced. This push towards process control requires the output to be designed so as to yield process decisions in an unimpeded manner. Design of the product so that it can be inspected easily is the natural result: Design for Inspectability (Drury, 1992a; Black, 1990). There is also a parallel use of the term 'Design for Inspectability' where inspection is part of the service life of the product rather than of its manufacture. Thus aircraft structures, bridges, automobiles and trucks must undergo periodic checks to determine their continuing fitness for use. These checks are as much inspections as statistical process control during manufacture. Design for Inspectability applies equally to these in-service inspections.

10.2 IMPACTS OF CURRENT CHANGES IN MANUFACTURING AND SERVICE

Global competition and new concepts of design, manufacturing and product use are forcing changes in the way items are manufactured, used and maintained. As a broad range of relevant changes are covered in detail elsewhere, for example Kidd and Karwowski (1994), only those most affecting the Design for Inspectability concept will be reviewed here.

As a structure for this review, consider the Product being designed / manufactured / serviced; the Process by which these actions are performed, and the People who perform the actions. Some important links between these three must be considered.

10.2.1 Product

Global competition has changed the nature of product design. In 1970, US companies had foreign competition in the USA for only 20% of their products. By 1990 this had risen to 80%. An enterprise cannot grow and prosper without extreme customer responsiveness. A whole engineering technique of Quality Function Deployment (QFD) has arisen to meet this need (Edmondson, 1992). The customer also expects the product to be free of defects in design and manufacture, leading to design changes aimed at improved quality and inspectability (Evans and Lindsay, 1993). Simultaneously, the pace of product design has changed. Automobiles are designed on a two-to-three year cycle instead of a four-to-five year cycle (Clark and Fujimoto, 1989).

Although products are reaching customers more rapidly, they are expected also to last longer. Complex products in particular can deteriorate during service with wear, cracking and corrosion all possible. Civil aircraft must undergo repeated inspections throughout their service life to check airframe, engines and systems (Goranson, 1993). Commercial trucks are subject to both periodic checks and random roadside inspections, covered under the North American Uniform Out-of-Service Criteria. Similar provisions exist for shipping, railroads, chemical plant equipment and bridges. Product design must consider these inspections just as much as it considers inspections during manufacture.

10.2.2 Process

Over the past ten years, methods of manufacture have changed considerably. The realization that efficient small-batch production can yield some of the benefits of mass production, without its high cost and inertia, has forced companies to innovate. Modern manufacturing is quality-driven (ISO-9000 series of standards; TQM philosophies such as Evans and Lindsay (1993), flexible (Hörte and Lindberg, 1991), leaner through JIT and low inventory production (Konz, 1990, p. 140), and measurement-driven (Drury, 1991; Compton, 1992).

In implementing strategies to reach these goals, industry has turned to two concepts in small-batch production: cellular manufacturing and Computer Integrated Manufacturing (CIM). Manufacturing cells typically have responsibility for their own inspection: incoming, in-process and outgoing. Cells can increase flexibility through worker cross-training, and be highly effective in terms of quality and cost as in Drury's 1991 example. CIM (Badham and Schallock, 1991) has brought the ability to use the new-found intelligent abilities of machines, and to integrate the information resources throughout the organization (Bullinger, Fähnrich and Niemeir, 1992). Inspection is a key information capture point for both cellular manufacturing and CIM.

10.2.3 People

Business cycles imposed upon a trend of company downsizing in response to global competition have created a new and less stable climate for the workforce. Typically, engineers and ergonomists have little to say about these broader structural changes in the workforce, concentrating instead on those still working. We are (at last) implementing some of the ideas of the socio-technical systems approach in modern manufacturing (Kidd and Karwowski, 1994). This can help modify the workplace and the working system to provide greater autonomy for the worker (Taylor and Felten, 1993). It can also provide healthier work (Karasek and Theorell, 1991). These trends towards a more autonomous workforce have been accentuated by moves towards multi-skilling in cellular manufacturing, which give more decision latitude in the minute-to-minute demands of manufacturing tasks.

Again, these changes are reflected in inspection. The inspector as such is required less in manufacturing: the activity of inspection becomes increasingly a part of every job. There are implications here for training (Drury and Kleiner, 1993) as more people will have to acquire inspection skills. Design for Inspectability should help to reduce at least some of these skill requirements.

10.2.4 Product, Process and People as a System

The Socio-Technical Systems (STS) discipline covers an important aspect of integration of product, process and people. It starts the design of the production system (process and people) by explicit consideration of the transformations required to go from raw material to finished product (Taylor and Felten, 1993). STS considers each of these transformations as a Unit Operation, and further bases the joint design of the technical and social systems on an enumeration of the quality characteristics at each step. Inspectability is a necessary requirement for control of these quality characteristics, and thus of the design of the social and technical aspects of the manufacturing system.

Other interactions have been noted, particularly in the integration of people into the manufacturing system. Siemieniuch (1992) and Sinclair (1992) describe a Design-to-Product initiative in which the technical support for design for manufacturability is considered from a human factors viewpoint. Wilson (1992) in the same volume casts a suspicious eye on the supposed improvement in human work as a result of Just-In-Time (JIT) and Total Quality Management (TQM) initiatives. Nagamachi (1994) integrates Design for Manufacturability concepts and customer satisfaction into *kansei* engineering. Finally, Brödner (1994) introduces the concept of anthropocentric design, meaning the use of allocation of function and job design based on a human-centered approach rather than a techno-centered approach. Throughout all of these is the implication of design to fit human needs: inspection design to meet human goals is a single extension.

10.3 STATUS OF DESIGN FOR INSPECTABILITY

Design of the product for ease of inspection is a necessity when the customer demands high quality and continuing safe service from a product. Inspectability in manufacture provides rapid and accurate feedback for process control. Inspectability in service allows rapid and accurate determination of deterioration in structure or function, ensuring safety of use. With defect rates measured in parts per million, and major in-service system failures (e.g. civil aircraft accidents) expected to occur at even lower rates, concurrent design of the product, process and people to ensure inspectability is increasingly important. While some bodies of literature exist in electronic component design and airworthiness inspection, more common examples are available. The examples below extend the listing in Drury (1992b):

1. To ensure that all colors have been printed on a product or label, a special color array is provided to the printer. An example from a sheet of postage stamps is given in Figure 10.1.
2. On automobiles, fluid containers such as coolant, brake fluid and washer fluid are now manufactured from transparent material to show levels easily (Figure 10.2). In addition, wear-prone items such as disc brake pads can be inspected through special holes in caliper castings (Figure 10.3).
3. A new use for smart materials could include constant, real-time crack and delamination detection in critical composite aircraft and spacecraft structures. The process incorporates micron size particles of Terfenol-D, a "giant magnetostrictive" material, into the composite. Sensing coils monitor internal stress based on the magnetic field generated by the particles. Stress concentration caused by a defect or delamination results in an abnormally high local magnetic field, according to Terfenol-D

manufacturer Etrema Products, Ames, Iowa. (Quoted from <u>Aviation Work and Space Technology</u>, February 27 1995, p. 15)

4. Almost any personal computer is expected to test itself whenever it is switched on. This power-on-self-test (POST) performs functional inspection of various subsystems and reports any discrepancies to the user for action.

5. A much older example is from the landing gear on the World War II Spitfire fighter, which when fully extended or retracted gave the pilot a visible indication on the upper surface of each wing.

Figure 10.1 Printing symbols (B3333) on edge of postage stamp sheet to help printer inspect for color and alignment.

In electronic equipment design, inspectability at the integrated circuit chip level and at the circuit board level are important design considerations. Design-for-Testability (DFT) adds a small amount of circuitry, typically only 5%, to allow test probes to isolate certain subsets of the circuit to find and diagnose faults (Markowitz, 1992). DFT can involve isolation of components, synchronous logic to allow timing of test signals (Novellino, 1991), removal of feedback paths (Black, 1990), and boundary scanning for boards (Venkat, 1993). There are standards which cover DFT procedures, such as IEEE Stand 1149.1. Note that in order to perform DFT, a list of the possible failures or defects is required to act as the challenges to the test procedure. Note also that the DFT procedures referenced above cover functional test rather than visual inspection.

Civil aircraft in-service inspection is a field where inspectability is important. Spectacular failures of the inspection system, such as the Aloha Airlines B-737 fuselage failure, have driven public demand for inspection system improvements. The design philosophy is Damage

Tolerance (Goranson, 1993), which allows structures to function safely until defects are detected and repaired. Different levels of inspection, from the daily walk-around, through general inspection of structural zones, to detailed inspection of small areas, have different capabilities of detecting failures such as cracks, corrosion or wear. As inspection intensity increases towards a detailed level, so smaller defects can be detected, but at increased cost. Knowledge of the probability of detection by inspection is used with the mechanics of crack growth to determine safe inspection intervals (Goranson, 1993; Finch, 1994).

Figure 10.2 Inspection hold in disc brake calipers to improve inspectability of pads and rotor by car owner or mechanic.

However, as many researchers have determined (Lock and Strutt, 1987; Spencer and Schurman, 1994; Drury, 1992a) inspection is a fallible procedure, whether performed by unaided humans, automated systems or hybrids (Hou, Lin and Drury, 1993). In airframe inspection this means that multiple inspections must be scheduled between a defect becoming visible and the same defect growing to an unsafe size. Any moves towards improved inspectability will allow more frequent inspection for the same cost. In fact the choice of inspection interval is highly dependent upon the inspection technique used. Visual inspection and non-destructive inspection (NDI) have widely different capability and cost characteristics. Indeed, the choice between NDI techniques themselves is a complex one (Roberge, 1995). Note that in order to specify inspection intervals and procedures, all possible failures/defects must be listed.

Considered in another way, lack of inspectability forces the designer to rely on a Safe Life concept to ensure structural integrity (Hagemaier, Skinner and Wikar, 1994). Here, a very conservative lifetime is selected for the component so that, despite variability arising from manufacture and service use, no component will fail before replacement. This is a costly strategy, as the necessary conservatism in safe life specification means that most components will be replaced long before their useful life has ended (Finch, 1994). Lack of inspectability can be very costly.

A final example of the literature available on inspectability also comes from aviation. Built-in test equipment (BITE) is an increasingly common means for conducting functional evaluations of avionics without removal from the aircraft. Similar in concept to a computer's POST mentioned earlier, a BITE system saves inspection time and cost while improving inspection accuracy (Goldsby, 1991). Successive generations have been increasing in sophistication and sub-system integration but the goal remains the same: improved inspectability. As with the other inspectability examples, the starting point must be a list of possible system failures.

Figure 10.3 Translucent hydraulic fluid reservoir to allow inspection of brake fluid level by car owner.

10.4 SYSTEMATIC PRODUCT DESIGN FOR INSPECTABILITY

All of the Design For Inspectability (DFI) procedures quoted so far have had a common starting point in the defect or failure list. At one time, the analyses necessary to produce such a list (such as Failure Modes and Effects Analysis, FMEA), were only used in complex military or aerospace products. With the current emphasis on quality in manufacturing, such techniques as FMEA are more widely practiced in consumer and industrial product design. Any DFI procedure implies use of such a technique to certain extent.

What is also required is a thorough knowledge of the inspection systems available, so that the system appropriate to the challenges defined by the failure list can be chosen. The detailed design of this inspection system must be concurrent with product design for inspectability, but a knowledge of the technological options is a prerequisite to design.

To develop a systematic procedure for inspection systems design, we propose the use of a generic function description of inspection, so that the impact of each function can be assessed systematically. The function analysis used in our original DFI procedure (Drury, 1992b) is expanded here to include one more function, based on extensive study of aircraft inspection tasks (Drury, Prabhu and Gramopadhye, 1990) but applicable to all inspection tasks. Table 10.1 shows this function list, adapted from Drury and Prabhu (1994).

Table 10.1 Generic Inspection Functions and Their Outcomes

Function	Outcome
Setup	Inspection system functional, correctly calibrated and capable
Present	Item (or process) presented to inspection system
Search	Indications of all possible non-conforities detected, located
Decision	All indications located by Search correctly measured and classified, correct outcome decision reached
Respond	Action specified by outcome decision taken correctly

The next step is to list those factors known to affect inspection performance within each function. A comprehensive discussion is given in Drury (1992a), so that only a listing of these factors is presented here, in the first column of Table 10.2. Also shown in Table 10.2 is how each of the factors is influenced by changes in the product, process and person (i.e. inspector). Note the number of primarily human factors in Table 10.2 which are affected by product design changes. Clearly there is scope for DFI.

Table 10.2 Seven different defects were listed by Drury (1992b)

Components		
Components	1.	Missing
	2.	Wrong
	3.	Damaged
	4.	Reversed
Solder Joints	1.	Missing
	2.	Inadequate
	3.	Excess

To use Table 10.2 in design, the impact of the factors listed must be found for each possible defect. This produces the Factor/Defect matrix, shown in generic form in Table 10.3. Entries in this matrix are specific statements of effects or impacts.

Table 10.3 Factors affecting performance, and how they are impacted by system changes

Task	Factors Affecting Performance	Possible Changes		
		Product	Process	Person
Setup	Job Instructions		X	X
	Accuracy of defect list	X	X	
	Calibration of equipment		X	X
	Equipment capability		X	
Present	Accessibility	X		
	Location of areas to inspect	X	X	
	Handleability of product	X		X
Search	Visual Lobe			
	• Size of defect set		X	
	• Defect/field contrast	X	X	X
	• Field complexity	X		
	• Defect size	X		
	• Illumination		X	
	• Peripheral acuity			
	Search strategy			
	• Random/systematic		X	X
	• Interfixation distance		X	X
	Timing			
	• Fixation duration		X	X
	• Time available		X	
Decision	Discriminability			
	• Defect versus standard	X	X	
	• Presence of standard		X	
	• System noise	X	X	
	• Human noise			X
	Criterion			
	• Defect probability		X	X
	• Cost/value of accept		X	
	• Cost/value of reject		X	
	• Perceived probabilities/costs		X	X
Action	Action complexity, e.g. defect counts	X	X	X
	Action convenience to operator		X	X
				X

Table 10.4 Generic factor/defect matrix

Task	Factors Affecting Performance	Possible Changes		
		Defect 1 ..... 2 .....3		
Setup	Job Instructions Accuracy of defect list Calibration of equipment Equipment capability			
Present	Accessibility Location of areas to inspect Handleability of product			
Search	Visual Lobe • Size of defect set • Defect/field contrast • Field complexity • Defect size • Illumination • Peripheral acuity Search strategy • Random/systematic • Interfixation distance Timing • Fixation duration • Time available			
Decision	Discriminability • Defect versus standard • Presence of standard • System noise • Human noise Criterion • Defect probability • Cost/value of accept • Cost/value of reject • Perceived probabilities/costs			
Action	Action complexity, e.g. defect counts Action convenience to operator			

Table 10.5 Factor/defect matrix and DFI changes

Factors Affecting Performance	Components				Solder Joints		
	Missing	Wrong	Damaged	Reversed	Missing	Inadequate	Excess
Calibration of equipment	1	1	1	1	1		
Accessibility	2	2	2	2			
Location of areas to inspect	3	3					
Handleability of board	4	4	4	4			
Defect/field contrast	5	6		7	10		8
Field complexity	9	9	9	10	10		
Defect size	11	12		10	10		
Defect versus standard		13	13,14	7		13	13
System noise		15		13	10	10	16
Action complexity	17	17	17	17	17	17	17

1. Add probe points for functional test.
2. Ensure that component is visible from a range of eye/board angles.
3. Subdivide board into visually-logical areas to simplify recognition of area which needs to be inspected.
4. Use automated insertion which crimps component to board so that components will not fall off if board is moved vigorously during inspection.
5. Place colored patch (of a color contrasting with the board top layer) behind component to increase discriminability of missing components.
6. If possible, code components to match identifiers on board.
7. Use obviously asymmetric components to detect reversals.
8. Use a dark-colored undersurface of board to provide good contrast with excess solder.
9. Subdivide board into visually-logical areas so that patterns of correct components may be easily recognized.
10. Use a regular grid of solder joints to simplify detection of missing or inadequate solder.
11. Use a large colored patch (#4 above) to increase conspicuity of missing components.
12. Use components with lettering or identifiers in large printing.
13. Provide comparison standards for all defects close to the line of sight.
14. Make colored patch same size and shape of component to simplify detection of damaged or misaligned components.
15. Reduce the number of different component types so that conspicuity of wrong components is increased.
16. Keep the design of the undersurface of the board as simple as possible so that bridges are not confused with legitimate connectors.
17. Reduce the number of defect types searched for at one time (and reported) by splitting task into populated and solder sides of board.

We are now ready to begin searching for solutions to the inspection system design requirements summarized in Table 10.3. To make the ideas easier to follow, a specific example will be used, that of inspecting printed circuit boards for defects. This continuing problem has been shown to be an example of error-prone inspection (Drury and Kleiner, 1984), and has been used as an example in both design for manufacturability (by Anderson, 1990) and design for inspectability (Drury, 1992b). Here, these earlier examples are expanded to include the setup function added in Tables 10.1, 10.2 and 10.3 of the current chapter.

Considering only the product changes, which are the essence of DFI, we must produce a matrix relating the factors impacting the product (from Table 10.2) and the list of defects given above. Table 10.5 gives the completed factor/defect matrix for inspection of both functional failures (by functional and test) and potential failures (by visual inspection). Each DFI change is numbered and listed below the table.

There is obviously considerable scope for DFI ideas even within such a simple task. In addition, by going back to Tables 10.3 and 10.4, it is possible to add similar changes for process design and person design. Both of these have well-known literatures and examples of improvements. For example, lighting changes are an obvious process improvement in inspection (Drury, 1992a) while personnel selection (Thackray, 1994) and training (Kleiner and Drury, 1993) can give beneficial process changes. There are also systematic procedures for considering particular aspects of system design in inspection, for example, automation (Drury, 1994a) and the effects of speed or inspection accuracy (Drury, 1994b).

10.5 SUMMARY

As manufacturing and maintenance activities move into a more quality-conscious environment, the importance of knowledge about the product must increase. This knowledge comes from inspection processes, which are themselves undergoing rapid change. Design for inspectability is one strategy by which quality and service life can be improved. It is also one that uses the classic engineering improvement approach of replacing continuing downstream costs with relatively small increases in initial cost. All of the data required to implement DFI is typically available at the design stage, so that systematic DFI procedures can be applied directly, as shown by the worked example in this chapter.

REFERENCES

Anderson, D.M. (1990) *Design for Manufacturability*, Lafayette, CA: CIM Press.

Aviation Work and Space Technology, Industry Outlook, February 27 1995, 15.

Badham, R.; Schallock, B. (1991) Human Factors in CIM: a human-centered perspective from Europe, *International Journal of Human Factors in Manufacturing*, **1**, 121-141.

Brödner, P. (1994) Design of work and technology, In: *Design of Work and Development of Personnel in Advanced Manufacturing*, (Edited by G. Salvendy. and W. Karowowski), New York: John Wiley & Sons, Inc., 125-157.

Black, S.L. (1990) Consider testability in your next design, *Electronic Design*, **39**(54), December 27, 1990, 54-59.

Bullinger, H.J.; Fähnrich, K.P.; Niemeier, J. (1992) Computer-integrated business systems, In: *Handbook of Industrial Engineering*, Second Edition, (Edited by G. Salvendy), New York: John Wiley & Sons, Inc., Chapter 10, 241-268.

Clark, K.B.; Fujimoto, T. (1989) Reducing the time to market: The case of the world auto industry, *Design Management Journal*, **1**, 49-59.

Compton, W.D. (1992) Benchmarking, In: *Manufacturing systems: Foundations of world-class practice*, (Edited by J.A. Heim and W.D. Compton), Washington, D.C.: National Academy Press, 101.

Drury, C.G. (1990) Design for inspectability, In: *Proceedings of the IEA Human Factors in Design for Manufacturabiilty and Process Planning*, Honolulu, Hawaii, 133-139.

Drury, C.G. (1991) Ergonomics practice in manufacturing, *Ergonomics*, **34**, 825-839.

Drury, C.G. (1992a) Inspection performance, In: *Handbook of Industrial Engineering, Second Edition*, (Edited by G. Salvendy), 88, New York: John Wiley and Sons, 2282-2314.

Drury, C.G. (1992b) Product design for inspectability: a systematic procedure, In: *Design for Manufacturability: A Systems Approach to Concurrent Engineering and Ergonomics*, (Edited by M. Helander and M. Nagamachi), London: Taylor and Francis, Ltd., 204-216.

Drury, C.G. (1994a) Function allocation in manufacturing. In: *Contemporary Ergonomics 1994*, (Edited by S.A. Robertson), Keynote address to the Ergonomics Society Meeting, Taylor & Francis, Canada, 2-16.

Drury, C.G. (1994b) The speed-accuracy trade-off in industry, *Ergonomics*, **37**, 747-763.

Drury, C.G.; Kleiner, B.M. (1984) A comparison of blink aided and manual inspection using laboratory and plant subjects, In: *Proceedings of 1984 International Conference on Occupational Ergonomics*, Rexdale, Ontario: Human Factors Association of Canada, 670-676.

Drury, C.G.; Kleiner, B.M. (1993) Design and evaluation of an inspection training programme, *Applied Ergonomics,* **24**(2), 75-82.

Drury, C.G.; Prabhu, P. (1994) Human factors in test and inspection, In: *Design of Work and Development of Personnel in Advanced Manufacturing*, (Edited by G. Salvendy and W. Karwowski), New York: John Wiley & Sons, Chapter 13, 355-402.

Drury, C.G.; Prabhu, P.; Gramopadhye, A. (1990) Task Analysis of Aircraft Inspection Activities: Methods and Findings, In: *Proceedings of the Human Factors Society 34th Annual Conference*, Volume 2, Santa Monica, California, 1181-1185.

Edmondson, H.E. (1992) Customer satisfaction, In: *Manufacturing systems: Foundations of world-class practice,* (Edited by J.A. Heim and W.D. Compton), Washington, DC: National Academy Press, 129.

Evans, J.R.; Lindsay, W.M. (1993) *The Management and Control of Quality,* Second Edition, West Publishing Company, Minneapolis/St. Paul, MN, 143.

Finch, D.V. (1994) The management of "on condition" aircraft maintenance, presented at Aerotech 1994, Birmingham, England.

Goldsby, R.P. (1991) Effects of automation in maintenance, In: *Proceedings of the Fifth FAA Meeting on Human Factors Issues in Aircraft Maintenance and Inspection,* Atlanta, GA, 103-123.

Goranson, U.G. (1993) *Damage Tolerance Factors and Fiction*, Preseneted at the 17th Symposium for the International Committee on Aeronautical Fatigue, Stockholm, Sweden, June 9, 1993, Seattle, WA: Boeing Commerical Airplane Group.

Hagemaier, D.; Skinner, R.; Wikar, D. (1994) Organizing NDT in the aerospace industry: the airworthiness limitations instructions (ALI) document, *Materials Evaluation*, April 1994, 470-473.

Helander, M.; Nagamachi, M. (1992) *Design for Manufacturability: A Systems Approach to Concurrent Engineering and Ergonomics,* London: Taylor and Francis, Ltd.

Hou, T-S; Lin, L.; Drury, C. G. (1993) An empirical study of hybrid inspection systems and allocation of inspection function, *International Journal of Human Factors in Manufacturing*, **3**, 351-367.

Hörte, S.A.; Lindberg, P. (1991) Implementation of advanced manufacturing technologies: Swedish FMS experiences, *Advanced Manufacturing Technologies*, New York: John Wiley & Sons, Inc., 55-73.

Karasek, R.; Theorell, T. (1991) *Healthy Work*, New York: Basic Books.

Kidd, P.T.; Karwowski, W. (1994) *Advances in agile manufacturing*, Amsterdam: ISO Press.

Konz, S. (1990) *Work design: Industrial ergonomics* (3rd ed.), Worthington, OH: Publishing Horizons, Inc., 140.

Lock, M.W.B.; Strutt, J.E. (1981) *Reliability of In-Service Inspection of Transport Aircraft Structures, CAA Paper 85013*, London.

Luggen, W.W. (1991) *Flexible Manufacturing Cells and Systems*, New Jersey: Prentice Hall.

Markowitz, M. (1992) Design for test (without really trying), *EDN*, **37**, February 17, 1992, 114-126.

Nagamachi, M. (1994) Kansei engineering: an ergonomic technology for a product development, In: *Proceedings of the 12th Triennial Congress of the International Ergonomics Association*, Volume 4, 120-121.

Niebel, B.W.; Liu, C.R. (1992) Designing for manufacturing, In: *Handbook of Industrial Engineering, Second Edition*, (Edited by G. Salvendy), New York: John Wiley and Sons, Chapter 13, 337-353.

Nevins, J. L.; Whitney, D.E. (1989) *Concurrent Design of Products and Processes*, New York: McGraw-Hill.

Novellino, J. (1991) Design teams evolve to face testability, *Electronic Design*, **39**, January 10, 1991, 56-76.

Roberge, P.R. (1995) A knowledge-based shell for selecting a nondestructive evaluation technique, *Materials Evaluation*, February 1995, 166-171.

Siemieniuch, C. (1992) Design to product, In: *Design for manufacturability: A systems approach to concurrent engineering and ergonomics*, (Edited by M. Helander and M. Nagamachi), London: Taylor & Francis, 35.

Sinclair, M.A. (1992) Human factors, design for manufacturability and the computer-integrated manufacturing enterprise, In: *Design for manufacturability: A systems approach to concurrent engineering and ergonomics*, (Edited by M. Helander and M. Nagamachi), London: Taylor & Francis, 127.

Spencer, F.; Schurman, D. (1994) *Reliability Assessment at Airline Inspection Facilities, Volume III: Results of an Eddy Current Inspection Reliability Experiment*, DOT/FAA/CT-92/12,III, Atlanta City, NJ: FAA Technical Center.

Taylor, J.C.; Felten, D.F. (1993) *Performance by Design*, NJ: Prentice Hall.

Thackray, R. (1994) Correlates of individual differences in non-destructive inspection performance, *Human Factors in Aviation Maintenance - Phase Four, Volume 1 Program Report*, DOT/FAA/Am-94/xx, Springfield, VA.: National Technical Information Service.

Venkat, K. (1993) Follow these guidelines to design testable ASICs, boards, and systems, *EDN-Design Feature*, **38**, August 19, 1993, 117-126.

Wilson, J.R. (1992) Human resource issues in manufacturing systems: The effects of JIT/TQM initiatives, In: *Design for manufacturability: A systems approach to concurrent engineering and ergonomics*, (Edited by M. Helander and M. Nagamachi), London: Taylor & Francis, 269.

DESIGN FOR STORAGE AND DISTRIBUTION

B. Gopalakrishnan; S. Chintala; S. Adhikari; G. Bhaskaran

This chapter analyzes the concepts related to product design for effective product storability and retrieval for distribution and describes the use of computer based systems for implementation in a concurrent engineering environment. The product design function and the manufacturing function are far apart from the storage, retrieval, and distribution functions in terms of the product life cycle. However, they have a significant impact on product storability and retrieval for distribution in terms of the costs attributable to the storage infrastructure as well as to materials handling activities. As products are often packaged as units for distribution, the product design parameters influence the unit design aspects.

11.1 OVERVIEW

Concurrent Engineering has become a very popular term amongst organizations attempting to improve productivity in manufacturing systems. The basis for this is the fact that substantial savings in resources can be obtained if several operations relating to preliminary design, detail design, manufacturing, assembly, support, and disposal, are done at the same time. This implies that some of the operations occur concurrently, while some other functions in an advisory mode. Early interactions between the operating work stations lead to the job being done "right the first time".

Concurrent engineering can be termed as the systematic approach to the concurrent design of products, their related processes and support systems, so that considerable reductions in time and cost may be achieved at both levels (Winner *et al.*, 1988). Concurrent engineering can also be defined as the merging of the efforts of product designers and other downstream specialists such as manufacturing engineers to improve products and their life cycle characteristics with concurrence, constraints, coordination, and consensus as the primary ingredients, as mentioned by Stauffer (1988). The practice of concurrent engineering principles, merely by the use of enhanced and sophisticated computer networks to share information between product design and other downstream functions, is not likely to deliver very many benefits it promises. This contrasts the use of integrated "Design for X" tools which are engineered to share information in a constructive manner.

The product life cycle is composed of a number of stages, namely preliminary design, detail design, manufacturing, assembly, storage and retrieval for distribution, and disposal. The extent of costs actually incurred is low during the initial stages of product development, while the costs committed are high, although "invisible" as far as the company's cost accounting system is concerned. Considering the issue of storage and retrieval, costs are actually being "committed" but not actually incurred. These are not recorded in the cost accounting system and therefore invisible to the management. If the product design is accomplished without considering storage and retrieval concerns, the product life cycle costs are likely to be high, especially when product storage and retrieval cannot be accomplished in a cost effective manner. To improve storability, the product design must be modified, thus resulting in the increase of the product development time and cost. In many cases, when concurrent engineering principles are not being adopted, expensive solutions to the storage and retrieval problem may be the only alternatives mainly because the product manufacturing activities may well be underway after the completion of the product design.

This chapter aims to provide design tools which can bring about the benefits promised by concurrent engineering. Product design for storability and effective retrieval for distribution can significantly reduce the above mentioned problems. This means that the product designer should be provided with tools which can evaluate the storability and effective retrieval or at least have the tools that function as a dialogue mechanism between the product design function and the storage design function. Efficient product retrieval from the storage areas provides for effective distribution. The designer should be in a position to alter the product design and observe its effects on storability and retrieval aspects. In this manner, the designer can iteratively try to improve the product design, thus reducing the product life cycle costs and enhance the product's storage and retrieval characteristics.

11.2 DESIGN FOR STORABILITY: DEFINITION AND IMPLICATIONS

Storability may be defined as the ability to implement an organized and efficient approach towards materials storage, its identification, and retrieval by means of effective materials handling equipment. The term "design for storability" implies the use of effective product design techniques so as to facilitate storability. This is especially important to consider because improper product design may often render the product storage function to be economically inefficient. For example, the design parameters may render a low utilization of warehouse cubic space. In some cases, an expensive solution to the order picking strategy required, including the materials handling equipment, may become the only available alternative due the nature of the product design. In general, high costs in product storage and retrieval may result if "design for storability" concepts are not considered. In addition the ability to deliver products on time may be seriously impaired, thus reducing a company's competitive edge.

11.2.1 Product Design Parameters Which Affect Storability

The product design phase may be divided into two distinct areas: preliminary design and detail design. During preliminary design, the overall product geometry, functionality oriented product characteristics, and other relevant parameters are specified, as mentioned by Hoover and Rinderle (1989). During the detail design, the product geometry specifications are more detailed, the material characteristics are well defined, and in general, the design of the product is fully completed, as found in Mistree and Muster (1990). The key parameters relating to preliminary and detail design include the overall product geometry and material characteristics. These parameters along with the product demand volume to be distributed per unit of time influence product storability in terms of factors such as pallet selection, unit load design, and rack design. They also affect factors such as the selection of the type of material handling equipment, the number of personnel involved in the material handling function, and the type of order picking scheme to be used.

For discrete structural products, there are numerous design oriented aspects which are important, but the major design characteristics considered important from the viewpoint of this chapter are the overall product geometry and the weight of the product as rendered by the material characteristics. The domain under consideration for emphasis in this chapter is one which pertains to the storage of discrete structural products packaged as prismatic units to comprise unit loads. A unit load may be defined as a collection of individual discrete products as units, packaged and arranged on a holding platform such as a pallet, a skid, or a bin. The holding platform is then stored in rack shelves to enable product retrieval along with the platform or as individual units. The product, if stacked on a platform, is limited to the height to which it can be stacked owing to safety regulations and product quality standards. The type of racks being considered is the standard pallet rack. Rack design for effective product storability and retrieval is an issue which will be analyzed later in this chapter.

11.2.2 Unit Load Design And Its Effect On Storability

A unit load may be defined as a collection of products which can be stored as a unit and moved from one location to another so as to save material handling costs, time, and product damage. The products may be "unitized" in numerous ways, such as in totes, pallets, cartons etc. The products may also be shrink-wrapped to conform as a unit load. In the context of this work, units contain one or more product(s) arranged on pallets and transported to their locations by using material handling equipment such as fork lift trucks, automated guided vehicles (AGVs), and automatic storage and retrieval systems (ASRS), as indicated by Tompkins and White (1984). Figure 11.1 shows a unit load resting on a pallet.

The size and configuration of pallets vary considerably. Pallets are usually made from wood, although metal and plastic pallets are not uncommon. The storage of the units on a pallet needs to be maximized so as to enable effective and efficient storage. This means that the maximum amount of units have to be arranged to fit on a pallet. The restriction on this effort will pertain to the height to which the units may be stacked. This is based on the weight and volume considerations which influence the characteristics of the storage system and the materials handling system (Apple, 1977; Tompkins and White, 1984). Other considerations include the door widths, turning angles of vehicles, and aisle widths.

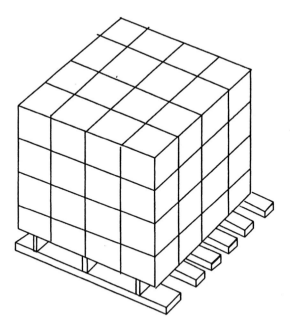

Figure 11.1 Unit Load on a Pallet.

11.2.3 Design For Maximum Pallet Space Utilization

The term "space utilization" refers to the percentage utilization of the available space on the surface of a pallet. The design of the product in terms of its dimensions significantly affects the space utilization, thus influencing storage and material handling costs. Space utilization close to 100 % are ideal, but not attainable in most cases in reality. However, an attempt should be made to increase the space utilization as much as possible by analyzing the product dimensions, product arrangement, and pallet dimensions. For given pallet dimensions, the space utilization depends entirely on the dimensions of the unit comprised of one or several products and the method of arrangement on the pallet. However, if the pallet dimensions can be held as variables, then the maximization of the space utilization would depend upon the pallet dimensions, the unit dimensions, and the method of arrangement.

Consider pallet dimensions as follows in inches: 36 x 48, 42 x 42, 32 x 40, 40 x 48, and 48 x 48. If the user is allowed to choose from amongst these pallets for purposes of storage, what impact would the unit dimensions have on space utilization ? Unit dimensions are directly dependent upon the overall product dimensions and the method of packaging. The question is how can one alter the product design dimensions in an attempt to increase space utilization on the pallet? For a pallet of size 36 x 48 inches, considering unit dimensions to be 14 x 7 x 4 in inches and allowing for no overhang, the best arrangement results in a space utilization of 85 % while unit dimensions of 10 x 12 x 3 result in a space utilization of 100 %. This illustrates the importance of product design dimensions as applied to product storability, irrespective of the method of storage. Low space utilization implies wasted storage space on the pallet and hence wasted resources. Changing product dimensions to improve storability is an intelligent decision capable of saving valuable plant resources. Figure 11.2 shows the unit arrangements on pallets of differing sizes.

11.3 KEY ISSUES IN DESIGN FOR STORABILITY

The key issues in design for storability relate to the unit load dimensions and weight as influenced by the product design, as well as on their impact on the type of order picking strategy to be used. The economics of storing and retrieving products or unit loads is a key issue in pursuing design for storability concepts. The volume and variety of products stored and retrieved are critical parameters to analyze, as will be the nature of the unit loads used.

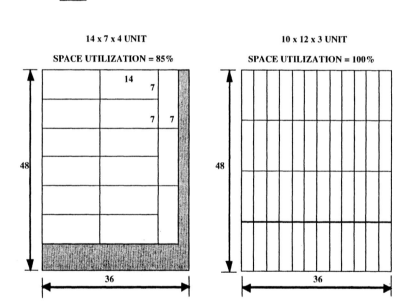

Figure 11.2 Unit arrangement on pallets.

11.3.1 Product Design and Order Picking

It is important, though difficult, to examine how the design of a product, specifically its dimensions and weight, influences order picking strategies and costs. A computer based software system such as the one described in this chapter is beneficial in informing product designers about design impact on storage and retrieval for distribution costs and function as a tool in the dialogue between product storage/material handing system design personnel and the product design function within the concurrent engineering domain. Product design influences not only the packaging for storage and distribution as described earlier but also on the type of order picking strategy used. This is again dependent upon the type of unit packaging and the demand volume to be delivered. If demand volumes are low and manual order picking may be sufficient, the product dimensions and weight have to facilitate the packaging of the products into integral units which are economical to be stored and convenient to be handled by operators. On the other hand, if the demand volumes are high, the product dimensions and weight have to be commensurate with the packaging strategies used so as to be handled by forklift trucks and similar equipment as material may have to be retrieved in large quantities in pallet loads for distribution. This aspect will be discussed in the "system performance analysis" section.

11.3.2 Product Variety

One important aspect which plays a key role in deciding upon the type of order picking is the variety of the products being stored and distributed. The variety, often depicted as "stock keeping units" (SKUs), significantly impacts the type of order picking to be used. When the product variety is large and demand volumes are low, it may be necessary and sufficient to employ people to pick the products for distribution. Such a situation may occur, for example, in a mail order facility of consumer goods. On the other hand, if the product variety is large and the demand volumes are high, fast moving equipment such as forklifts and AGVs may not be appropriate for retrieving material in large pallet loads as individual orders for products may not be substantial. It is in these instances that the product design dimensions play a more critical role in influencing storage and distribution costs more than in other situations. Low product variety and high demand volume may require people to pick items while riding on AGVs or similar fast moving equipment and the nature of the unit shape and its weight then will significantly influence the order picking efficiency and costs.

11.3.3 Relevant Literature

Unit load design considerations may be found in (Apple, 1977; Tompkins and White, 1984) including the steps for designing unit loads and the selection of the platforms on which they may be placed. The different types of material handling equipment and their uses are well documented in (Apple, 1977; Konz, 1994; Tompkins and White, 1984). The importance of considering materials handling system design principles early in the product design stage is emphasized in (AGVS, 1991; Apple, 1977; Apple and McGinnis, 1987; Bolz and Hageman, 1958; Francis and White, 1992; Tompkins and White, 1984). The importance of analyzing product, process, and schedule design for facilities planning effectiveness is also discussed in these references.

Design for efficient packaging is illustrated through case studies pertaining to industrial equipment by Vittal (1993). The importance of value engineering with respect to the product design function so as to improve storability and packaging effectiveness is analyzed. Industrial applications and case studies for product storage and distribution may be found in (Apple and McGinnis, 1987; Kulwiecz, 1992), Material Handling Reprint series (MHTI, 1993). Building and planning aspects for industrial storage and distribution are addressed in Falconer (1975). A quantitative study on serving a multi-aisle system by a single ASRS (Automated Storage and Retrieval Systems) is discussed by Hwang and Ko (1988). An ASRS may be described as a large carousal traveling at high speeds in X, Y, and Z directions so as to retrieve items from racking systems. A detailed presentation on warehousing aspects may be found in Ozden (1988) and NAVSUP (1978). As observed in the presentation of the literature herein, much work has been done in plant layout and materials handling but there is a need to integrate them to product design in the concurrent engineering environment.

11.4 SYSTEMATIC PROCEDURE OF DFS&D

This section describes the computer based systems software STORE with its components PALLET and SIMPICK. The unit design and the pallet design information are required from the user. PALLET can then be used to determine the most effective arrangement for maximizing the space utilization by changing pallet sizes and unit dimensions. Once the optimal arrangement has been achieved for the unit load, this information is channeled into a system which focuses on rack design for storage. Thus, product design dimensions could have an impact on a far removed entity such as warehouse storage, especially when the company's storage space is limited.

The determination of warehouse storage feasibility leads to the analysis of order picking strategies for product distribution. The term "order picking" refers to the methods employed for removing units from the racks so that they may be consolidated as required for distribution purposes. The removed units may either be "palletized", or packaged into small containers for shipping. The decisions made at the stage of order picking have a significant influence on the nature of packaging and distribution costs. Since people and/or equipment will have to be used to retrieve the units from the storage racks, the type of order picking used will depend upon the type of demand volumes which are to be met. For example, the type of order picking needed to deliver 1000 units per day will be quite different from the type of order picking needed to deliver 100,000 units per day. In the former situation, people may be utilized to pick items at a slower pace whereas in the latter case, forklifts, automated guided vehicles or such means must be used to deliver the high demand volume required.

The system STORE to be discussed later in this chapter will consider the following modes of order picking: people, forklift trucks, automated guided vehicles (AGVs), people riding on automated guided vehicles, and automated storage and retrieval systems (ASRS). The computer system STORE contains within it a simulation system, SIMPICK. Its inputs include the size of the warehouse, the type of order picking strategy used and their characteristics. The system simulates order picking to arrive at the rate of demand volume delivery possible with a chosen order picking strategy. This leads to the determination of the number of order picking entities required to satisfy the demand volumes, and the costs incurred in their use and maintenance. Since a significant amount of investment is often involved in implementing storage and order picking schemes to deliver large demand volumes, the payback on investment is determined by the system as a ratio between the costs incurred and the profits obtained through distribution. Figure 11.3 shows the systems diagram for STORE.

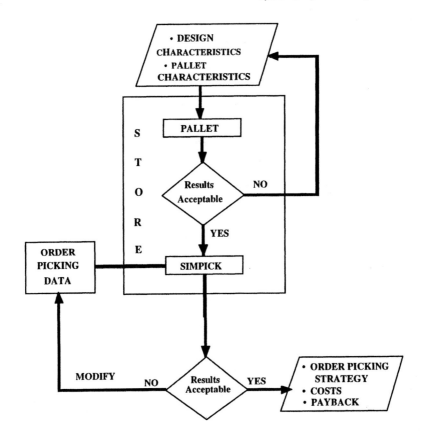

Figure 11.3 Systems diagram for STORE.

11.4.1 The PALLET System

The system PALLET is used to maximize pallet space utilization. It is an integral part of STORE, the overall system to function in the design for storability and retrieval for distribution in concurrent engineering. The inputs to the system are the length and width of the pallet (allowance for any feasible overhang), the maximum height to which the pallet may be stacked with units, the weight of each unit to be stored on the pallet, the maximum load which a unit may be able to withstand without being damaged, and the demand volume to be met. The system generates as output the best arrangement of the units for maximum space utilization. The system considers the weight of each unit and the maximum weight which each unit can bear for arriving at the number of units which can be stored on each pallet. If the units may be damaged on account of considerable load on them, then the pallet may be stacked only a few levels high, meaning a general loss in storage capacity and resources. Thus, the weight of the product comprising the unit, which is dependent mainly upon the material, a design parameter, considerably influences its storability. Finally, the system outputs the total number of pallets required to satisfy the demand volume.

11.4.2 Simulation of Order Picking: SIMPICK

In order to determine the costs and efficiency of a chosen order picking strategy, STORE uses a simulation module. Once the user has executed the PALLET system and obtained a satisfactory arrangement of units on the pallet, the SIMPICK system may be executed. This system requires the following inputs the length and width of the warehouse facility, the length of a rack, the length of an aisle, the demand volume to be delivered in a day, the number of hours per shift and the number of shifts per day in which order picking is accomplished, and the profit on each unit distributed. The weight of a unit and the number of units on a pallet are passed on from the PALLET system. Once all the inputs for the system have been obtained, the user may select through a menu, the type of order picking strategy. When AGVs, ASRS, or forklifts are used to pick units by the pallet load, the cost of the equipment, its maintenance costs, the picking time for a pallet, and the speed are required as inputs. When manual order picking is used and in cases where the above mentioned equipment have to be operated by people, the labor rate, the time to pick units, the capacity of pickers to carry units, and the speed are required as inputs.

The SIMPICK system then begins to simulate the order picking effort by accomplishing the following, namely, racking system design, and tracking the movement of order picking entities through the storage area. The system designs the racks by using the information on the warehouse dimensions, rack width, aisle width, and the number of pallets to be stored. The racks are identified and subdivided into identifiable locations where the order pickers will stop for picking the units by the pallet load or as individual units. Using the speed of movement, the system simulates and obtains the delivery rates for the order picking entity chosen. The system is thus capable of determining the number of order picking entities required, the costs involved, and the estimated payback on investment.

The user may change unit dimensions and weights as key parameters in the product design function and observe the changes to the storage and material handling systems design function and obtain an estimate of the costs involved. Thus the user may iteratively go back and forth between the product design dimensions and weight and STORE to optimize on the design parameters for the lowest possible storage and material handing distribution costs. The algorithms for STORE were coded in C language.

11.5 SYSTEM PERFORMANCE ANALYSIS

The STORE system creates a concurrent engineering environment to allow a dialogue between the design function and other downstream functions such as product storage and packaging for distribution. The computer system STORE, with its components PALLET and SIMPICK are meant to be advisors to the product design function so that designers may have a flavor of the storage and material handling costs when the design dimensions and material specifications are being made. This section tests the system performance and analyzes the results from STORE. It should be understood that the data generated from the execution of STORE and reported in the following sections are best estimates and are intended to be used as such in assisting the facilities planners and product designers. In addition, the data have been generated under several assumptions as outlined earlier. It is the intent to use STORE to underline the importance of design for product storability so that interested readers may attempt to develop similar systems according to the operating conditions at hand.

11.5.1 Optimizing unit dimensions

In this section, a number of examples will be used to illustrate the system performance of STORE. Consider a unit with dimensions of 11 x 11 x 7 inches. For a pallet size of 48 x 40 inches, allowing for no overhang, PALLET delivers an optimal unit arrangement resulting in a pallet space utilization of 80.2 %. This means that 19.8 % of the space on the pallet is wasted and not being used for storage. When the unit dimensions were changed to 7 x 9 x 9 inches, the pallet space utilization improved to 82.03 %. A further change in unit dimensions to 7 x 5 x 11 gave a space utilization of 87.5 %. Finally with a unit dimension of 8 x 8 x 8, the pallet space utilization of 100 % is obtained. Since the 48 x 40 pallet is the most popular size in pallets, this analysis leads to advantageous space utilization. Similar analysis with other pallet sizes could be accomplished to optimize not only on the unit dimensions but also on the pallet dimensions. Table 11.1 shows the unit dimensions, and their corresponding pallet space utilization, obtained from the executions of PALLET.

Table 11.1 Pallet space utilization for varying unit dimensions on 48 x 40 pallet

UNIT DIMENSION	SPACE UTILIZATION
11 x 11 x 7	80.20%
7 x 9 x 9	82.03%
7 x 5 x 11	87.50%
8 x 8 x 8	100.00%

11.5.2 Influence of unit dimensions and weight on storage

Consider a warehouse with a width of 400 feet. Allowing for a aisle width of 7 feet on either side, as shown in Figure 11.4, the racks are to be placed parallel to the width of the warehouse, their length being 386 feet. If the clear ceiling height is 20 feet (maximum allowable height of the racks), and the clearances between the pallet and the edges of the rack shelf are 2 inches, calculations may be made on the number of shelves which are in each rack, depending upon the dimensions of the pallet and the unit load on it. Consider the unit with dimensions of 11 x 11 x 7 inches. On one layer of the pallet (48 x 40), 20 units may be stored for a space utilization of 80.2 %. If each unit weighs 5 lbs, and the maximum allowable weight on each unit is 25 lbs with one less layer being used for safety reasons, then the pallet may only be stacked 5 high, leading to the height of the unit load on the pallet being 55 inches. If the height of the pallet is 10 inches, then the total dimensions of the pallet will be 40 x 48 x 65, with the effective pallet height being 67 inches including the 2 inch clearance between the unit load and the rack shelf.

Allowing for clearances and rack bar widths of 2 inches as appropriate, 100 shelves may be present horizontally on any rack, and 3 shelves may be present vertically, leading to a total of 300 shelves on any rack. Please note that the pallets are lifted by the forklift trucks by the 40 inch side. If 2 million units have to be stored in the warehouse, a total of 20000 pallets and 67 racks are required. Changing the unit dimensions will have a significant effect on these numbers, thus providing the designer with a means to link design parameters and warehouse space requirements. Table 11.2 shows the varying unit dimensions, the unit weights, and the requirement on the number of pallets and racks for a maximum weight restriction of 25 lbs on the units on the pallet as obtained from the execution of STORE. The input data used to generate the results on Table 11.2 are the height of the pallet of 10 inches; required clearances on the shelves and rack beam widths of 2 inches; clear ceiling height of 240 inches; maximum weight permitted on any unit of 25 lbs based on the sum of the weights of the units vertically above it (allowing for one less layer); rack width of 4 feet; aisle width of 7 feet; total required storage of 2 million units; and no overhang permitted on the sides of the pallet.

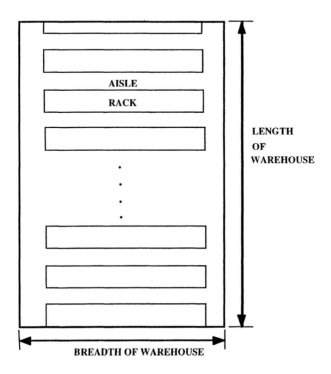

Figure11.4 Warehouse layout and rack arrangement.

For each unit, the weight was varied from 5 through 7 to 10 lbs. The unit weight influences the height to which the pallet can be stacked and this in turn has a significant effect on the total number of pallets and racks required to store a required quantity of 2 million units assuming the constraint on the availability of warehouse space. It can be observed from Table 11.2 that for any given unit, the number of pallets and the number of racks required to store 2 million unitsincreases with the increase in unit weight and unit volume. It can also be understood from Table 11.2 that units with poor pallet space utilization tend to require larger number of pallets and racks, especially when the unit weight becomes higher.

It can be observed from Table 11.2 that the maximum number of units that can be arranged on a pallet is found with respect to the unit which does not have the best pallet space utilization. It is clear that maximizing the number of units on a pallet plays a key role in storage and retrieval for distribution costs as demonstrated in the case of the 7 x 5 x 11 unit. This unit can be stacked 240 units to a pallet when each unit weighs 5 lbs, while the 8 x 8 x 8 unit having a 100 % pallet space utilization can be stacked only 150 units per pallet. The main reason for this is the dimension of the unit which contributes to the height of the pallet, after maximizing pallet space utilization. For the 7 x 5 x 11 unit, its height of 11 inches when stacked 5 high contributes to a larger effective pallet height than for the 8 x 8 x 8 unit.

Table 11.2 Variations in unit dimensions and corresponding storage requirements

Unit Dimensions (inches)	Unit Weight (lbs)	Effective pallet height (inches)	Number of pallets required& (units/pallet)	Number of racks required
11x11x7	5	67	20000 (100)	67
11 x 11 x 7	7	45	33300 (60)	67
11 x 11 x 7	10	34	50000 (40)	72
7 x 9 x 9	5	57	16000 (125)	40
7 x 9 x 9	7	39	26670 (75)	54
7 x 9 x 9	10	30	40000 (50)	58
7 x 5 x 11	5	67	8330 (240)	28
7 x 5 x 11	7	45	13890 (144)	28
7 x 5 x 11	10	34	20834 (96)	30
8 x 8 x 8	5	52	13340 (150)	34
8 x 8 x 8	7	36	22220 (90)	37
8 x 8 x 8	10	28	33340 (60)	42

It is interesting to note that the lowest number of pallets and racks are found for the 7 x 5 x 11 unit with a space utilization of 87.5 % while the 8 x 8 x 8 unit shows 100 % space utilization but requires a larger number of racks and pallets. The understanding of this is complicated by the fact that the effective height of the pallet on the shelf including clearances for the 7 x 5 x 11 unit with a unit weight of 5 lbs is 67 inches while the effective height for the 8 x 8 x 8 unit under the same circumstances is only 52 inches. This leads to the logical conclusion that the 8 x 8 x 8 unit should have the larger storage effectiveness. However, the key aspect here is that the pallet containing the 7 x 5 x 11 unit has 48 units per layer for 5 layers while the pallet containing the 8 x 8 x 8 unit has 50 units per layer for 3 layers only. This show that good pallet space utilization alone cannot guarantee maximum storage on the pallet, but that the unit dimensions and the arrangement pattern play important roles.

In general, the rate of increase of storage requirements in terms of the racks for a given unit is minimal with increase in unit weight, except in the case of the 7 x 9 x 9 unit which shows a relatively high rate of increase. This can be attributed to the relationship between the dimensions of the unit and the dimensions of the pallet which affect the type of unit arrangement and hence the effective pallet height, especially since the case study assumes no overhang on the sides of the pallet. As far as Table 11.2 is concerned, it can be concluded that the product design dimensions and weight impact the unit characteristics which in turn has shown remarkable effect on the storage characteristics. Hence, in the concurrent engineering environment, it pays to inform the product design engineers about information regarding "design for storability".

11.5.3 Influence of unit dimensions and weight on order picking details

Consider that the demand volume is 20,000 units per day of 5000 stock keeping units. Since the volume requirement from the warehouse is very high, and so is the variety of the products, the best order picking strategy would be to have workers riding on automated guided vehicles to pick the different units. Table 11.3 shows the unit dimensions and the variations in unit weight, the number of automated guided vehicles required in each case, the costs determined for the purchase of the AGVs, their maintenance, and the labor required for operating them, and the estimated payback on investment as related to the obtained profit as acquired from the execution of STORE. The total costs of order picking are determined as the costs of the AGVs, their maintenance costs per year, and the labor costs involved. The profits obtained from distribution are determined as the profit per unit times the number of units distributed per year. The payback is estimated as the ratio between the costs incurred and the profits. The data used to generate the information on the Table 11.3 are the demand volume of 20000 units in 2 shifts; number of shifts being 2; number of hours per shift of 8; labor cost of $ 25 per hour; maintenance costs of $ 1000 per vehicle per year; cost of each vehicle being $ 100000; speed of vehicle of 250 feet per minute; time for picking units being 30 seconds; capacity of each vehicle of 300 lbs; profit on distributing each unit of $ 1, and profit per year being $ 7300000.

The information in Table 11.3 provides for some interesting results. The estimated payback on investment generally increases for any given unit size as the unit weight increases. This is because as the unit weight increases, the pallet height reduces, thus requiring more storage space, and a larger number of AGVs are required to retrieve the units. Also, the poorest paybacks are associated with the unit sizes having the least pallet space utilization. It is interesting to note that the unit weight has the most significant impact on the number of AGVs required to retrieve the units, and not the total number of pallets. This is because the unit weight influences the pallet height. The height of each pallet influences the rack design aspects considerably, especially in terms of the number of racks required and the height of the racks. The rack design parameters then have a direct effect on the number of AGVs required for product retrieval.

Consider the 11 x 11 x 7 unit with weights of 5 and 7 lbs respectively. For a unit weight of 5 lbs, the number of pallets required is significantly lower than for the unit weight of 7 lbs. However, the number of AGVs required for unit retrieval are the same. This is because the effective pallet height for the 5 lbs unit is 67 inches while the effective pallet height for the 10 lbs unit is only 45 inches. Thus the allowable height of the pallet has been a factor in making the number of racks required in each case to be the same, thus impacting the number of AGVs required. A similar comparison can be made on the 7 x 5 x 11 unit between unit weights of 5 and 7 lbs.

The information on Table 11.3 has shown that the unit dimensions and weight as resulting from the product design function plays a key role in storage costs. The estimated paybacks are strictly related to the number of AGVs required which is tied in with the product design parameters. The low paybacks reported in Table 11.3 illustrate the fact that although investment costs may run into millions, the profits obtained by delivering a large demand volume every day can justify the investment. A system such as STORE can thus be a valuable dialogue generator between the product design and facilities planning functions in the concurrent engineering domain offering various choices in unit packaging and pallet selection so as to minimize storage and distribution costs.

Table 11.3 Unit dimensions, unit weights and corresponding order picking details

Unit Dimensions (inches)	Unit Weight (lbs)	Number of AGVs Required	Total Costs ($)	Estimated Payback (years)
11 x 11 x 7	5	30	7410000	1.02
11 x 11 x 7	7	30	7410000	1.02
11 x 11 x 7	10	33	8151000	1.12
7 x 9 x 9	5	17	4199000	0.58
7 x 9 x 9	7	23	5681000	0.78
7 x 9 x 9	10	25	6175000	0.85
7 x 5 x 11	5	12	2964000	0.41
7 x 5 x 11	7	12	2964000	0.41
7 x 5 x 11	10	14	3458000	0.47
8 x 8 x 8	5	14	3458000	0.47
8 x 8 x 8	7	18	4446000	0.61
8 x 8 x 8	10	20	4940000	0.68

11.6 SUMMARY

The computer systems described in this chapter have been found to demonstrate effectively a concurrent engineering principle regarding design for product storability and distribution. They also have the potential to expand their current capabilities in terms of incorporating other types of storage and retrieval methods. Other concepts such as product design to minimize the costs of in-process storage entities such as jigs and fixtures ought to be explored. Design for product storability and distribution concepts are particularly useful for designers to appreciate the effect of product design decisions on a function such as product storage and retrieval. Products well designed, not only in terms of manufacturing and assembly, but also for product storability, offer low product life cycle costs and increase the competitive edge of industrial organizations.

REFERENCES

AGVS Application Profiles (1991) The Material Handling Institute, Charlotte, NC.

Apple, J.M. (1977) *Plant Layout and Material Handling*, John Wiley.

Apple, J.M., McGinnis, L.F. (1987) Innovations in Facilities and Material Handling Systems An Introduction, *Industrial Engineering,* **19** (3), 33-38.

Bolz, H.A., Hageman, G.E. (1958) *Material Handling Handbook*, Ronald Press.

Falconer, P. (1975) *Building and Planning for Industrial Storage and Distribution,* John Wiley and Sons.

Francis, R.L., White, J.A. (1992) *Facility Layout and Location: An Analytical Approach*, Prentice Hall.

Hoover, S.P., Rinderle, J.R. (1989) A Synthesis Strategy for Mechanical Devices, *Research in Engineering Design*, 87-103.

Hwang, H., Ko., C.S. (1988) A Study on Multi-Aisle System Served by a Single Storage/Retrieval Machine, *International Journal of Production Research*, **26** (11), 1727-1737.

Konz, S. (1994) *Facility design*, Publishing Horizons Inc.

Kulwiec, R.A. (1992) Material Handling: Hero of the Nineties for Manufacturing and Warehousing, *Modern Materials Handling*, **47**, 37-42.

"Material Handling: Storage Systems", *Plant Engineering*, Reprint Series 45-5.

Material Handling Teacher's Institute Training Program, Training Materials, (1993), *CIC-MHE*, Troy, NY.

"Material Handling: Warehousing", *Plant Engineering*, Reprint Series 45-3.

Mistree, F., Muster, G. (1990) Conceptual Models for Decision-Based Concurrent Engineering Design for the Life Cycle, In: *Proceedings of the Second National Symposium on Concurrent Engineering*, Morgantown, WV, 443-468.

Ozden, M. (1988) A Simulation Study of Multiple Load Carrying AGVs in an FMS, *International Journal of Production Research*, **26** (8), 1353-1366.

Stauffer, R.N. (1988) Simultaneous Engineering: Beyond a Question of Mere Balance, *Manufacturing Engineering*, 43-46.

Tompkins, J.A., White, J.A. (1984) *Facilities Planning*, John Wiley and Sons.

Vittal, M.S. (1993) *Value Engineering for Cost reduction and Product Improvement*, Systems Consultancy Services, Bangalore, India.

Warehouse Modernization and Layout Planning Guide, (1978), NAVSUP Publication 529, Naval Supply Systems Command, Department of the Navy, Washington, D.C.,

Winner, R.I., Pennel, J.P., Bertrand, H.E., Slusarczuk, M.M.G. (1988) The Role of Concurrent Engineering in Weapons Systems Acquisition, *Report R-338*, Institute for Defense Analysis.

12

DESIGN FOR RELIABILITY FOR MECHANISMS

John A. Stephenson; Ken M. Wallace

This chapter describes a design for reliability (DFR) method for assessing which design configuration has the greatest potential for reliability at the early stages of the design process. The method has been developed through analysing case histories of the reliability of a series of mechanisms. From these analyses the key design factors that influence reliability were identified. Using the method, designers can model their designs quickly and simply to identify areas which are most likely to cause reliability problems. DFR enables designers to address potential failure areas and to produce a reliable design configuration quickly and cheaply.

Product reliability is of key importance to customers. Improvements in reliability are driven by developments in design, manufacturing, and maintenance, as well as by customer expectations, and the performance of competing products. As design is the first factor to shape a product it has a major influence on its reliability. The ability to "design in" reliability is therefore an important capability that will help companies increase the market share of their products. Failure to achieve sufficient levels of reliability during a product's life cycle can have several effects. If testing of a prototype during product development shows that the expected level of reliability is not being met, then redesign may be required which will delay the product's introduction to the market. Time to market is a key factor in product profitability, and delays can prove costly (Nichols, 1992). Failures of a product, once sold, may lead to the company facing warranty claims. In safety critical areas, e.g. aero engines, reliability is of crucial importance to customers. In non critical areas product failures will create dissatisfied customers, which will lead to a negative image for both the product and the company. Future sales for the company will be in turn reduced.

12.1 REVIEW OF CURRENT APPROACHES

At present there are three main approaches to ensuring that a design will be reliable: reliability prediction, design techniques such as Failure Mode Effects and Criticality Analysis (FMECA) or Fault Tree Analysis (FTA) and development or pre-production reliability testing. Each approach has a different basis and each is used at different stages of the design process, or during the early stages of manufacture. In order to achieve the required level of reliability a combination of these approaches tends to be used to create the most reliable product.

12.1.1 Reliability Prediction

Reliability prediction tends to be used earliest in the design process, after component selection has been completed. Reliability block diagrams are used to identify the dependencies between components and this enables component failure probabilities to be combined to make a numerical prediction of the system reliability. Commonly failure probabilities are calculated using failure rates from databases which are based on tests of standard components (USAF). In some cases (US MIL-HBK-217), mathematical models are available for taking the effect of different factors into account (e.g. part quality for electronic components).

Reliability prediction has the benefit that it gives a quantifiable estimate of the likely reliability which can then be assessed to see if this is appropriate for the market. There is however, debate over the validity and accuracy of reliability prediction for mechanical design (O'Connor, 1991; Carter, 1986). Reliability prediction is not based on a theory of how a mechanical design works and so it provides the designer with little help in understanding what is wrong with an unreliable design. Such an understanding is an essential first step, in order to be able to deal with a redesign to improve reliability.

12.1.2 FMECA and FTA

FMECA and FTA are techniques which can be used at the detailed stage of the design process. FMECA is based on analysing the design and asking "what if?" questions about individual parts to assess potential failures. FTA is similar to FMECA, but starts at a system level and creates a logical structure of possible failure causes.

The principal benefit of using FMECA or FTA is that they both increase insight into a design and its possible operation, enabling designers to identify where failures may occur. The use of a systematic approach is especially important as it reduces the number of possible failure modes that are overlooked. However FTA and FMECA tend to be fairly expensive exercises due to the time involved in a detailed analysis. Designers often consider these techniques tedious to apply, and the conclusions often appear obvious. The relative lateness of the use of these techniques in the design process means that if a redesign is required then time and cost penalties are then incurred.

12.1.3 Reliability Testing

Reliability testing tends to be used towards the end of the design process and is often seen as part of the development phase of a product. Typically the product will be tested under a range of different conditions such as temperature, vibration, shock, humidity, dirt, and power input or output variations. Reliability testing can be used in two main ways. Firstly it can be used to identify all the actual failure modes that will be encountered. If a redesign is required as a result of testing, then the new design can be tested in a similar way. Secondly reliability testing can be used to predict the reliability of a design when it is in service. Generally testing comes towards the end of the design process, but if a significant technological risk exists then testing can be brought forward to reduce the level of uncertainty.

If testing can establish that a product is sufficiently reliable before production, then it is a cost effective method of reducing warranty claims, or the possibility of product recalls. However in terms of the design process, the majority of design decisions will have been made and the costs associated with them committed prior to reliability testing. Hence the time delay and cost of redesign can be considerable if the testing highlights problems. Late changes need to be minimised.

Each of the three different approaches has its strengths and its weaknesses in terms of "designing in" reliability. As a result the best programmes for achieving high levels of product reliability tend to incorporate all three approaches. However, as a "design for reliability" method which allows designers to assess the effect of the decisions of layout and part selection, which have the greatest influence on reliability, all three fall short to some degree. There is therefore a need for a new design for reliability method, to support mechanical designers during the earlier stages of the design process.

12.2 A DFR METHOD FOR MECHANISMS

It is the earliest decisions made in the design process, over what type of design solution and how to pursue it, which are the ones that have the biggest influence on the final outcome of the design process. An example of this is the growth of costs committed during the design process which shows that after the conceptual stage around 75% of the cost has been committed (Ullman, 1992). If this is true for costs, then does it follow for reliability? The research work that this method is based on, confirms this view that it is the decisions of concept selection and how to embody the concept that have the most profound effect on the subsequent product reliability. Clearly then, what is needed is a design approach which can assess the consequences of the decisions that are made at the earliest stages, in terms of their likely effect on reliability.

12.2.1 Aims and Scope

During conceptual and embodiment stages a mechanical design will consist of a partially developed design concept or design configuration. This could be, for example a solution principle, such as a lever linkage, and a general spatial layout of the parts proposed. This section describes a DFR method which quickly and easily identifies and assesses potential failure areas of a design configuration. The method is not a numerically predictive as such techniques are often inaccurate and require too much of the designer's time to be used regularly in the early stages of the design process, even if appropriate reliability data is available. Instead, the method concentrates on identifying the interfaces and components that may cause failure. By providing an underlying theory for avoiding such failures, the method aims to increase insight into the operation of a design configuration so that the appropriate steps can be taken to "design in" reliability from the start.

The development of this DFR method is based on the research findings from a number of case histories that investigated the failure data from backhoe loaders in use. The method focuses on mechanisms and specifically those that perform positioning and load bearing functions. After further research and development it is possible that the method will be extended to other mechanical domains.

Aguirre (1990) analysed 3500 design guidelines from engineering design literature. He identified three overall design principles: simplicity, clarity and unity. Aguirre proposed that simplicity, clarity and unity can be viewed as internal properties of a mechanical (technical) system and that an appropriate mix of these internal properties determines the external properties of performance, economy and reliability (see Figure 12.1). Thus when a designer is working on a proposed design and makes a change that effects the internal properties then this will result in a change to the external properties (e.g. simplifying a design will make it cheaper to produce and so effect the economy of the design). Consequently if we can relate these internal properties to reliability and measure them for a new design configuration it should be possible to determine the underlying reliability of a design configuration.

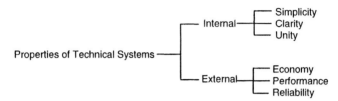

Figure 12.1 Link between external and internal properties of technical systems.

The concepts of simplicity, clarity and unity are at their most powerful when making the mapping between the functions required to the chosen physical embodiment. Hence these properties are appropriate for evaluating the decisions made at the earliest stages of the design process. Evidence from the case histories has supported the theory that simplicity, clarity and unity all contribute to reliable design and can be used to explain many of the failures that have occurred.

12.2.2 Concepts of Design for Reliability

Work on the case histories has established which aspects of each property relates to reliability, and the method focuses on these aspects. For example, simplicity relates to many issues such as number of parts, part shapes, manufacturing and assembly operations, etc. Within a DFR context it may be that not all of these issues are of importance. Thus the definitions for simplicity, clarity and unity, used here correspond to the most relevant elements of each total concept.

Simplicity

Aguirre (1990) defined the principle of simplicity as "the number of elements in a technical system should be the minimum necessary for its correct operation". Simplicity is often linked to reliability through the concept of "what isn't there can't go wrong". Further to explaining the role of simplicity in reliability, the case histories have indicated that it is the active interfaces (i.e. moving surfaces) that are the most common regions for failures to occur. This suggests that the most relevant way of describing a design configuration is to model these active interfaces and the elements or "components" (either single parts or groups of parts fastened together) that the interfaces link. Hence simplicity within a DFR context corresponds to *minimising the number of interfaces and components*. Suh's information axiom has a parallel with the principle of simplicity but uses a more complex definition of simplicity based around information content, in order to identify the best design (Suh, 1990).

Clarity

Pahl and Beitz (1984) state that clarity of function relates "to the lack of ambiguity of a design (and) facilitates reliable prediction of the end product". Aguirre expanded upon this by defining the principle of clarity "as the degree of independence between the physical and the functional relationships defining the configuration of a technical system should be the minimum necessary for its correct operation". Again Suh's other axiom, the independence axiom provides a parallel here to the concept of clarity. The independence axiom states that good designs are those that maintain independence of their functional requirements. This is equivalent to saying that functions within a design should be independent of each other.

Within a mechanism, functions are carried out by forces being transferred at active interfaces in order to position the components. Functions relate to sub-elements of the overall task being carried out in the mechanism. Clarity suggests that the operation of a mechanism should be a single, unambiguous action. In looking at an individual interface within a mechanism, if its operation is unclear (i.e. ambiguous) then in reliability terms it means that it can have two states, it can either operate or fail. For instance, imagine an interface which performs two functions, the first carried out by a large force, and the second by a small force. If the design of the interface allows the possibility that the larger force can overcome the smaller then the functions are not independent and so they are not clear. If this occurs then operation of the second function has the possibility of failing. This does not mean that interfaces performing single functions are automatically clear. In order for such an interface to be clear its function must be independent of the side effects related to operation (e.g. restraint forces) and the external conditions (e.g. corrosion).

If we consider a function being carried out in an ideal, reliable way at an interface, then there will be a constant force level available to carry out the function despite the influence of any other factors (e.g. other functions or the environment). Hence there will always be sufficient force to carry out the function. Any departure from the ideal of a constant force level available will suggest some loss of clarity, i.e. any variation in the force transfer due to other effects. Obviously some force variations may be too small to affect the operation of the mechanism. In such cases the interface will remain clear, but moves away from the ideal. A lack of clarity will occur if the forces performing a function acting on the interface varies so much that its unable to operate satisfactorily, i.e. a failure occurs.

Thus clarity is the guideline that describes the actual operation of a mechanism and links directly to the concept of failure, and hence to reliability. Within a DFR context, the definition of a clear interface is one in which each function is able to operate independently of all other effects; and this is achieved if variation of the resultant force performing each function at an active interface is minimised. A definition of clarity based around minimising variations has similarities with the concept of quality loss in Taguchi methods (Clausing, 1994).

Unity

Unity is Aguirre's third principle and is defined as "the relative contribution that each element in the configuration of a technical system makes to the correct operation of a technical system should be equal". Strength is an important factor in reliability and so a DFR interpretation unity is that a mechanical system which has unity is one that has components of equal strength, and so each component is equally likely to fail. However the ideal of an equal strength can be difficult to achieve when components perform tasks.

Maintainability is also an issue which can have a major effect on reliability. Often sacrificial components that are cheap and easy to replace are used as part of a maintainability strategy and this contradicts, to some extent, the principle that each component should be equally likely to fail. Sufficient strength is a vital to reliability and it is the strength aspect of unity that appears to relate most closely to reliability. A sacrificial component is strong enough for its normal loadings, but will be the first to fail if an overload occurs. Hence in the DFR context unity is limited to the concept that components should be strong enough for the static and dynamic loads a component has to carry. This involves both loading and environmental considerations as these can cause a change in strength across a component's life-cycle (e.g. wear, fatigue, corrosion, creep).

Interactions between Simplicity, Clarity and Unity

Research into the interactions between simplicity, clarity and unity suggests that clarity has the biggest influence on a mechanism's reliability. Mechanisms fail if they cannot carry out their functions. Functions are carried out by forces being transferred at active interfaces and then being passed on to other interfaces by the components that join them. If a mechanism consists of clear interfaces then its functions will be performed independent of events or effects around them. Thus a "clear" mechanism rather than an "unclear" one will tend to have better reliability.

Simplicity has often been linked to reliability and the industrial case histories have illustrated that the best design configurations, i.e. those with high reliability are both simple and clear. However, with increasing levels of performance products become more complicated and simplicity decreases (Glegg, 1972). In these cases obtaining the simplest design configuration possible is important but if this is done at the expense of a loss of clarity then the result will be less reliable than a more complicated configuration that is clear.

Unity as a strength concept, i.e. a component has to be strong enough to bear the forces required, interacts very closely with clarity. The clarity of an interface deals with how forces are transferred between components. Unity deals with how components transfer these forces to the other interfaces. If a component lacks unity, i.e. it is not strong enough for the forces of the function passed through it then it will break and cause a failure (i.e. it is a weak link). Unity and clarity can interact, e.g. though minor deformation of a component (such as wear) affecting how functions are performed at an interface.

Thus in overall terms, clarity and unity can be seen as contributing directly to the likelihood of failure through the interface interaction and component strength. These features are usually developed during the embodiment stage of the design process. The overall level of simplicity, however relates more directly to the solution principle chosen at the conceptual stage, rather than the types of interfaces used. It appears then that simplicity tends to apply earlier in the design process than clarity and unity.

12.2.3 Evaluation of Simplicity, Clarity and Unity

The DFR method aims to provide a technique by which the levels of simplicity, clarity and unity can be assessed in a proposed mechanism design configuration or for a series of alternative proposed configurations in order to establish which has the highest potential for reliability:

- *Simplicity.* Simplicity is assessed by counting the number of components and interfaces in the model. This assessment is not complete as simplicity relates to other concepts such as ease of manufacture. However it is easy to perform and provides a rough comparison of configurations. In most cases only significant differences in the overall numbers of components and interfaces between configurations impact on reliability, provided that the level of clarity is maintained.

- *Clarity.* Clarity is assessed using a ranking system to indicate interfaces with different levels of clarity. The ranking system is equivalent to a "penalty point" system, where a "1" is the clearest type of interface and a "3" is the least clear. The assessment of clarity is based on how the forces performing each function are transferred at an interface. Thus a particular type of interface is not intrinsically "clear" or "unclear", it depends on the different functions that the interface is required to carry out and so the forces it has to transfer.

- *Unity.* As unity relates to strength, conventional means for assessing strength across the life-cycle can be used, such as modelling stresses using Finite Element Analysis (FEA), or estimating the strength loss due to the effects of corrosion.

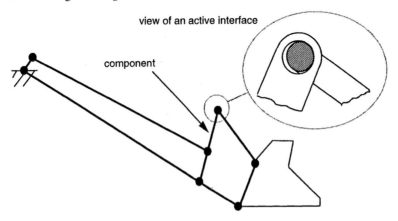

Figure 12.2 An example of a generalised pin joint interface from a backhoe loader linkage.

Modelling of Mechanisms

The approach to evaluating the levels of simplicity and clarity in a design configuration starts with modelling the configuration as a series of components linked together by interfaces. Conventionally the components are shown as blocks and the interfaces as pairs of double lines (an example model is shown in Figure 12.6).

12.2.4 Evaluation of Clarity

The forces involved in performing a function vary from one application of an interface to another. Hence one type of interface is used to describe each category and it is the different force cases that are the key to assessing the level of clarity. The interface analysed is a general pin joint from a four bar mechanism, and a typical application it could be used is the loader linkage from a backhoe as shown in Figure 12.2. In the case shown, the interface forms part of a larger configuration which is carrying out the task of carrying the forces from the bucket or loader. However as the pin joint is a general interface it could be used in another application performing other functions such as accurate positioning of components, or some combination of load bearing and accurate positioning.

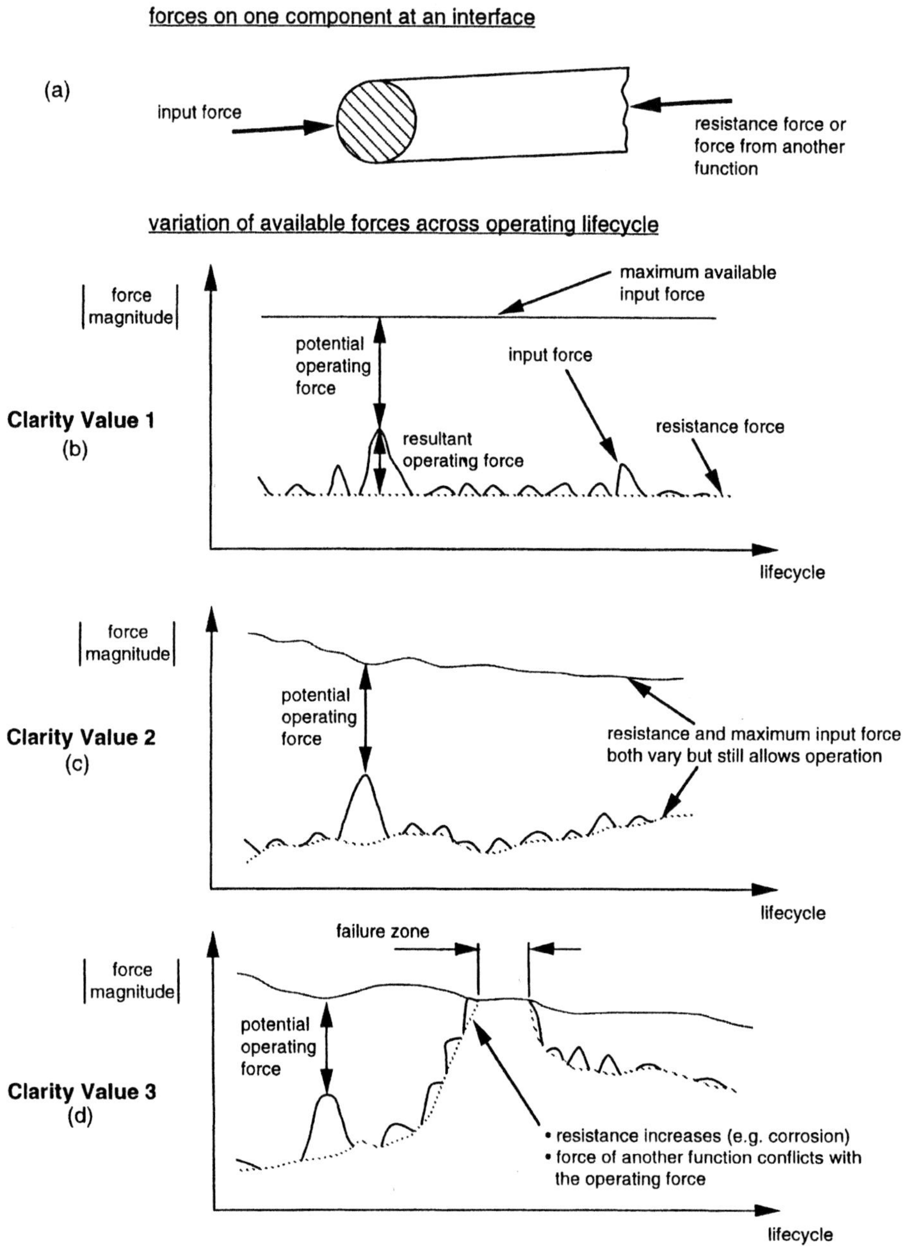

forces on one component at an interface

(a)

input force

resistance force or
force from another
function

variation of available forces across operating lifecycle

force
magnitude

maximum available
input force

Clarity Value 1
(b)

potential
operating
force

input force

resultant
operating force

resistance force

lifecycle

force
magnitude

Clarity Value 2
(c)

potential
operating
force

resistance and maximum input force
both vary but still allows operation

lifecycle

force
magnitude

failure zone

Clarity Value 3
(d)

potential
operating
force

• resistance increases (e.g. corrosion)
• force of another function conflicts with
the operating force

lifecycle

Figure 12.3 Illustrative examples of clarity values.

Initially the interface will be assumed to be carrying out a single function to provide an introduction to the assessment of clarity. If an interface carries out several functions then the force that relates to each function must be analysed separately to assess the sources of possible force variation. Such interfaces introduce an extra source of force variation through the possibility of the interaction with the other functional forces.

Clarity Value 1

Figure 12.3(a), shows a single bar from the pin joint to show the forces acting on it. The other connecting bar could just as equally have been analysed. Figure 12.3(b) shows the forces involved in the clearest type of interaction, where the pin joint is subject just to a single input force to position it, and this is countered by a single resistance force.

The two forces acting on the interface are the input force, and the resistance force. The acting force represents the force carrying out the positioning or load transferring function and the resistance force represents the counter forces to this motion, such as drag or bearing stiffness. The difference between the input force and the resistance is the resultant operating force which can change the motion of the interface by causing acceleration.

The actual size of the input force actually applied to the interface will depend on the motion of the interface. If the interface is stationary or moving at a constant speed it will be in equilibrium and the input force will equal the resistance force. However, if the resistance increases then the size of the acting force will have to increase to counter the resistance up until the maximum available input force is reached. When this happens the interface is no longer under the control of the input force and so the function the input force is performing will fail. The amount of reserve force available to operate the interface at any time is represented by the potential operating force.

In the example shown the four bar linkage might be used as part of a mechanism whose function is to accurately transfer position and so is subject to forces which are just large enough to move the components. Alternatively the linkage might be part of a mechanism that has to transfer large forces. If this interface has a clarity value of "1" this could be achieved through restricting the operation of the linkage so that the available input force and the resistance force are both constant across the linkage's life-cycle.

As there are no other aspects or features such as auxiliary forces (which will be discussed later) to affect the potential input force or the resistance force there will always be a sufficient resultant operating force to carry out the interface's function and so failures are extremely unlikely. Thus an interface assessed as a "1" denotes a potentially reliable interface.

Whilst it is difficult to find real interfaces that match the ideal of constant force levels, a practical assignment of the "1" ranking would apply to situations of very minor force variations, or to situations where an interface's function is to prevent a motion being transferred, and this can be ensured despite the influence of surrounding effects, e.g. at a stop or a rotating bearing. In these cases such interfaces can be treated as being close to this ideal.

Clarity Value 2

Figure 12.3(c) illustrates an example for clarity value "2". It shows a similar loading case to the clarity value "1" case, but neither the potential input force to position the components nor the resistance are constant during the interface's lifetime. Variations in both forces may be due to many effects, either related to the operation of components within the whole mechanism (an internal variation, e.g. due to geometry changes as the mechanism moves) or due to other effects such as environmental changes (an external variation, e.g. corrosion). In this case the four bar linkage might be part of a larger mechanism which is performing either a positioning or a load transferring function, such as the loader linkage in Figure 12.2 which encounters different loads, e.g. through digging different quantities of earth. Clarity value "2" is a more usual case than "1", as the forces are is not assumed to be constant.

The key point about a clarity value 2 interface is that whilst there is a variation in the level of the maximum potential acting force and the resistance force, this is not enough to increase significantly the likelihood of failure at this interface. Hence whilst this case moves away from the ideal of no variation during the lifetime, it is still a clear interface as it is still likely to operate reliably across its lifetime despite these variations.

Clarity Value 3

Figure 12.3(d) shows the final clarity ranking which identifies an unclear interface and so the possibility of failure. In the case of a single function being carried out by this interface, failure of the function will occur if the maximum potential acting force is insufficient to overcome the resistance force. The increase shown in the graph of the resistance force may come from an external source, such as an environmental change (e.g. freezing of components together) or it could come from an internal source (e.g. wear of components). There can also be a loss of the maximum available input force, e.g. gasket leaks in an internal combustion engine can result in a loss of pressure on the piston.

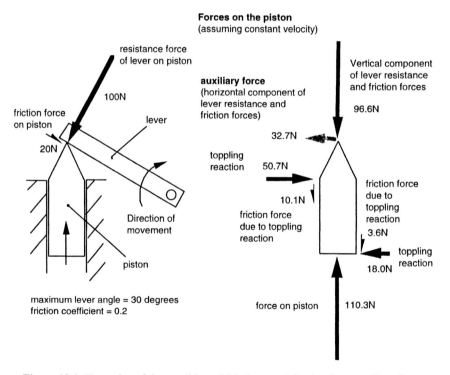

Figure 12.4 Illustration of the possible variable forces originating from auxiliary forces.

In the case of an interface performing several functions, it would be possible for the resistance force illustrated to be replaced by a force performing another function, with much higher forces associated with it than the potential acting force. This could occur if the four bar linkage was used primarily to perform a positioning function involving small forces, but on occasions was also required to support a large force. In such a case the positioning force would be overridden by the load bearing force and so the positioning function would fail. In this example the two functions being carried out are not independent of each other and so failure could occur as a direct result of the lack of clarity.

In order to carry out a clarity assessment all the sources and levels of the variations in the forces have to be identified. This involves a degree of subjectivity because these assessments have to be made using engineering judgement. The degree of subjectivity can be reduced if the force levels can be predicted accurately. However the aim of this method is to increase insight into the type of problems that can cause unreliability, and so the main benefit comes from a rapid analysis of force cases that highlights the relevant factors.

The possible sources of variations in the force performing a function frequently come from the side effects of related to the transferring of the function across the interface. Typically these side effects or auxiliary forces result from the geometry chosen for an interface. One example of an auxiliary force is the axial separating force created in helical gears (Pahl and Beitz, 1984). As pin joints do not create auxiliary forces, another example from mechanism design will be used to illustrate the point.

Figure 12.4 shows a lever which can rotate and is positioned by a linearly acting piston. Due to the geometry chosen, a side force will occur at the interface, as the components slide, and this force will vary during operation. The side force represents an auxiliary force and comprises of the horizontal components of the lever resistance and the friction force.

The auxiliary force creates toppling reaction forces on the piston and thus additional frictional force on the piston. The resistance forces (including the additional frictional forces) will act against the pressure force on the piston which is trying to position the lever. Figure 12.4 shows the size of the additional friction forces generated if the lever is at 30 degrees to horizontal, with a lever resistance force of 100N at the point of contact. The friction coefficient is assumed to be 0.2. For this lever angle the friction forces add around 10% extra to the force on the piston above the 100N required to move the lever from a horizontal position. If the input force on the piston can increase to overcome this, regardless of changes across the life-cycle then no failures would be anticipated. In this case these interfaces would be classified as a "2".

However if the force on the piston is restricted and so cannot increase to overcome further variations (e.g. an increase in friction coefficient due to corrosion or lack of lubrication) then this could prevent operation and cause a failure. If so these interfaces would be classified as a "3". This is an example illustrates how an auxiliary force can have a direct influence on the level of clarity at an interface. Hence identifying the action of auxiliary forces is an important part of looking for potential failure areas.

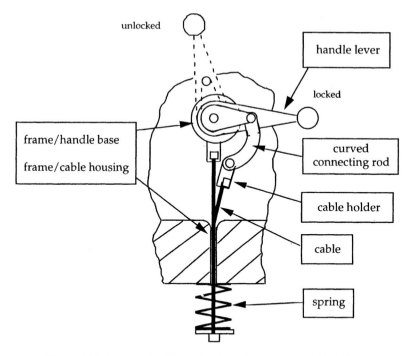

Figure 12.5 Lever and cable mechanism with components identified.

12.3 PROCEDURE OF DESIGN FOR RELIABILITY

In order to explain the application of the DFR method, a model of the lever and cable mechanism shown in Figure 12.5 will be built and analysed. The lever has two positions and be used to actuate a locking mechanism.

Step 1 - Evaluation of Simplicity

In the first step the assembly is divided into components and active interfaces. The components are shown in Figure 12.5, and the model is shown in Figure 12.6. A count of the number of components and interfaces gives a rough basis for assessing the level of simplicity relative to an alternative design configuration. As the cable housing and the handle base are both fixed to the frame, these are considered to be one component and are thus joined together in the model.

Step 2 - Evaluation of Clarity

The second step is to assess each active interface using the three categories described previously. The results are shown in Figure 12.6.

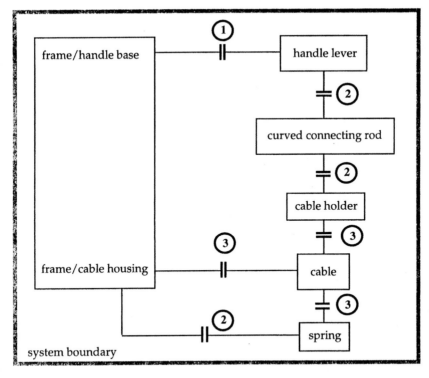

Figure 12.6 A completed model of the lever mechanism.

<u>Interfaces assessed as clarity value 1</u>

- **Handle base (frame)/handle lever interface**

This interface is the bearing for the handle lever. As the interface's function is to constrain the handle lever's motion and the performance of this function would not be expected to vary considerably with time (e.g. little change to the torque to move the lever) it is classified as category "1".

<u>Interfaces assessed as clarity value 2</u>

- **Handle lever/curved connecting rod interface**
- **Curved connecting rod/cable holder interface**

These interfaces each have a single function which is to transfer the motion of the handle lever (i.e. a positioning function). The maximum available forces transferred by the interfaces between the handle lever, curved connecting rod and the cable holder vary due to changes in the angle of pull of the cable. This variation in force would not be expected to interfere with the operation of these interfaces (e.g. little change in joint resistance). Due to the force variations of these interfaces they are assessed as "2"s.

- **Spring/frame interface**

The spring is required to be in contact with the frame to provide a reaction force so that the spring can act on the cable. Whilst the force in this interface will change with spring extension, if the spring is always in compression it will remain in contact with the frame. Continuous contact can be ensured through the geometry of the spring dimensions and the dimensions of the cable movement. Hence this is a clear interface as it is able to carry out its function despite changes in the force transferred. Hence it is classified as a "2".

Interfaces assessed as clarity value 3

- **Cable/cable housing interface**
- **Cable holder/cable interface**
- **Cable/spring interface**

Each of these interfaces carries out a single function to transfer the movement of the lever. If these interfaces are to be reliable the input force acting on the cable must be able to overcome the resistance force. The input force comes from the difference between the force on the cable holder and the tension generated in the cable by the spring. The resistance comes from two sources. Firstly there is the friction force of pulling the cable through its housing. Secondly there is the additional friction caused by the cable being pulled against its housing, as shown in Figure 12.7. Due to the cable not being pulled coaxially with the cable housing by the cable holder a side reaction force is created. The side reaction force represents an auxiliary force and has a friction component associated with it that varies as the lever moves.

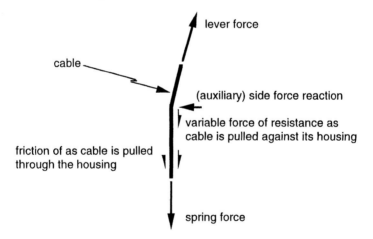

Figure 12.7 Forces acting on the cable.

The input force is also subject to variations. At the cable and spring interface the force of the spring will change with its extension. If a constant force is put on the lever by the operator the force transferred to the cable holder will vary as the lever moves due to the geometry of the lever arrangement. Whilst both of these variations are predictable it may be that the variations in the input force combined with those of the resistance force may be sufficient for the resistance to overcome the input force (e.g. high friction from the cable housing) and so a failure will occur. The point at which the auxiliary resistance force is the greatest is also the point at which the force of the spring is the lowest. Hence it would appear that the most vulnerable point in the operation of the mechanism occurs when the lever is being moved to or from its locked position. Thus each of these interfaces plays a role in creating a potential failure and so each has been classified as a "3". These interfaces will require further analysis to establish the likelihood of failure. If a redesign is required the level of clarity should be re-evaluated.

Step 3 - Evaluation of Unity

The third step is to estimate the peak loads to ensure that the components are strong enough. If there are components that might fail under load, then the appropriate component box in the model should be shaded. In the lever mechanism example considered, each component is strong enough.

Step 4 - Use of Results

Clarity is the main issue in this example as simplicity is only relevant when several design solutions with different working principles are being compared. Unity is not an issue as no weak components were highlighted. From the assessment of clarity, the cable's interfaces are ranked as "3"s, i.e. they are unclear interfaces with the potential to be a failure area. These areas therefore need to be reviewed carefully before proceeding to detail design. As discussed dividing functions does not guarantee clarity and this is an example of where a single function carrier (the cable) has unclear interfaces.

The problem highlighted relates to whether the varying tension provided by the lever or spring is sufficient to overcome the varying resistance created by pulling the cable against its housing during operation. Hence the designer needs to investigate these forces to assess how much they will interfere with each other and lead to possible failures. This could be done using calculations, tests and engineering judgement. If after analysis the variations in the force levels are considered unacceptable then the clarity of the cable's interfaces can be increased by redesign in the following areas:

- Increase the force of the positioning function
  - longer lever
  - stiffer spring, or a pre loaded spring.
- Reduce the resistance force
  - reduce the auxiliary force by pulling the cable coaxial with its housing
  - change cable materials to reduce frictional forces.

Thus applying the DFR method to analyse a mechanism can help to identify potential failure areas and provides insights that can assist the redesign process.

12.4 CASE HISTORY

This section presents another example to demonstrate the application of the method. It is based on the case history of the return-to-dig mechanism on a backhoe loader. The mechanism is part of the control system for positioning the front bucket, or loader. In total, five different return-to-dig configurations have been analysed in the research into the internal properties, and two of these will be described here. The components relevant to this case history are illustrated in Figure 12.8.

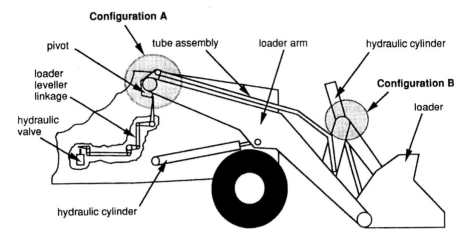

Figure 12.8 Components in the return-to-dig mechanism.

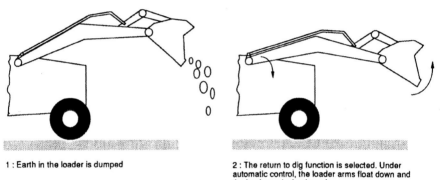

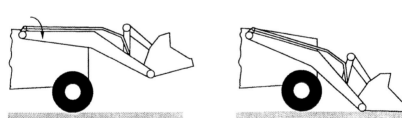

1 : Earth in the loader is dumped

2 : The return to dig function is selected. Under automatic control, the loader arms float down and the loader racks backward.

3 : When the loader reaches a set angle the microswitch is triggered, and the loader stops racking back, whilsts the arms continue to float down.

4 : The loader arms drop until they reach the ground. The loader is now positioned ready to dig again; hence return to dig.

Figure 12.9 Operation of the return-to-dig mechanism.

Exploded view of components

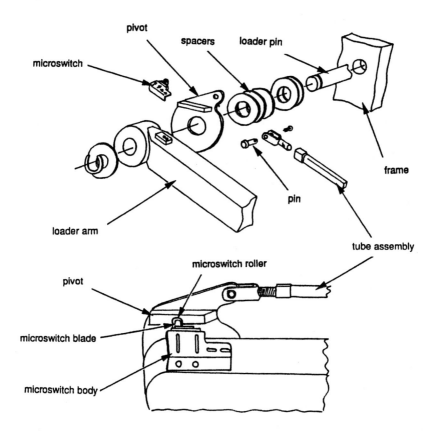

Assembled view of components

Figure 12.10 Exploded view of return-to-dig mechanism: Configuration A.

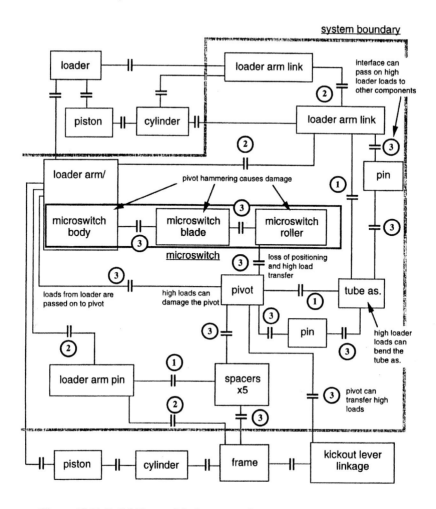

Figure 12.11 Reliability model of return-to-dig mechanism: Configuration A.

The positioning system controls two loader functions: loader levelling function and the return-to-dig function. Loader levelling keeps the loader at the same angle to the ground as the loader arms rise or fall and is achieved by a mechanical linkage feeding back the position of the loader to a hydraulic valve mounted on the frame. The hydraulic valve then controls the position of the loader through a hydraulic cylinder. The return-to-dig function utilises some of the loader levelling linkage and it is this function that is the focus of the case history. The purpose of the return-to-dig function is to speed up the digging cycle and thus increase machine productivity. Its operation is shown in Figure 12.9.

12.4.1 Configuration A

Configuration A is shown in Figure 12.10 with a close-up of the components in the microswitch area. The tube assembly transfers the movement of the loader, relative to the loader arms to the pivot, which then actuates the microswitch. A component/interface model of the configuration is shown in Figure 12.11.

- *Analysis of Simplicity.* (within the system boundary for the return-to-dig function)
  Number of interfaces ˙ = 18
  Number of components = 11
- *Analysis of Clarity.* The key point about clarity to emerge from the model is that the loader arms share the same bearing (the loader pin) with the pivot, and this results in a mismatch between the functional requirements of these components. The loader arms lift the loader and so they move with large forces over a large distance and are relatively insensitive to their positioning at the bearing. The pivot's function is however to actuate the microswitch and so it moves over a small distance with small forces. Consequently there is a loss of clarity when these two different functions are brought together on the same bearing and can interfere with each other. The interference can occur in two ways:
  a) *Positioning of the pivot.* If the loader arms move axially along the loader pin so that some slop is created between the machine frame and the loader arms, this will leave the pivot free to move axially. Failures can thus occur if the pivot is too far from the microswitch to actuate it. Thus the function of positioning the pivot fails through its restraining forces no longer being able to act. Thus there is a lack of clarity through this variation in the force levels.
  b) *Load transfer to the pivot.* Alternatively if the loader arms move axially in the other direction so that the pivot becomes trapped in between the loader arm and the frame, then the loads of the loader arm can be passed through the pivot and will effectively swamp the small, positioning loads acting on the pivot. Again a lack of clarity arises as the functions of the loader arm and the pivot are no longer independent. Failures will also occur if the loader arms move while the pivot is trapped because this will result in the transfer of high loads generated by the loader through the tube assembly to the pivot. The forces transferred through the tube assembly thus varies greatly in size, indicating a further lack of clarity. The end result can be the bending or breaking of either component.
- *Analysis of Unity.* Within the system boundary of the return to dig mechanism there are a number of components which have the potential to fail through deformation because they are exposed to high peak loads created by the loader. If, for instance, high loads are transferred through the pivot (i.e. due to the movement of loader arms) then either the pivot or the tube assembly will be damaged. Similarly, if the pivot is free to vibrate axially the hammering action may damage the elements of the microswitch (i.e. the body, blade and roller). As these peak loads may cause damage, the affected components are shaded in the model.

This example demonstrates a close interlink between clarity and unity. It is the lack of clarity at the interfaces that allows high loads to be transferred and a lack of unity (i.e. strength in a DFR context) that results in the final failure mode. If these loads have to be carried, then one solution is to strengthen the affected components. An alternative is to protect these components from the high loads by providing an alternative force transmission paths.

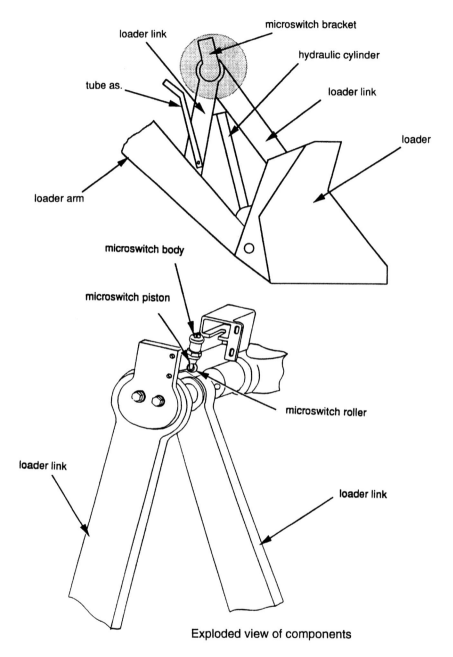

loader link

microswitch bracket

hydraulic cylinder

tube as.

loader link

loader

loader arm

microswitch body

microswitch piston

microswitch roller

loader link

loader link

Exploded view of components

Figure 12.12 Exploded view of return-to-dig mechanism: Configuration B.

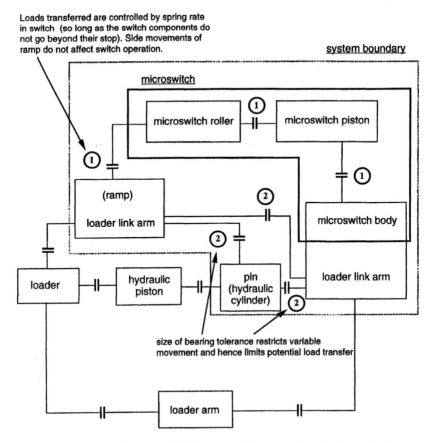

Loads transferred are controlled by spring rate in switch (so long as the switch components do not go beyond their stop). Side movements of ramp do not affect switch operation.

size of bearing tolerance restricts variable movement and hence limits potential load transfer

Figure 12.13 Reliability model of return-to-dig mechanism: Configuration B.

12.4.2 Configuration B

Configuration B of the return-to-dig mechanism is shown in Figure 12.12. The microswitch is located much closer to the loader and shares only the loader linkage with the loader levelling function. A component interface model is shown in Figure 12.13. The loader levelling function is still performed as before using the tube assembly and pivot as part of a mechanical feedback linkage to the loader valve. However these components are no longer involved in performing the return-to-dig function.

• *Analysis of Simplicity.* (within the system boundary for the return-to-dig function)

		Configuration A	Configuration B
Number of interfaces	=	18	6
Number of components	=	11	5

This configuration is much simpler than configuration A, involving a much shorter distance and a more direct path for the information about the loader position to travel to the microswitch. Further simplifications have been made by incorporating the ramp triggering the microswitch into the loader link, removing the need for a separate component such as the pivot to perform this task.

• *Analysis of Clarity.* The model shows that this design has a higher level of clarity of the two, implying that it is likely to be more reliable. In particular the components that interface with the microswitch do so in a clearer way. Any small variations in their positions do not affect the operation of the microswitch. For instance, the loader link has only limited movement in two directions, neither of which will affect microswitch operation. Any axial movement is not in the direction of operation of the microswitch and any radial movement is easily limited, by controlling the dimension of the bearing surfaces, to a level that will not damage the microswitch.

• *Analysis of Unity.* Due to the design of the interface it is not possible for large forces from the loader to be transferred to the microswitch, so the unity issue of weak components bearing large loads does not apply. Other components, such as the loader links, are strong enough to bear the highest loads likely to be encountered.

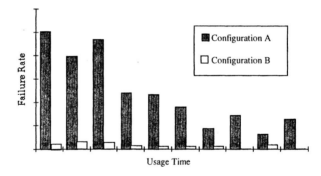

Figure 12.14 Field Failure Rates for Configurations A and B.

A comparison of the reliability data for the two configurations from backhoe loaders that have been manufactured and sold over a year is shown in Figure 12.14. As the models suggest the simpler and clearer design has a significantly lower failure rate.

12.5 SUMMARY

This chapter describes a new design for reliability (DFR) method for analysing mechanisms at the early stages of the design process. The method gives designers a greater insight into how and why a proposed design may fail and identifies the aspects of the design that may need to be improved. As the method addresses these issues at an early stage in the design process, redesign time and time to market can be reduced. Although the method is based on evidence from mechanism design, the overall approach should be capable of being extended to all other types of mechanical design.

The method is based on an understanding of the link between the internal properties of simplicity, clarity and unity and one of the external properties, i.e. reliability. The link has been validated using a series of case histories of different subsystem designs used on backhoe loaders. By evaluating the levels of simplicity, clarity and unity in a design configuration, the method aims to increase insight into the issues behind configuration selection quickly and easily. It complements previous techniques to support those decisions that have been shown to have the greatest influence on product reliability.

REFERENCES

Nichols, K. (1992) Better, Faster, Cheaper Products - by Design, *Journal of Engineering Design*, **3** (3), 217-228.

Ullman, D.G. (1992) *The Mechanical Design Process*, McGraw-Hill, Inc.

Non Electronic Parts Reliability Data, USAF Rome Air Development Center.

US MIL-HBK-217, *Reliability Prediction for Electronic Systems*, National Technical Information Service, Springfield, Virginia.

O'Connor, P.D.T. (1991) *Practical Reliability Engineering*, 3rd. ed. Wiley & Sons.

Carter, A.D.S. (1986) *Mechanical Reliability*, Macmillan Education Ltd.

Aguirre, G.J. (1990) *Evaluation of Technical Systems at the Design Stage*, Cambridge University Engineering Department, Cambridge.

Suh, N.P. (1990) *Principles of Design*, Oxford University Press.

Pahl, G., Beitz, W. (1984) *Engineering Design*, (English edition by K.M. Wallace), Design Council / Springer-Verlag.

Clausing, D. (1994) *Total Quality Development*, ASME Press.

Glegg, G.L. (1972) *The Selection of Design*, Cambridge University Press.

13

DESIGN FOR ELECTROMAGNETIC COMPATIBILITY

J. F. Dawson; M. D. Ganley; M. P. Robinson; A. C. Marvin; S. J. Porter

This chapter discusses design for electromagnetic compatibility (EMC) in the design of electrical and electronic equipment. We first describe how the subject of EMC divides into 'emissions' and 'immunity', discuss the implications of the European EMC Directive, and stress the importance of considering EMC at all stages of the design of equipment. The next section gives practical design advice on system level design, partitioning, shielding of enclosures, cables, printed circuit boards, use of filters, circuit design (analogue, digital and power) and software. Finally we discuss the use of computer-aided engineering tools for EMC, explaining the different types, where they can be applied and what they cost.

13.1 EMC: ELECTROMAGNETIC COMPATIBILITY

13.1.1 What is EMC?

Electromagnetic compatibility (EMC) is the ability of electronic equipment to function without either causing electromagnetic interference (EMI) or being disrupted by interference. Many people will be aware of everyday examples of EMC problems such as a radio receiver not operating when placed on top of a microwave oven, or a car radio giving poor reception when the windscreen washers are working. Air travellers are banned from using electronic equipment at certain times during the flight; this is to prevent any risk of dangerous interference to the plane's navigation and communication systems. Although the effect of poor electromagnetic compatibility is often only a minor nuisance, there have been cases of it leading to disruption of emergency broadcast services or to fatal industrial accidents.

EMC problems have been reported since at least 1890 (Holliday, 1995). During the Second World War, EMC problems resulted from electronic systems being mounted close together on ships or aircraft. The first legislation in the UK concerning EMC requirements of consumer products was introduced in 1952 to restrict the electromagnetic emissions from the ignition systems of motor cars (Noble, 1992). The growth of electronic equipment since then has made EMC an important consideration in many other areas. Legislation to control EMI was passed in many European countries, the most stringent regulations being in Germany. In order to harmonise EMC regulations throughout the European Community (now European Union), the European Commission adopted the EMC directive 89/336/EEC in 1992. This is very wide ranging in scope and restricts both emission of and susceptibility to electromagnetic interference.

The subject of EMC can be conveniently divided into two parts:

1. Electromagnetic emissions, where the equipment is the source of electromagnetic interference
2. Electromagnetic immunity, where the equipment is disrupted by electromagnetic energy in its environment.

Many types of equipment may be either a source or a victim of interference. EMC is also classified according to the path that the interference takes. Conducted interference travels along cables, while radiated interference travels through the space between source and victim. Conducted interference tends to be more important at lower frequencies (below about 30 MHz), and radiated interference at higher frequencies.

Both the emissions and immunity aspects of EMC have become increasingly important, because of the increased use of electronics in all areas of life, and because of changes in the type of electronics being used. For example, the large number of communications systems now in use (e.g. portable phones) makes immunity important, while the use of fast digital circuitry in computers and other equipment leads to high radiated emissions if not designed well. EMC is now important in all areas of electrical and electronic design, including information technology (White, Atkinson, and Osburn, 1992), automotive electronics (Noble, 1992), industrial controllers (DTI and New Electronics, 1994) and medical equipment (Collier, 1994).

13.1.2 The European EMC directive

In 1992 the European Single Market was created, the aim being to remove trade barriers throughout the European Union (EU). One obstacle to this was the difference in EMC standards throughout the member countries. In order to harmonise the standards, Directive 89/336/EEC was agreed in May 1989. It was adopted in 1992, and after a transitional period becomes legally binding from 1st January 1996.

There is space here for only a brief discussion of the EMC Directive. More detail is given by Marshman (1992). Copies of the Directive are available from European Information Centres and European Documentation Centres located throughout the UK.

The EMC Directive requires that apparatus shall be so constructed that:

• the electromagnetic disturbance it generates does not exceed a level allowing radio and telecommunications equipment and other relevant apparatus to operate as intended;
• it has a level of intrinsic immunity which is adequate to enable it to operate as intended when it is properly installed and maintained, and used for the purpose intended.

The Directive thus covers both emissions and immunity. The EMC Directive has a very wide ranging scope. However a few types of equipment are not covered by the directive, including equipment for which there is provision in other directives (e.g. medical equipment), second hand equipment, and equipment intended only for military use or for export outside the EU (There is currently some dispute as to whether military equipment is included within the scope of the directive or not).

There are three ways to demonstrate that equipment complies with the Directive.

1. The first and most straightforward is by 'self-certification'. This involves satisfying the relevant EN (Euro Norm) standards, either by performing the tests oneself or by contracting them to an independent test house. Harmonised standards are produced by the European electrical standards committee, CENELEC are listed in the Official Journal of the European Community.

2. The second way is to keep a 'technical construction file,' which is a record showing that appropriate design procedures and tests have been followed to ensure compliance with the Directive. The file must include a technical report by a competent body appointed by the Department of Trade and Industry (DTI).

3. The third way is to obtain a 'type-examination certificate.' This applies only to radio-communications transmitters and transceivers.

Having demonstrated compliance with the Directive, the manufacturer makes a 'declaration of conformity' and can affix the 'CE' mark to the product (assuming that it also conforms to any other relevant directives). The product is now free from technical restrictions on its sale throughout the EU.

In the UK the regulations will be enforced by the trading standards departments of local authorities, with enforcement being 'complaint driven'. Penalties for contravening the regulations include fines and imprisonment, and the authorities will have the power to seize and destroy apparatus.

Medical equipment is covered by separate directives: 90/385/EEC for active implantable medical devices (e.g. pacemakers), proposed directive 91/C237/03 for medical devices generally (under preparation) and a proposed directive for in vitro diagnostic medical devices (under negotiation) (Nensi, 1994).

13.2 EMC AND THE DESIGN PROCESS

There are three reasons why EMC is important to designers and manufacturers of electrical and electronic products:

1. Legal - products that do not comply with the EMC Directive will effectively be banned throughout the European Union. There is currently some dispute as to whether military equipment is included within the scope of the directive or not.

2. Social - poor equipment can cause a nuisance or be a hazard to health and safety.

3. Commercial - products that suffer from EMC problems will have a reputation for unreliability and be less competitive.

With a few obvious exceptions, such as battery operated torches or domestic electric heaters, products designed without EMC in mind stand a good chance of suffering from EMC problems. These problems are often complicated and do not have cheap, simple solutions. However, by good design practice it is often possible to prevent the problems ever arising.

13.2.1 Design strategies for EMC

A common strategy at present is simply to do nothing about EMC until the product is ready for manufacture, then test it and fix any EMC problems that are found. Although the time and effort of proper EMC design is saved if the product passes first time, expensive re-design and re-testing will be needed if the product fails. Unfortunately most electronic equipment has the potential to cause EMC problems if not carefully designed, and it has been estimated that only 15% of products not designed for EMC pass the tests first time (Williams, 1991). Moreover, at a late stage in development the options available for controlling the problems will be much restricted. It may be necessary to adopt expensive solutions like metal shielding, where the problem could have been avoided by laying out the PCB correctly earlier in the design process. The 'do nothing' strategy is not recommended.

One reason that current design practice frequently ignores EMC may simply be lack of awareness of the subject. EMC is regarded as a 'black art,' only comprehensible after many years of experience. In fact, the physical principles are well understood and can be described by standard electromagnetic theory.

Electromagnetic phenomena are explained by equations formulated by Maxwell at the end of the last century. Maxwell's equations relate electric and magnetic fields to charges and currents. Given sufficient time and computing power, it should theoretically be possible to predict the EMC performance of any equipment. However, a typical electronic product may contain hundreds or even thousands of components and conductors, whose positions are not always well defined (for example wiring looms in motor vehicles). The complexity of the problem means that accurate prediction is generally not possible. EMC performance cannot at present be predicted in the way that the total weight can be calculated by summing the weights of the components.

Quantifying EMC is an important area of research, and the possibilities of prediction are likely to improve. Computer aided engineering (CAE) tools are now available that incorporate EMC considerations, and these are discussed in Section 13.5. Since the 'do nothing' strategy leads to expensive fixes, and complete prediction is not feasible, the best strategy is to apply good practice at each stage in the design process. EMC is affected by many electrical and mechanical aspects of the complete design, some of which have little or no effect on the intended electronic function of the equipment.

Some companies issue guidelines to ensure good design practice for EMC. For example, Butcher and Withnall (1993) describe how a large electronics company (Racal-Comsec) ensures that EMC is considered in the design of its products. The company has developed checklists which assist designers in avoiding possible EMC problems throughout the development of a product. Development is divided into three phases, referred to as 'design philosophy,' 'equipment design' and 'measurement'. The checklist for 'design philosophy' raises such questions as 'is the case metal,' 'can linear power supplies be used' and 'can a ground plane be used on PCBs and motherboards.' The 'equipment design' checklist covers power supplies and PCB design in greater detail, and 'measurement' covers preparatory in-house testing.

Guidelines and design rules are based on principles of electromagnetic theory, and an understanding of these principles will lead to better application of the rules. If designers and manufacturers can improve their understanding of EMC phenomena, then not only will good design be made easier, but also any problems that do arise will be easier to trace and to fix.

A useful method in EMC design is to allow for possible modifications at a later design stage. For example, unless the available space is very limited, space can be left for a mains filter. It is possible to design a circuit board with more EMC protection than may be necessary, then remove components if they are found to be unnecessary at the prototype stage. This is likely to be cheaper than redesigning the circuit board to allow for additional components.

13.2.2 Factors affecting EMC

EMC depends on factors that may not affect the 'normal' operation of the equipment. EMC performance cannot be evaluated simply by looking at the circuit diagram (although this may be helpful). This is partly because components that can be treated as ideal for the intended operation of a circuit may behave quite differently at frequencies outside the operating bandwidth. For example at radio frequencies (RF), cables and PCB tracks can act as antennas, capacitors can act as inductors and wires to ground can act as high impedance loads. This difference between the ideal and the true nature of components and conductors is especially relevant to fast digital circuits. For software design, only the logic functions of a circuit are important. however from the point of view of an EMC engineer, the integrated circuits are RF signal generators and the connecting tracks are antennas, emitting RF radiation.

We can consider the general EMC problem as that of unwanted electromagnetic energy entering or leaving an electronic system. Figure 13.1 shows radiated energy entering equipment, and possibly disrupting its operation. Whether this disruption happens depends on five factors influencing the path of the unwanted energy. These are

1. The threat: its amplitude, frequency, polarisation, and its low-frequency modulation.
2. The path of entry into the enclosure: the size and shape of the enclosure, its conductivity, the thickness of the material from which it is made, the size, shape and position of apertures in its walls, the type and length of cables connected to it, the cable entry, and filters used at cable entries.
3. The circuit layout: the length and width of PCB tracks, loop areas, position of components, whether there is a ground plane.
4. The components: if digital, the logic family; if analogue the bandwidth and non-linearity.
5. The signal: if digital, the clock speed and software; if analogue the signal levels and operating bandwidth.

For radiated emissions the same factors apply in reverse. For conducted immunity or emissions the figure would be similar except that the energy would enter (or exit) by conduction along cables and could be controlled by filters at the ports. Cables are also important to radiated EMC because they can act as antennas.

Of the five factors above, all but the first are influenced by the design of the equipment. Mechanical design is important as well as electrical. To avoid problems it is best to consider EMC at all stages of the design process including the conceptual stage, circuit capture, physical layout and mechanical construction.

An important part of electromagnetic theory is the reciprocity theorem, which generally applies to factors 2 and 3 above. This means that if the enclosure and the circuit layout are designed to reduce the emissions, then the immunity will also improve. However, the reciprocity theorem does not apply to factors 4 and 5. For instance, low emissions can be achieved by using a low power logic family (which emits less unwanted energy), but this can result in poor immunity (because less energy is needed to produce a spurious transition).

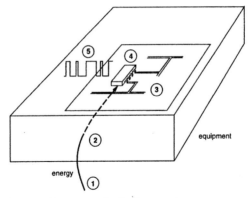

1 - external threat; 2 - entry path; 3 - layout; 4 - components; 5 - signal

Figure 13.1 Factors influencing radiated immunity.

13.2.3 Considering EMC at different design stages

Let us consider how EMC may be important at the various stages of electronic design. At each stage, the sort of questions that should be asked are

1. Conceptual design. How noisy will the electromagnetic environment be? What signal levels and bandwidths are appropriate?
2. System partitioning. Can the system be partitioned into critical and non-critical subsystems?
3. Circuit capture. What logic families are used? Can filters and decoupling capacitors be applied?
4. PCB layout. Are loop areas minimised for noisy or sensitive tracks? Are decoupling capacitors placed near enough to ICs?
5. Power supply. Is it linear, switched mode or a battery?
6. Cables. Are they shielded? How are they terminated?
7. Shielding. Is the case made of metal, insulator or coated polymer? How large are the apertures? How is the lid attached?

In a large organisation, many people will be involved in the design process. It is therefore important that there is either a high awareness of EMC throughout the organisation, or that EMC experts are able to participate at each stage. Communication and documentation are important, and the trend towards concurrent design environments is likely to help with this. EMC should be considered in choosing CAD/CAE tools if these are used in design process.

Installation and maintenance procedures may also affect EMC. One way to ensure that EMC performance is not degraded is to document critical features which should not be altered. Such features include the position of cables, connection of cables to equipment, bonding between metal cases and ground straps for connecting equipment to earth.

Many companies are adopting quality control systems such as BS5750. Armstrong (1994) suggests that three modifications are necessary to include EMC in quality control: identification of critical parts, methods and processes; clear and obvious marking and documentation; and appropriate in-house testing.

13.2.4 Evaluating EMC options

Taking insufficient measures to prevent EMC problems will lead to products failing to comply with legislation and requiring expensive redesign and re-testing. However applying more protection than is necessary could also be expensive. The ideal is to apply only what is necessary, allowing for a margin of error.

In both formal and informal design procedures, there must be a stage at which a selection is made from a number of options for achieving a particular function (Cross, 1989). For a given piece of equipment there will be a number of ways to improve its immunity or reduce its emissions. Some of these, such as the position of PCB tracks, will have little effect on the cost of the equipment. Others, such as the use of metal enclosures for shielding, can be expensive, especially for large production runs.

To decide which options are best to ensure adequate EMC performance, we can consider the benefits and costs of each option.

The benefit is simply the reduction in emissions or the improvement in immunity. While exact prediction of the EMC performance of equipment is difficult, there are formulae describing the behaviour of various EMC measures such as shielding of enclosures and cables or radiation from loops. Some of these may be found in the following section, which gives practical advice on EMC design. Computer aided design (CAD) and computer aided engineering are becoming increasingly important in electronic design. Section 13.5 covers the use of CAD/CAE tools in assessing the benefits of EMC design options.

A case study of cost-effective design is given by Upton (1992). Discussing the design of a radio teleswitch, he divides the financial costs into design, material and production costs.

The intended market and the likely size of production run will influence the choice of EMC options. For example, if only small quantities are made then shielding may be an appropriate solution. For a large production run, or where the price of the product must be as low as possible, it may be better to concentrate effort on correct layout and choice of components. This would often apply, for example, to the automobile industry.

As well as financial cost, there may be conflicting requirements between EMC and other constraints. These could include weight, safety, size, appearance or ease of use. Some computers have a microprocessor that generates heat and must be situated next to a cooling fan at the edge of the case, although good EMC practice would be to place it at the centre of the PCB (DTI and New Electronics, 1994, pp 38-40). Such conflicting requirements may depend on the particular product and the earlier EMC is considered, the more easily they can be resolved.

It is impossible to give simple rules for decision making, as each situation must be considered separately. Take for instance the costs and benefits of shielding. If a product is small and the case does not need many holes then a metal box may give good shielding at reasonable cost. However if the product is larger and requires the case to have large holes for displays, then one either has to accept reduced shielding effectiveness or to use a transparent screening material over the apertures, thus adding to material and production costs.

Some interesting case studies of EMC design are described in a DTI technical report (DTI and New Electronics, 1994).

These include controller boards for domestic and industrial gas boilers designed by IMI Pactrol, an air powered mattress system for hospital beds designed by Huntleigh, and a personal computer (using a DEC Alpha microprocessor clocked at 300 MHz) designed by Digital Equipment.

13.3 PRINCIPLES OF EMC

This section aims to explain the principles of design for EMC in more detail, providing some practical design advice.

13.3.1 System level design

Thoughts about EMC should start at the same time as the conceptual design of the system. Ask yourself, "Do I really need a sensor with a micro-volt output when a version with an output of volts is available at only a marginal increase in cost?", when you know that the system will be installed in an electrically noisy environment. Also the way a system is partitioned can have an impact upon its EMC performance. Don't put noisy circuits next to sensitive ones!

The next step is at the circuit design stage: you should endeavour to use circuits that produce as little electrical noise as possible and that are robust in the presence of noise. Provision should also be made for additional filter components now. The design of software can also affect the EMC performance of a system. Finally, the physical implementation of the system is critical to ensure a well-behaved system. This includes enclosure design, correct use of filters and isolating components, and control of interconnecting cables.

13.3.2 Partitioning the system

As the system block diagram is being developed, the implications, on the EMC performance of the system, of any decisions taken should be considered. The type of subsystem chosen and how it will be connected to other sub-systems should be examined with EMC in mind.

Consider an industrial control system with sensors and power actuators. When considering a position sensor for a servo-system driven by a high-power switching controller you must decide on a sensor which will not be sensitive to interference from the switching controller on the drive, or perhaps separate trunking for the motor and sensor cables should be used. There is no one correct solution but an individual solution must be decided upon for each particular system. Perhaps, if other sensitive circuits are near the servo-system, a linear motor drive or improved filtering on the switching drive is the best solution. On the other hand, if the servo mechanism is controlling a robot arc welder, then there may be little point in providing a high level of screening and filtering to stop motor drive interference when the arc welder is the largest source of interference: the sensor signals must be robust and probably screened too!

When partitioning the system between circuit cards, between racks, or even on a single circuit card, thought must be given to the proximity of sensitive and noisy sub-systems. The paths of signal connections and their ground returns should be considered now, as should the grounding and power supply system.

13.3.3 Shielding

Enclosures, cables and PCB layout all affect the EMC performance.

13.3.3.1 Enclosures

A common starting point in the quantitative comparison of screening enclosures is to define the Screening Effectiveness (SE) of the enclosure. The screening effectiveness is defined separately for electric and magnetic fields. Figure 13.2 illustrates the point for the electric field shielding effectiveness of an enclosure. The SE is usually defined as the ratio of the field at a point P, some distance (r) from the source of radiation, with (E_s) and without (E_0) the shield in place. The electric field screening effectiveness in decibels, illustrated in Figure 13.2 is given by:

$$SE = 20 \log(E_0/E_s) \qquad dB \qquad (1)$$

The magnetic field screening effectiveness is determined similarly for the ratio of magnetic fields with and without the enclosure.

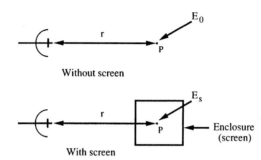

Figure 13.2 An illustration of the definition of Screening Effectiveness.

The SE of an aircraft body, for example, may vary from 20-100 dB which, for some applications, may be inadequate and consequently drives us to consider methods for the optimisation and improvement of screening effectiveness. Note that the SE is a fairly simplistic measure as it ignores several important features, namely

1. The vector nature of electromagnetic (EM) waves. In general, E_s will have a different polarisation to E_0 and both will be complex numbers relating amplitude and phase shift due to the presence of the screen.
2. Usually the SE of a volume is specified by a single number (typically a worst case figure) but this ignores any spatial variation there may be throughout the shielded enclosure.
3. Inherent in our definition of SE is the fact that the introduction of the shield does not change the source characteristics, an assumption only valid if the distance between the source and measurement point (P) is large (in terms of wavelengths).

Finally, note that the reciprocity theorem mentioned in Section 13.2.2, and well known from antenna engineering, may be used to prove that the enclosure reduces internally generated fields measured externally by the same ratio as for internally measured fields generated externally. Hence the same figure of merit may be used for source locations inside or outside the screening enclosure.

Electric field shielding. Consider a spherical shell of conducting material with an applied static electric field E as shown in Figure 13.3.

Charges will distribute themselves over the surface of the shell to produce an internal field which exactly cancels the applied field. Thus the SE is infinite and a closed metallic enclosure provides perfect isolation from static electric fields. However, as an alternating field of increasing frequency is applied, this surface charge moves over the shell in phase with the incident field causing a surface current $I = dq/dt = j\omega q$ which is in phase quadrature with E. This current increases with frequency (at 6 dB per octave) and, if the shell has finite conductivity, currents will then appear on the inside of the shell and re-radiate to generate a non-zero electric field in the interior. This causes the SE to decrease with frequency (again at a rate of 6 dB/octave). This decrease is halted at a frequency where the thickness equals the skin depth (δ) of the material. For frequencies higher than this critical value, the internal currents are reduced due to the skin effect and hence the SE starts to increase again. This increase occurs (initially) at 6 dB per octave for SE above the critical frequency.

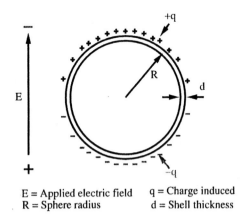

E = Applied electric field   q = Charge induced
R = Sphere radius   d = Shell thickness

Figure 13.3 Static electric Field on a spherical shell.

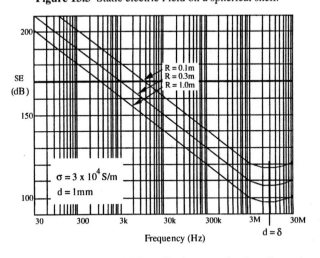

Figure 13.4 Electric field shielding effectiveness of carbon fibre spheres.

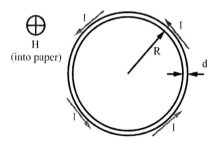

H = Applied magnetic field   I = Current induced
(uniform across sphere)
R = Sphere radius   d = Shell thickness

Figure 13.5 Eddy Currents on sphere surface that are necessary for magnetic field screening to occur.

For example, Figure 13.4 shows the SE of carbon fibre reinforced plastic (CFRP) spheres of differing radii. We notice that the SE for fixed frequency decreases with increasing radius and all curves attain a minimum value when the skin depth is equal to the thickness of the material (d = δ) at around 8.5 MHz.

Magnetic field shielding. The case of magnetic field shielding is very different. The basic shielding mechanism now relies on induced eddy current flow in the shield to generate a magnetic field to oppose the incident field as illustrated in Figure 13.5.

Since the induced current flow is proportional to the frequency of the incident magnetic field, it follows that there is no screening effect at dc and so the SE is zero. Note that this may not be the case in practice since many systems contain ferrous materials which have a relative permeability greater than one and hence provide some screening even at dc. In fact, if low frequency magnetic fields are to be screened then the only effective method is via the use of high permeability shield materials.

However, as the frequency increases so the magnitude of the induced current increases and there is partial cancellation of the incident field. As a result, the SE increases at 6 dB per octave as frequency increases. The skin effect again causes a modification of this behaviour when d = δ, after which the SE increases at a faster rate due to exponential attenuation of surface current. Note however, that this increase is offset slightly by an increase in the series impedance of the surface, again due to the skin effect. An example of a typical SE variation for CFRP spheres is shown in Figure 13.6. Note that the SE increases with increasing sphere radius, the opposite effect from that found for E field screening. We can see the initial 6 dB per octave increase and then, at the point when the skin depth is equal to the thickness of the material (d = δ), we observe an increased rate of (initially) 12 dB per octave. The main conclusion of this is that at low frequencies it is more difficult to screen magnetic than electric fields.

The screening effectiveness of ideal enclosures at frequencies below resonance (i.e. when the enclosure is small compared to the wavelength) is discussed by Field (1983), and by Hill, Ma, Ondrejka, Riddle, and Crawford (1994) when enclosure resonances are significant.

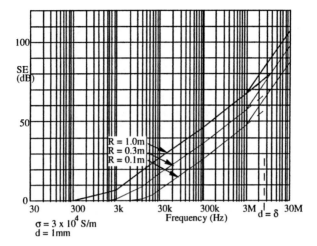

Figure 13.6 Magnetic shielding effectiveness of carbon fibre spheres.

Apertures and joints. Whilst it is simple to achieve a screening effectiveness of 100 dB or more from an ideal enclosure, the main factor limiting the SE of real enclosures is the presence of imperfections such as apertures and joints. Ott (1988 , pp188-190) suggests the following approximation for the shielding effectiveness of an enclosure with apertures of less than half a wavelength in size:

$$SE = 20 \log(\lambda/2l) - 10\log(n) \qquad dB \qquad\qquad (2)$$

where λ is the wavelength under consideration, l is the largest dimension of the aperture and n is the number of apertures. This is very approximate but serves to indicate that a relatively small aperture can dominate the SE of an enclosure and that a number of small apertures (the largest dimension is the significant one) perform better than a single large one.

A joint behaves the same way as an aperture if a gap (i.e. poor electrical contact) is present between points of contact. The quality of the electrical connection between two materials depends on the fasteners and on the surface finish of the materials to be joined. Many common surface finishes (e.g. anodising) have a high surface resistance leading to poor shielding effectiveness.

In order to achieve a low contact resistance at a joint a gas tight contact must be made (The pressure of contact between the metal surfaces must be such that oxidation can not occur, it does not imply that the enclosure (or the joint as a whole) must be gas-tight). This can be achieved by the use of a conductive gasket. A wide variety are available - Weston (1991, pp340-353) gives a good overview. In many circumstances the use of gaskets is too expensive. When this is the case the main source of contact is likely to be at each fastener. The distance between the fasteners determines the maximum dimension of a thin slot, so the constraint imposed by equation 2 determines how the joint limits the overall SE of the enclosure.

Screening material. Traditional electromagnetic screening is achieved with metal, but an alternative is to use a plastic case with a conductive coating (Molyneux-Child, 1992). This gives the advantages of cheapness and lightness, although the SE tends to be worse than with a solid metal shield.

The methods of applying a conductive coating to a plastic surface are paint spraying (nickel, copper or silver paint), arc spraying (zinc), electroless painting (copper and nickel) and vacuum deposition (aluminium). Arc spraying is being superseded by paint spraying, which is the cheapest technique. Electroless plating and vacuum deposition are more expensive but are claimed to give a better finish and higher SE.

It should be noted that quoted values of the SE of conductive coatings apply to a flat sheet illuminated by a plane wave, and will not be the same as the SE of a coated plastic enclosure. The quality of the application (especially into corners) and the contact resistance across joints are also important.

13.3.3.2 External cables

The major sources of radiated electromagnetic emissions are the interconnecting cables between equipment. They are also the major antennas by which EMI enters a system. Conducted interference also enters and leaves the system by means of the cables. The main principle of good physical design is to minimise this coupling. If unscreened cables are used then the coupling is dependent upon the filters used at the cable entries. For screened cables the screen must be connected to the enclosure walls at the cable entry. Ideally the screen should appear to be an extension of the enclosure and therefore a pig-tail should not be used. A connector which allows 360 degree termination of the screen to the enclosure should be used so that no open loop is formed. The screen should not be taken inside the enclosure without termination to the enclosure walls.

Coupling of external fields to cables. Apart from crosstalk between cables in close proximity, a second important interference mechanism is due to the coupling of external RF and pulse waveforms to electronic systems via interconnecting cables. Shown in Figure 13.7 is a schematic of a typical system under study in which two units are connected by means of a screened cable.

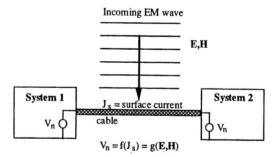

Figure 13.7 EM wave striking screened interconnect cable.

The incident field (which may be a high or low impedance source or a plane wave) generates a surface current J_s on the cable screen, the magnitude of which depends on the cable geometry and tangential electric (E) and magnetic (H) fields. These fields can then produce two kinds of interference sources on the internal signal conductor:

- The electric field can generate a current (I_{sc}) on the central conductor via the transfer admittance Y_t per unit length, such that

$$Y_t = \frac{1}{V_o} \frac{dI_{sc}}{z} = j\omega C_t \qquad \text{S/m} \qquad (3)$$

where V_o is the voltage on the screen (with respect to a reference conductor), C_t is the transfer capacitance between the outer shield and inner conductor, and z is the cable length. The transfer admittance is usually very small because of two reasons:

- If the cable is electrically short and the shield grounded, then the voltage will be very small;
- For cables with large optical coverage, C_t will be very small.
- The tangential magnetic field generates a noise voltage source V_n via the transfer impedance Z_t such that

$$Z_t = \frac{1}{I_o} \frac{dV}{dz} \approx \frac{V_{oc}}{I_o z} \qquad \Omega/\text{m} \qquad (4)$$

where V_{oc} is the open circuit voltage on the inside of the shield, I_o is the current flowing on the shield and z is the cable length.

The transfer impedance tends to be the more important parameter at low frequencies and here we study methods for its prediction and measurement. Typical values are of the order of a 10 mΩ/m at low frequencies falling to less than 1 mΩ/m at around 1 MHz and rising to over 100 mΩ/m at 100 MHz.

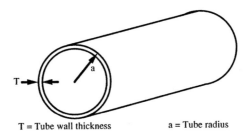

T = Tube wall thickness   a = Tube radius

Figure 13.8 A solid tube as a screen for a cable where is the permeability of the screen.

Transfer impedance. At low frequencies the transfer impedance is given by the dc resistance per unit length of the cable. For example, a solid cylindrical shell as shown in Figure 13.8 would have a transfer impedance (Z_t):

$$Z_t = R = \frac{1}{2\pi\sigma aT} \qquad \Omega/\text{m} \qquad (5)$$

where R is the resistance of the tube (screen) per unit length, a is the radius of the tube, σ is the conductivity of the tube, and T is the tube thickness.

Thus at low frequencies, Z_t is frequency independent. This behaviour changes when the period of the incoming wave is of the order of the diffusion time through the shield. The exact expression for Z_t then becomes

$$Z_t = R\frac{(1+j)T/\delta}{\sinh(1+j)T/\delta} \qquad \Omega/m \qquad (6)$$

where δ is the skin depth of the shield material. For frequencies above a break frequency f_{crit} the transfer impedance falls off rapidly due to the skin effect. The frequency fcrit is given by:

$$f_{crit} = \frac{1}{\pi\sigma\mu T^2} \qquad Hz \qquad (7)$$

where μ is the permeability of the screen.

In practice, shields are seldom solid and are more usually braided in some way. Consequently, this fall of Z_t with frequency does not continue at high frequencies, but rises again due to shield imperfections. The effect of cable braiding is modelled as a mutual inductance M, such that a more accurate expression for Z_t is

$$Z_t = R\frac{(1+j)T/\delta}{\sinh(1+j)T/\delta} + j\omega M \qquad \Omega/m \qquad (8)$$

The mutual inductance is due to two effects: aperture coupling and porpoising. The net result of these variations is a typical variation of Z_t with frequency as shown in Figure 13.9.

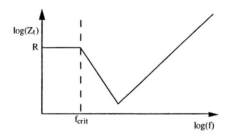

Figure 13.9 Typical variation of Z_t with frequency for a braided cable shield.

Note that f_{crit} (Equation 7) may be around 10 kHz and shield imperfections are generally only important for frequencies above 1 MHz. For very high frequencies Z_t may be much larger than R.

13.3.4 Physical layout

The physical layout of a system has a profound effect on its electromagnetic performance. It is possible to significantly improve the performance of a system by careful physical layout at no extra material cost to the production item.

13.3.4.1 PCB layout

A good PCB design can reduce the radiated emissions from an equipment by 10 to 20 dB. The system immunity may be increased by a similar amount. The main aspects of good PCB design are the careful partitioning of circuits and the reduction of loop areas. Correct partitioning of circuits reduces the noise levels due to self interference and improves the immunity of the system to EMI. Circuits should be categorised as noisy or quiet for emissions, and robust or sensitive for immunity. Sensitive and noisy circuits should also be categorised as high- or low-impedance. High impedance circuits are more likely to couple to electric fields and low impedance circuits are more likely to couple to magnetic fields.

The circuit board should be partitioned so that noisy circuits (including interconnecting busses) are well separated from sensitive circuits. Robust circuits and quiet circuits can be used to obtain separation between noisy and sensitive circuits. Power supplies and ground connections can be partitioned so that only robust circuits share power and ground connections with noisy circuits, and sensitive circuits only share power supplies and ground connections with quiet circuits.

Input/output (I/O) circuits should be placed near the edge of the board along with filter and suppression components grounded to their own ground plane which is bonded to the chassis, ideally at the panel through which I/O connections occur, so that any incoming interference does not flow to earth across the circuit card and interference generated on-card does not flow on external connections.

Reduction in circuit loop areas is the key to minimising on-card noise generation, emissions of radiation from the PCB and reduction of pick-up from external electromagnetic fields. The ideal is that every signal or power supply track has a return conductor which takes the same path. This can be achieved in practice if ground and power supply planes (separated into quiet and noisy sections) are used on the circuit board. Whilst ground planes can usually be incorporated into double sided circuit boards for analogue circuits, the track density on digital circuit cards precludes this unless a multi-layer board is used. Even in digital circuits a good approximation to ground and power supply planes can be achieved on double sided boards if the connections are gridded.

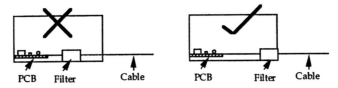

Figure 13.10 Cables must be filtered as they enter the enclosure.

13.3.4.2 Cable looms

Internal cable looms are often the means by which interference is coupled between the system and the outside world. To avoid this it is necessary to ensure that filtering occurs where cables enter and leave the enclosure and not on the circuit card or inside the enclosure (Figure 13.10). Cables from circuit cards to front panel mounted controls or displays should not drape over a PCB. Ideally they should leave the edge of the circuit card and run close to the metal chassis (if used) until they reach the control/display. This reduces the loop area of the circuit and hence the coupling between the loom and other circuits in the enclosure. Front panel items are likely to be subject to electrostatic discharge (ESD). Many items (e.g. keypads) can be purchased with an earthed screen designed to accept the discharge current and pass it to earth via the front panel. When a front panel item is subject to ESD, large currents may flow in the loom back to the PCB. This can be reduced by filtering or suppression at the front panel. If this is not possible a separate ground plane on the PCB can be used for filter and suppression components. It should be bonded to the chassis in a way which minimises the loop area formed by the circuit which returns the discharge to chassis ground.

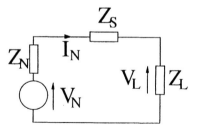

Figure 13.11 Series impedance filter.

13.3.5 Filters

We start with a disclaimer! Most filter theory is based on designing filters with a well defined frequency response on the assumption that they operate with a well defined (resistive), linear source and load. Most filters in EMC applications do not work under these conditions; it is interesting to note that most EMC filters are specified for a fixed linear load so don't believe everything that you read in the specification. This is especially true for power supply filters where the load may vary greatly depending on the operating conditions of the equipment.

13.3.5.1 How filters work

Filters are used to pass or block a range of frequencies by the use of frequency dependent impedances. Here we will consider how we can block a signal. Figure 13.11 shows how a series impedance can be used to reduce the noise voltage at the load. The load voltage is given by

$$V_L = \frac{V_N Z_L}{Z_N + Z_S + Z_L} \tag{9}$$

and it can be seen that the noise voltage can only be reduced if

$$Z_S \gg Z_L \tag{10}$$

and

$$Z_S \gg Z_N \tag{11}$$

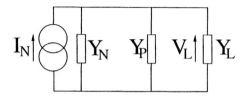

Figure 13.12 Use of a parallel admittance.

The technique can only work well for low load and source impedances. A ferrite bead would be useful as a filter in such an application.

The alternative approach is to divert the noise current away from the load by putting a low impedance in parallel with the load. Here it is more convenient if we work in admittances. Figure 13.12 shows the use of a parallel admittance as a filter to reduce the noise current in the load. The load voltage is given by

$$V_L = \frac{I_N}{Z_N + Z_P + Z_L} \tag{12}$$

and it can be seen that the noise voltage can only be reduced if

$$Z_P \gg Z_L \tag{13}$$

and

$$Z_P \gg Z_N \tag{14}$$

The technique can only work well for high load and source impedances (low admittances). An application of this technique is the use of decoupling capacitors on a power supply rail. The use of simple series or parallel elements alone has two disadvantages:

- they only work when both source and load have similar high or low impedances;
- the slope of the filter frequency response cannot be more than 6 dB per octave (assuming the use of a single inductor/capacitor and a resistive load and source).

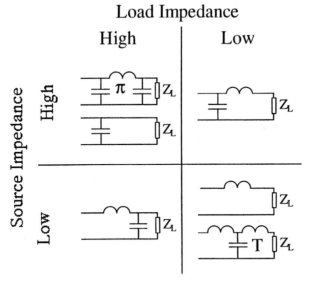

Figure 13.13 Use of π, T, L sections, and single element filters for various source and load impedances.

Both limitations can be overcome, at the expense of extra complexity, by the use of additional elements. This leads us on to L, π, and T section filters and their multiple stage versions. Figure 13.13 illustrates some of the possibilities with optimum source and load characteristics.

13.3.6 Circuit design

Whilst there is no rigid border between analogue, digital, and power switching circuits it is convenient to categorise circuits in this manner to consider EMI effects. Those circuits which straddle the boundaries set here, such as analogue to digital converters and discrete-time continuous amplitude circuits (e.g. switched capacitor filters), may exhibit the (worst) characteristics of both analogue and digital circuits.

1. remember that EMI can enter through outputs and escape via inputs. Equal attention must be applied in controlling EMI at all system interconnections.
2. try to identify sensitive and noisy circuits, then design in protection to the circuit. It is often cheaper and easier to remove protection that is not required than to add protection at the prototype stage.
3. be aware that the EMI protection and reduction measures designed into the circuit will only work if the physical layout is correct.

13.3.6.1 Analogue circuits

Analogue circuits tend to suffer from susceptibility to electromagnetic interference (EMI) because they often deal with low level signals. Interference in analogue circuits can be broadly classified into two types: interference within the circuit bandwidth and interference outside the circuit bandwidth.

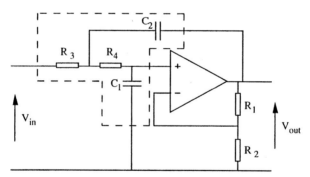

Figure 13.14 A two-pole, low-pass Sallen-Key filter built on to a non-inverting op-amp circuit.

Once interference within the circuit bandwidth enters an analogue circuit it cannot be distinguished from the wanted signal. This interference must be kept out by screening of the circuit and interconnections. However, it can often pay to remember that there is no need to make the circuit bandwidth wider than necessary to accommodate the signal. The circuit bandwidth can easily be restricted by using a simple RC filter or turning an op-amp circuit into a Sallen-Key filter as shown in Figure 13.14 (Horowitz and Hill, 1980, pp148-162).

Sometimes it is not possible to prevent the entry of interference but even then its effect can be minimised by defensive design. For example a pair of diodes can be used to limit pulse interference levels and minimise the time a circuit takes to recover from impulsive interference.

Interference outside the circuit bandwidth can still cause problems. Almost all electronic circuits contain non-linear elements; all active devices are non-linear. This means that most circuits are capable of acting as a mixer and producing additional frequency components that are not in the original signals or interference present in the circuit. The effect of non-linearities manifests itself as the rectification of high frequency interference which causes DC level shifts in circuits; if the interference is amplitude modulated this will result in the modulation being superimposed on the normal circuit voltages. If you have experienced the local taxi firm coming over loud and clear on your hi-fi then you know how annoying demodulation effects can be. The problem can be overcome either by filtering the inputs to a circuit and ensuring that the circuit is well screened or by providing small high frequency bypass capacitors across op-amp input and transistor base emitter junctions near the inputs to a circuit (see Figures 13.15 and 16).

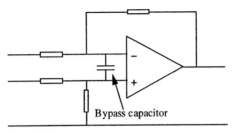

Figure 13.15 High frequency bypass capacitor in an op-amp circuit to prevent demodulation effects.

13.3.6.2 Digital circuits

Digital circuits tend to be a significant source of EMI but have much better immunity than analogue circuits.

EMI is generated by the high rates of change of current and voltage present as the logic gates switch; therefore the slowest possible logic family should be chosen. The EMI from digital circuits can only be reduced by limiting the propagation of the EMI generated to other sensitive circuits or the outside world. This is achieved by the use of high quality power supply decoupling, along with filtering or isolation of circuit interconnections - this means inputs and power supply connections as well as outputs.

Semiconductor manufacturers are now producing logic devices with output circuits which control the signal switching times and, by rounding the pulse edges, reduce the harmonic content of the signals. This greatly reduces the noise generated by the circuit. One of the most useful characteristics of digital circuits is their ability to completely regenerate a noisy signal provided the level of the noise is below a certain threshold - the noise margin of the circuit. Once the noise level exceeds the circuit noise margin, it is impossible to regenerate the signal.

Many factors may contribute to the noise levels within a digital circuit including: power rail noise generated by the logic circuit itself and from other circuits; noise due to ringing and reflections; and electromagnetically coupled noise both from within the system and from external sources. The first step in ensuring good immunity to EMI is to ensure that the internal noise generated by the circuit is minimised - this is dependent upon the circuit layout and power supply decoupling. The lower the internal noise the larger the proportion of the noise margin which is available for immunity to external sources. If the digital circuit is to operate in a particularly noisy environment then isolation and filtering of circuit interconnections will considerably improve the circuit immunity.

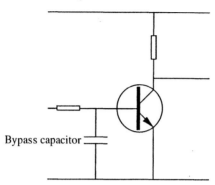

Figure 13.16 High frequency bypass capacitors in a discrete amplifier circuit to prevent demodulation effects.

In microprocessor based systems it is often worth having a separate low-speed bus to operate I/O devices connected to the outside world. This can be achieved by using an I/O port, rather than the processor bus, to communicate with A/D converters, D/A converters, and digital I/O. The signals in the I/O circuit change only when an I/O device is being accessed and at a lower speed than the processor bus. This results in reduced emissions from the I/O leads. The isolation of the I/O devices from the main processor bus also increases the immunity of the processor circuit to EMI.

In all sequential digital systems (computers and sequential logic circuits) provision should be made to restart the system in a known, safe state if a failure due to noise or interference occurs. This can be achieved in many systems by the use of a watchdog timer. This is a timer which should be periodically reset by the system as part of its normal operation. If the system crashes the timer times-out and resets the system to a known state from where normal operation can resume.

13.3.6.3 High power switching circuits

High power switching circuits are considered only as a source of EMI here.

High power switching circuits such as switched-mode power supplies (SMPS), stepper motor drives, and thyristor power controllers are used because of their high efficiency. In order to achieve high efficiency it is desirable that the switching process occurs as quickly as possible - a high power is dissipated in the switch whilst it is switching and the less time spent in the switching process the less power dissipated. Where magnetic components are involved, their size decreases with increasing switching frequency. These two factors mean that high power circuits are continually moving to higher switching frequencies and reduced switching times. It is precisely these features that contribute to the high levels of EMI generated by these circuits.

If you are a designer of high power switching circuits then you should seek ways of reducing the rates of change of current and voltage within the circuit. This can sometimes be achieved without loss of efficiency, if natural commutation of the currents is used, such as in the pulse resonant converter.

If you are a user of high power switching circuit modules then you should ensure that: the noise generated by the circuit is adequately filtered within the module and that the module itself is adequately screened. Filters external to the module are often ineffectual as the radiation from the connecting leads is sufficient to cause problems. Direct radiation from the module can be a problem if it is not contained within its own screened enclosure.

13.3.6.4 Power supply noise

Power supply noise can be due to the power supply itself, generated by the circuits attached to it, or from an external source. The level of power supply generated noise affects the circuit immunity by adding to the total noise present. External noise reaches the power supply rails by conduction. Internally generated power rail noise passes to external connections where it may become conducted or radiated interference to another system. The answers to these problems lie in choosing electrically quiet systems where possible and in adequate filtering, including power supply decoupling.

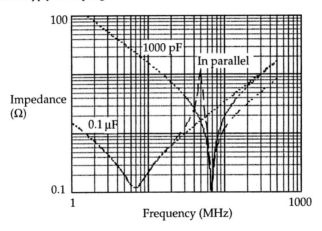

Figure 13.17 Typical impedance of different value ceramic capacitors, individually, and in parallel, showing parallel resonance.

Switched mode power supplies and voltage converters generate significant noise on both input and output connections. Most modern SMPSs have adequate built-in filtering, but many dc to dc voltage converters generate enough noise on their outputs to cause problems in sensitive analogue circuits and A/D or D/A converters. Despite claims of large power supply rejection ratios (PSRR) by manufacturers, most analogue and digital circuits pass high frequency power supply noise directly to their outputs. If you look closely on the data sheet the quoted PSRR only applies at very low frequencies.

Linear power supplies are electrically quiet but usually have no ability to filter the passage of high frequency noise. Voltage regulator circuits are feedback amplifiers with a limited bandwidth and are unable to prevent high frequency components from passing between the regulated and unregulated power rails.

The type and size of decoupling capacitors to be used is a subject of much debate. The basic problem is that all capacitors have a parasitic inductance due to their leads and internal construction. This is increased by the inductance of the PCB tracks connecting the capacitor to the circuit being decoupled. A decoupling capacitor will therefore form a series resonant circuit with these stray inductances at some frequency, and thereafter its impedance will rise as the inductance becomes the dominant effect. Smaller capacitors have a higher self resonant frequency, but a higher impedance at any given frequency.

The practice of putting small and large capacitors in parallel causes the additional problem of a parallel resonance between the small capacitor and the lead inductance of the large capacitor which can increase the impedance of the combination by a factor of 10 at the resonance (Figure 13.17). The combined impedance of the pair at high frequencies is only marginally lower than the large capacitor on its own. This is most noticeable for capacitors with low series resistance (e.g. small ceramic capacitors). The best solution would appear to be to use a large value (e.g. 100 µF) capacitor for good low frequency performance with a group of equal value small capacitors (e.g. 0.1 µF) spread around the board. The relatively large series resistance of the large value capacitor damps the parallel resonance (Figure 13.18). The impedance of the power supply rails at high frequencies is the parallel combination of the inductances of the capacitors and supply rails - ultimate values depend on the transmission line characteristics of the supply rails. Careful attention to layout is therefore important. For logic boards, and high frequency analogue boards, each IC should have its own decoupling capacitor (0.01 to 0.1 µF) placed as close to the chip as possible with larger electrolytic capacitor(s) also available on-board to maintain the low supply impedance at the lower frequencies.

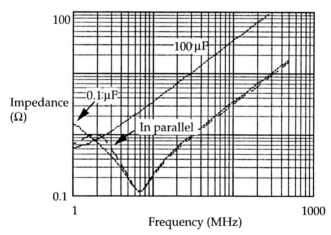

Figure 13.18 Typical impedance of an electrolytic and ceramic capacitor including the effect of parallel combination.

13.3.7 Software considerations

The software controlling an electronic system can have a significant effect on its EMC performance.

13.3.7.1 Emissions

Continual polling of peripheral circuits generates higher circuit noise levels, and hence emissions, than interrupt, timer, or demand driven access. Peripherals should only be accessed when absolutely necessary. The processor should be halted when its activity is not required rather than sitting in a loop polling a peripheral.

13.3.7.2 Immunity

Well-written software can significantly enhance the immunity of a system to EMI. Simple precautions like ensuring that all unused interrupts are trapped by an error handling routine can prevent unpredictable behaviour in the presence of noise. Impulsive noise on inputs can be ameliorated by comparing several samples of an input rather than just taking a single sample. If a number of samples taken over a period of time are identical (within limits) then the input has been read correctly. If samples differ, then the input can be re-read until the noise ceases to have an effect. The ability of a system to recover from a processor crash smoothly and quickly is also important in many systems and is critically dependent upon the design of the control software.

13.4 COMPUTER-AIDED ENGINEERING (CAE) TOOLS

Waiting until late stages of design to fix EMC problems is not only likely to be costly in terms of the expense of 'band-aids' but may also lead to the need to re-design the circuits. EMC therefore needs to be considered as early as possible, i.e. at the 'front-end' of the design process, and we must predict how decisions made at these early stages will affect the EMC of the finished design. This requires predictive methods such as CAE tools. CAE tools are also ideal for coping with complex problems and automating certain tasks, e.g. auto-routing PCB tracks.

The need for 'concurrent' design, which accounts for all disciplines (thermal, manufacturing, EMC, signal integrity, reliability, testability etc.) in an integrated manner, also encourages the use of computers. The computer tools should work within a framework which allows engineers from all disciplines to access the same data effectively.

The use of CAE tools is growing rapidly in all disciplines, driven by pressures such as time to market of the product, increased circuit density and speeds of operation, and more specifically related to EMC, regulations governing emissions and immunity. Extensive use of CAE tools has been made possible with the advances in computers.

Although these factors show the possible benefit of CAE tools for EMC, the availability of published performance data for most commercially available EMC CAE tools is very limited.

The problem of EMC appears at many levels of design from integrated-circuits(ICs) through multi-chip modules (MCMs) and PCBs, to subsystems which have shielding and external cables, and systems of interconnected subsystems. It is not possible at present to have a single CAE tool which can account for EMC at all of these levels, and the trend is towards specific tools for specific applications.

For CAE tools to be effective they must take into account both the engineer's EMC experience and familiarity with tools, especially with respect to interpreting the results of numerical tools. It is doubtful that the engineer will have experience of the latter, and the EMC experience may also be limited. Results must therefore be presented in a useful form that the engineer recognises, and the engineer should not be expected to interpret the limitations of the tools.

The tools should be able to account for all important EMC effects including radiation from both common-mode and differential mode currents (Paul and Bush, 1987), and immunity. However, because immunity is more difficult to predict, no tools exist which consider it.

13.4.1 Analytical Methods

In order to use analytical methods to solve most real problems it is necessary to make simplifying assumptions. However analytical solutions can give us significant insight into the interactions which may occur and their likely effects.

For example if the currents on a cable are fully known it is possible to estimate (analytically) with reasonable accuracy the radiation from the cable. However, whilst the differential mode currents are predictable from simple circuit analysis the dominant radiation effect comes from the common-mode currents which depend upon the physical layout of the circuit (Paul and Bush, 1987) and are not amenable to simple calculation.

13.4.2 Numerical Methods

Numerical methods solve Maxwell's equations in their complete form, expressed either as integral or differential equations, in the time or frequency domain, and in two-dimensions(2D) or three-dimensions (3D). General reviews of the methods applied to electromagnetics are given in (Hubing, 1991).

The methods can be divided up into those based on the differential equations called finite methods, and those based on the integral equations called the integral methods. Finite methods include the Finite Difference (FD) method, the Finite Element Method (FEM), and the Transmission Line Matrix (TLM) method, Integral methods include the Boundary Element Method (BEM), the Method of Moments (MOM), and the Time-Domain Integral Equation (TDIE) method.

Finite methods (see (Porter and Dawson, 1994)) 'discretise' the simulation space, i.e. divide it up into a grid. This grid is made up of uniform or nonuniform quadrilaterals (for 2D) or hexahedrals (for 3D) for the FD method. For FEM triangles (for 2D), or tetrahedrons (for 3D) are normally used. These discretised spaces are referred to as 'cells'.

In reality the fields would extend far beyond the simulation space that we are interested in, but of course the simulation space is limited by computer memory and speed. Radiation boundary conditions must be applied to the finite simulation space to approximate the effect of infinite space (Mur, 1981).

Integral methods do not need to discretise the whole simulation space as do the differential methods, but in general, just where current flows. Therefore just the surfaces or the volumes of the model are discretised, e.g. no discretisation is necessary in free space. This means that the simulation space can be considerably reduced and artificial boundary conditions do not need to be introduced, as exact boundary conditions are implicitly incorporated within the model. Green's functions are used to derive the integral equations which are then solved to calculate the current and charge distribution on the object. The far zone fields are then calculated from the charge and current distribution.

For numerical electromagnetic tools to be effective in EMC design they need to automatically import the geometry from the CAD layout tool and create the grid required for simulation. They may also need to interface with circuit simulation tools to determine the source currents and voltages used in the electromagnetic simulation.

Practically these methods are limited by available computer memory and time. The full detail of most real systems can not be considered, so simplifying approximations must be made.

No one method can be said to be better than another in general. Finite methods tend to perform well for enclosures and where dielectric bodies are present whilst integral techniques are best for wire-grid structures and open spaces.

Advances in the numerical methods include 'hybrid' methods, which combine different methods and solve them simultaneously, as opposed to using several methods serially as below (practical approaches to radiated field prediction). An example of a hybrid method which has been used is a combination of FEM and analytical solutions. Other improvements include variations of the above methods to increase accuracy and decrease demands on computer time and memory.

Practical approaches to radiated field prediction. In an attempt to simulate realistically complex circuits, more approximation is needed. This has resulted in the combination of several methods. First of all a 3D numerical solver is used to extract the circuit parasitic parameters. These are the resistance, inductance, conductance, and capacitance, referred to as the RLGC parameters. These parameters are used to create lumped element models of, for example, a transmission line, connector, or via, which are then input into a circuit simulator. The circuit simulation calculates the current distributions in the time domain by using these models and the device models included in the CAD libraries. Using the time domain allows non-linearities to be accounted for. The final part is the use of a 3D numerical solver to calculate the radiated fields, including at this point the shields and external cables. Approximations in creating the lumped element models restrict the frequency ranges applicable. The advantage is that complex PCBs can be modelled within reasonable time limits.

Identification of critical nets reduces the problem and may allow direct calculation of the fields. Many traces, by nature of the signals that they carry, their geometry, or the function that they perform, will not present an EMC problem. This may well include the majority of the traces on the board. There are tools available whose aim is to identify the traces that need further investigation, allowing the problem to be reduced considerably. This is usually performed by using first-order approximations. This may reduce the problem to a level where direct calculation of the fields can be done. How realistic the results are is not yet known.

13.4.3 Knowledge-based/expert systems

The terms knowledge-based systems and expert systems seem to be used interchangeably to denote an area of artificial intelligence that attempts to harness the knowledge and reasoning, borne out of experience, that an expert engineer possesses, and make it available to a wide number of engineers. They also tend to enhance the engineer's design environment by providing support in the form of data, background information, connections to analysis tools, and results interpretation tools.

The many types of tools that fit into this category overlap in their approaches and it is often difficult to clearly classify them. However, they can include some of the following capabilities.

Reasoning

1. Design guidelines and advice (Williams, 1991; Butcher and Withnall, 1993).
2. Identification of relevant EMC regulations and test procedures.
3. Assistance in interpreting the results of numerical simulations, or at least knowing their limitations.

Analysis Tools

1. Analytical expressions - closed-form expressions that are the rules-of-thumb used by engineers.
2. Numerical modelling tools.
3. Analysis tools such as Fourier analysis for manipulation of output data.

Support

1. Databases of material and component data.
2. Extensive background on EMC.
3. Specific application tools - e.g. filter design.

Design rules. Design rules based tools are a particular area of a knowledge-based system that are used extensively. Tools of this type either indicate where the rules have been violated, perform post-layout checking, or place constraints on the design which are automatically implemented as the design progresses. This latter approach 'guides' the designer and is preferred because decisions are made as early as possible. Applying the design rules as the design is progressing leads to a 'correct-by-design' methodology.

The rules have been developed from experience and are the sort of design checklists that companies develop to ensure EMC (Williams, 1991; Butcher and Withnall, 1993). Companies that produce these tools have verified the rules but further work is needed. The rules need to be developed and refined in a complementary environment of post-layout tools and test measurements. A methodology for verifying design rules is discussed in (Daijavad *et al.*, 1992). There must also be concurrent consideration of design rules and prioritisation for instances of conflicting constraints.

If the tool indicates violations, it does not at present offer advice on which design changes to implement. Alternatively, if the rules are automatically implemented as constraints, there may be instances where it would be better for the engineer to be offered alternatives, again with advice. Providing advice with the design rule tools would extend their expert system capability. This capability does not exist at present. Design rule checkers for EMC include Zuken-Redac's EMC Advisor, Altium's EMC Consultant, Cadence's DF/EMControl, and Harris' EDA Navigator.

13.4.4 How early can the tools be applied?

Knowledge-based, expert system methods can be applied at any stage as long as the relevant information can be developed. For example information could be supplied at the concept design stage to indicate the likelihood of problem areas such that EMC can be taken into account from the earliest possible stage. As the system level design progresses and more information is available the quality of possible advice is likely to improve. Advice on system partitioning, signal design and alternative design approaches may be possible. For example whether a particular type of sub-system is likely to require screening might be indicated; a choice could then be made as to:

1. look for an alternative sub-system;
2. screen the system locally;
3. use a screened enclosure.

At the schematic entry stage, detailed advice could be made available. This could include advice on component types, connectivity strategies, termination strategies, filtering and power supplies etc. At the layout stage simple analytical expressions or design rules can be used to estimate radiated fields. Advice and expressions for cables, shields, and systems can be developed. And finally they can be used to identify the particular EMC regulations that apply and advising on test procedures.

It must be stressed that these methods are only as good as the rules, advice, expressions etc. that they contain.

Limited numerical methods can first begin to be applied when components have been placed on the PCB, e.g. calculating PCB trace characteristics assuming Manhattan distances. This can lead to better placement before the optimal routing paths are decided. As the board is being routed, numerical methods can be applied to analyse specific areas which are considered critical.

It is possible that numerical methods can then be applied to the fully routed board, or even a complete system (although some simplification would be inevitable). This is likely to be necessary because rules can only give general solutions and indicate bestpractice. They can not predict whether a system will pass or fail specific tests.

13.4.5 Cost of the tools

Purchase cost of the software. The numerical tools are currently the most expensive, in the region of tens of thousands of pounds. This has been due to their very small market and most were designed as research tools and sold to a few companies where the in-house expertise was able to make use of the tools and large mainframe facilities were available. There is currently a trend to make such tools more readily available as EMC prediction tools which may herald significant increase in the possible market and reduction in cost. Knowledge based systems are currently available at costs in the region hundreds to several thousand pounds.

Computing requirements. Most of the tools are now available on engineering workstations. For the numerical codes high performance dedicated workstations are likely to be required (e.g. 100 MB memory, 20 Mflops processor performance, 2 GB disk).

This is likely to be close to the specification of existing CAE work-stations. Learning curve. The cost of gaining EMC expertise is inevitable for any company making electronic systems. Some software vendors offer the hope of achieving EMC (by means of their software) without the requirement of skilled EMC engineers. We think this is as unlikely as removing human skill and experience from any other part of the design process.

Integration with existing CAE. Compatibility between the ranges of existing CAE tools is generally poor. This seems particularly so in the EMC area.

13.5 SUMMARY

1. Electrical and electronic equipment should be designed to avoid
    *   excessive emission of electromagnetic interference
    *   excessive susceptibility to interference.
2. From 1996, the European EMC Directive will make this a requirement for most types of equipment.
3. Designers should consider EMC as early as possible because
    *   if EMC is ignored, products are likely to need expensive re-design and re-testing
    *   conflicts with other design constraints can be resolved more cost-effectively
    *   provision can be made for later changes to the design.
4. EMC is hard to predict, so good practice should be followed throughout the design process. Many electrical and mechanical aspects of a design affect EMC.
    *   When partitioning the system, sensitive subsystems should be separated from noisy ones.
    *   Metal or conductive enclosures provide shielding against electromagnetic fields. Magnetic fields are harder to shield against than electric. Apertures should be short and joints should make good electrical contact.
    *   Cables can act as antennas. Screened cables and good connectors can help.
    *   Good circuit board layout is very important. Keep noisy and sensitive circuits apart, put input/output at the edge of the board, reduce loop areas (a ground plane or multi-layer board helps), do not drape cables over circuit boards, suppress electrostatic discharge at the front panel.
    *   EMC filters can help, but are less simple than sometimes suggested.
    *   Immunity of analogue circuits can be improved with filters and bypass capacitors.
    *   Emissions from digital circuits can be reduced by choosing slower logic and decoupling power supplies.
    *   High power switching circuits can be a source of interference.
    *   Well written software can reduce emissions and improve immunity.
5. Computer aided engineering (CAE) tools for EMC design are either:
    *   Numerical, including field prediction tools. These have large computing requirements and cost tens of thousands of pounds.
    *   Knowledge-based tools, including design rule checkers. These cost hundreds to several thousand pounds.
6. Different tools are suitable at each design stage.

REFERENCES

Holliday, J. (1995) Of trams and telephones, *EMC Journal*, **1** (5).

Noble, I.E. (1992) EMC and the automotive industry, *Electronics and Communication Engineering Journal*, **4**, 263-271.

White, D.R.J., Atkinson, K., Osburn, J.D.M. (1992) Taming EMI in microprocessor systems, *IEEE Spectrum*, **22**, 30-37.

Department of Trade and Industry and New Electronics (1994), EMC Technical Report.

Collier, R.J. (1994) An introduction to EM interference in hospitals, In: *IEE Colloquium on Electromagnetic Interference in Hospitals*, October, 1/1-1/3.

Marshman, C. (1992) The Guide to the EMC Directive, 89/336/EEC, EPA.

Nensi, S. (1994) HTM 2014, Abatement of electrical interference, In: *IEE Colloquium on Electromagnetic Interference in Hospitals*, October, 2/1-2/4

Williams, T. (1991) Design for EMC and avoid problems, *Test*, **13** (7), 13-15.

Butcher, P.M., Withnall, S.G. (1993) A practical strategy for EMC design, *Electronics and Communication Engineering Journal*, **5**, 257-264.

Armstrong, K. (1994) EMC procedures in a BS5750 environment, In: *Euro-EMC Conference Proceedings*, October, 84-92.

Cross, N. (1989) *Engineering Design Methods*, Wiley.

Upton, M. (1992) Cost effective EMC design, In: *Euro-EMC Conference Proceedings*, October.

Field, J.C.G. (1983) An introduction to electromagnetic screening theory, In: *IEE Colloquium on Screening and Shielding*, November, 1/1-1/15.

Hill, D.A., Ma, M.T., Ondrejka, A.R., Riddle, B.F., Crawford, M.L. (1994) Aperture excitation of electrically large, lossy cavities, *IEEE Transactions on Electromagnetic Compatibility*, **36** (3), 169-178.

Ott, H.W. (1988) *Noise Reduction Techniques in Electronic Systems*, Wiley Interscience, 2nd edition.

Weston, D.A. (1991) *Electromagnetic Compatibility Principles and Applications*, Marcel Dekker.

Molyneux-Child, J.W. (1992) *RFI/EMI Shielding Materials: A Designers Guide*, Woodhead Publishing.

Horowitz, P., Hill, W. (1980) *The Art of Electronics*, Cambridge University Press.

Paul, C.R., Bush, D.R. (1987) Radiated emissions from common-mode currents, In: *IEEE International Symposium on Electromagnetic Compatibility*, Atlanta, GA, USA, August, 197-203.

Hubing, T.H (1991) A survey of numerical electromagnetic modelling techniques, *ITEM Update*, 17-30, 60-62.

Porter, S.J., Dawson, J.F. (1994) Electromagnetic modelling for EMC using finite methods, *IEE Proceedings*, **141** (4), 303-309.

Mur, G. (1981) Absorbing boundary conditions for the finitedifference approximation of the time-domain electromagneticfield equations, *IEEE Transactions on Electromagnetic Compatibility*, **23** (4), 377-382.

Daijavad, S., Pence, W., Rubin, B., Heeb, H., Ponnapalli, S., Ruehli, A. (1992) Methodology for evaluating practical EMI design guidelines using EM analysis programs, In: *IEEE International Symposium on Electromagnetic Compatibility*, Anaheim, CA, USA, August, 30-34.

DESIGN FOR SERVICE

Peter Dewhurst; Nicholas Abbatiello

This chapter presents a Design for Service (DFS) tool developed at the University of Rhode Island (URI) to help design teams address the serviceability of their designs at the same time as the important decisions are being made for ease of initial assembly. The DFS analysis procedure covers the range of disassembly and reassembly operations commonly carried out in service work (Whyland, 1993; Subramani and Dewhurst, 1994; Dewhurst, 1993).

Henry Ford is quoted with the remark that "in the Ford Motor Company we emphasize service equal with sales." However for the early pioneers of mass production, service meant little more than the availability of replacement parts used for repairs at service shops widely spread across the country. For the customers of those mass produced automobiles, ownership was viewed as a luxury which could be suspended occasionally for repair work. For today's consumers, however, most mass-produced appliances, including automobiles, have become a necessity of everyday life. For this reason, quality measured by reliability standards is now the most important attribute for success in the marketplace. In addition, customers expect service procedures to be carried out with the absolute minimum disruption of product use.

Reliability and serviceability are linked in the minds of both the manufacturer and the owner. For the manufacturer, they jointly determine the cost of the product warranty, while for the owner they define part of the continued cost of ownership. For example, for US automakers, annual warranty costs are now measured in billions of dollars (greatly in excess of profits) and approximately half this amount is for service labor. For the owner, the high maintenance costs as a product gets older translates into a too-rapid loss of value and dissatisfaction with the rate of depreciation. For low cost appliances, this often means early disposal in the municipal landfill when the likely cost of repair is felt to exceed the perceived product value.

14.1 INTRODUCTION

A major improvement in product serviceability would be beneficial to both the manufacturer and their customers. However, it appears that this is one aspect of product design that has not been improving. An extreme example is the use of spot welding in automobile body construction, coupled with the move to seamless body designs. This makes the repair of even minor body damage prohibitively expensive, even though the event is an anticipated part of normal product use.

It is clear that significant improvements in the serviceability of products will only occur if there are changes in the way in which products are designed. The current approach in most companies, of carrying out service reviews only when the design has been fully executed, only serves to avoid those service tasks which would be considered totally unacceptable. It is usually too late in the process to make changes to reduce long service procedures which can, nevertheless, be carried out in a routine manner. It is the belief of the authors that the analysis of important service tasks on new products should be carried out concurrently with the earliest design for assembly studies, and that these should take the place at the early concept-layout stage of product design. The goals of ease of initial assembly and that of subsequent service tasks, can be closely aligned provided that they are considered together by the development team. When they are separated, however, decisions about part locations and securing methods may be made with little consideration of later disassembly.

One important aspect of design for assembly which can have a positive influence on serviceability is the goal of reducing both part count and the use of separate fasteners. A review of the literature on DFA shows that in 74 case studies, which have been published on the results of using the Boothroyd Dewhurst DFA software (Boothroyd Dewhurst Inc., 1994), the average reduction in the number of parts to be assembled is 56 percent and the average reduction in the number of separate fasteners is 72 percent. These new designs are not necessarily easier to service. However, when designs are simplified in this way the potential for substantially easier service tasks clearly exists. Consider, for example, the service procedure of the replacement of a headlamp bulb shown in Figure 14.1. This illustration, taken from the owner's manual, shows that much of the front trim of the vehicle has to be removed to access the headlamp assembly; Figure 14.1(a). Then additional screws, trim, glass and the seal must be removed to uncover the bulb; Figure 14.1(b). In total, 32 items are removed and then reassembled in order to replace a relatively inexpensive item, which has a high likelihood of failure. In contrast, Table 14.1 shows the results of the proposed redesign of the GM-10 Headlamps and Panel assembly resulting from a DFA analysis of the previous design. The point of showing the GM-10 statistics in the present context is not that the new headlamp will necessarily be easier to service. With poor access or inappropriate securing methods it could even be more difficult to service. However, it is clear, that with fewer assembly operations, the new design has the potential to be easier to service than the older model. The challenge is to empower design teams to achieve that potential.

At the present time, concurrent engineering product development teams are driven by engineering designers together with manufacturing and industrial engineers. At the earliest concept stages, in some companies, marketing will also be involved in the identification of the concept layout of the new product. However, service engineers presently play a minor and somewhat negative role. At one large company, their participation in early design was described recently as pouring water on the campfire while the rest of the team were trying to keep the flames going. The main reason for this attitude is that service engineers have not had the tools needed to be pro-active in the design process. Typically, they have to wait until the design solidifies and then use historical company data to identify areas of service difficulty and to suggest possible changes.

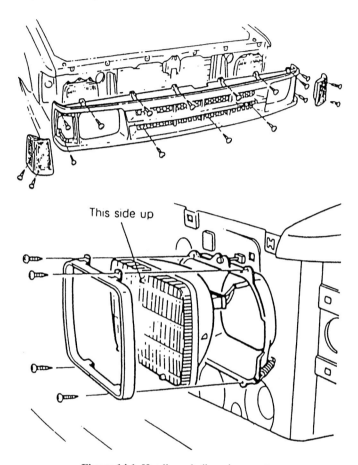

This side up

Figure 14.1 Headlamp bulb replacement.

Table 14.1 GM-10 Chevrolet headlamps and panel assembly: Impact summary

	Current ('90 Design)	DFM Proposal	
Parts	56	8	86% fewer
Operations	28	4	86% fewer
Assembly time	8.6	2.5	71% fewer
DFM savings per year: $3.7 Million			

Table 14.2 DFS Time-standards database charts

Item Insertion Tables	Table code
Item Insertion Times for Unsecured Items when Item Does Not Need Supporting During Insertion	00
Item Insertion Times for Unsecured Item when Heavy Item Requires Support During Insertion but Part is Easy to Hold, Handle or Control	01
Item Insertion Times for Unsecured Items when Heavy Item Requires Support and Part is Difficult to Hold, Handleor Control Due to Size	02
Item Insertion Times for Screws or Nuts Using a Power Tool	03
Item Insertion Times for Screws or Nuts Using a Screwdriver, Nut Driver, Rachet Wrench or Manual Fastening	04
Item Insertion Times for Screws or Nuts or Screw Fastened Items Using an Open-end Wrench or Box-end Wrench	05
Item InsertionTimes for Snap Fit Items	06
Item InsertionTimes for Push Fit Items	07
Item InsertionTimes for Interference or Press-fit Items	08
Item InsertionTimes for Rivets or Items Riveted on Insertion	09
Item InsertionTimes for Self-Stick Items	10
Item InsertionTimes for Items Secured Immediately by Bending or Bend Tab	11
Item InsertionTimes for Items Secured Immediately by Twisting or Twist Tab	12
Item InsertionTimes for Items Secured Immediately by Crimping	13
Item InsertionTimes for Items Secured Immediately by Staking	14
Item Removal Tables	
Item Removal Times for Unsecured Items	15
Item Removal Times for Screws or Nuts Using a Power Tool	16
Item Removal Times for Screws or Nuts using a Screwdriver, Nut Driver, Ratchet Wrench or Manual Unfastening	17
Item Removal Times for Screw Fastened Items Using an Open-end or Box-end Wrench	18
Item Removal Times for Snap Fit Items	19
Item Removal Times for Push Fit Items	20
Item Removal Times for Press or Interference Fit Items	21
Item Removal Times for Rivets	22
Item Removal Times for Self-stick Items	23
Item Removal Times for Items Unsecured during Removal by Bending	24
Item Removal Times for Items Unsecured during Removal by Twisting	25

Table 14.2 DFS Time-standards database charts *(Continued)*

Securing Operation Charts	
Securing Operation Times for Screw Fastening Using a Power Tool	26
Securing Operation Times for Screw Fastening Using a Screwdriver, Nutdriver, Rachet Wrench, or Manual Fastening	27
Securing Operation Times for Screw Fastening Using an Open-end or Box-end Wrench	28
Securing Operation Times for Snap Fits	29
Securing Operation Times for Push Fits	30
Securing Operation Times for Interference or Press Fits	31
Securing Operation Times for Riveting	32
Securing Operation Times for Self Stick	33
Securing Operation Times for Bending of Bend Tabs	34
Securing Operation Times for Crimping	35
Securing Operation Times for Twisting	36
Securing Operation Times for Staking	37
Securing Operation Times for Spot Re-welding	38
Securing Operation Times for Soldering	39
Unsecuring Operation Tables	
Unsecuring Operation Times for Screw Fastening Using a Power Tool	40
Unsecuring Operation Times for Screw Unfastening Using a Screwdriver, Nutdriver, Ratchet Wrench or Manual Unfastened	41
Unsecuring Operation Times for Screw Unfastening Using an Open-end or Box-end Wrench	42
Unsecuring Operation Times for Snap Fit Unfastening	43
Unsecuring Operation Times for Push Fit Unfastening	44
Unsecuring Operation Times for Interference or Press Fit Unfastening	45
Unsecuring Operation Times for Rivet Removal	46
Unsecuring Operation Times for Self Stick Unfastening	47
Unsecuring Operation Times for Bend Tab Unfasten	48
Unsecuring Operation Times for Crimp Unfasten	49
Unsecuring Operation Times for Twist Tab Unfasten	50
Unsecuring Operation Times for Stake Unfasten	51
Unsecuring Operation Times for Spot Weld Unfasten	52
Unsecuring Operation Times for Solder Unfasten	53
Unsecuring Operation Times for Pry Bar Unfasten	54
Miscellaneous Tables	
Item Set Aside Times	55
Item Acquisition Times	56
Tool Acquisition Times	57
User Defined Times for Frequently Performed Opretions	58

14.2 THE DESIGN FOR SERVICE EVALUATION PROCEDURE

In this section the URI Design for Service procedure is described for use in estimating the cost of servicing an item that has either stopped functioning correctly or is being replaced as part of routine maintenance. The time standard databases utilized in this procedure are a result of continuing research at the University of Rhode Island (Abbatiello, 1995).

The times represented in the databases were obtained from timed videotapes of work at service centers, and from time standard systems such as MOST (Zandin, 1990). Product case studies were used for validation of the times given in the data charts. The times are represented in a series of charts. A separate chart exists for common occurrences of item removal operations, item insertion operations, separate detaching or unsecuring operations, separate attaching or securing operations, and for part and tool acquisition and set-aside. Tool and item acquisition and set aside charts are included in a miscellaneous category so that a total of five categories of charts exist as listed in Table 14.2.

The structure of each of the database charts is arranged so that the time for the 'ideal' operation conditions are placed in the upper left hand corner with worsening conditions occurring along a diagonal line towards the lower right corner.

Table 14.3 Item removal times for screws or nuts using a screwdriver, nut driver, ratchet wrench or manual fastening

Chart Code 17		Easy to remove	Not easy to remove	Severe removal difficulties	Added time/rev
		0	1	2	3
No access or vision difficulties	0	8	14.3	28.2	1.2
Obstructed access or restricted vision	1	11.3	20.2	39.8	1.9
Obstructed access and restricted vision	2	14.8	26.5	52.1	2.6
Severe access obstructions	3	19.9	35.7	70.3	3.6

Note: Add extra time penalty from column 3 for total number of revolutions N > 5
Total time = Time from table + ((N - 5) × added time per revolution).

The database charts are each assigned a two-digit code number starting with "00." In addition the row on each chart is assigned a third digit and the column is assigned a fourth. Thus, for example, code number 1721 refers to row 2, column 1 on chart 17, and is for the removal of a screw with a manual screwdriver where there is obstructed access and restricted vision and the screw is difficult to remove (because of slight corrosion for example); see example chart in Table 14.3.

The following example of replacing the printed circuit board in the Pressure Recorder assembly, illustrated in Figure 14.2, is used to demonstrate the worksheet DFS method. The example is taken from the DFA handbook (Boothroyd and Dewhurst, 1990). The method simulates the disassembly and reassembly processes by considering the individual steps of tool acquisition, part removal, part set-aside, and later part acquisition (pick-up and orient) and part reinsertion. Worksheets were developed to allow for efficient organization of the data obtained from utilizing the serviceability time databases.

The first part of the DFS procedure is to complete a Disassembly Worksheet in Figure 14.3. Take the product assembly apart or imagine taking it apart, to access the service location, recording each operation as it is performed. Complete a row on the worksheet for each part or operation as you disassemble items from the main assembly. If the assembly contains sub-assemblies which must be removed to carry out the service task then treat them as 'parts.' If a subassembly must be disassembled for the service work then simply continue to enter lines on the disassembly worksheet for removing lower-level parts or sub-assemblies.

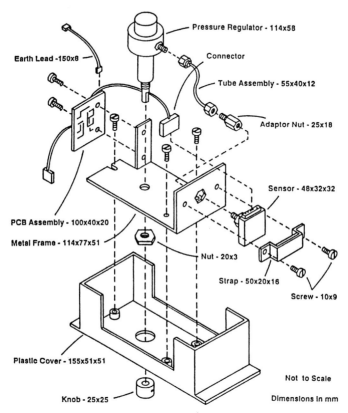

Figure 14.2 Exploded view of the pressure recorder assembly.

References to the relevant database charts are given in columns 3,5, and 7. For example, Chart number 57, referenced in Column 3, contains the data for tool acquisition. Row 0, column 0 of this chart contains the time for acquisition of a tool which is within easy reach. The time of 4.2 seconds given in the column 4 is an average value taken from hours of videotaped service work and includes a proportion of time for replacement of tools at the end of the service task. The estimated time for removal of the PCB is 104.3 seconds given by the sum of Column 9. If desired the estimated times in Column 9 can be converted to service labor cost by multiplying by the service technician labor rate. The division by 36 shown in the calculation at the top of Column 10 converts dollars to cents and hours to seconds.

Assembly Name: Pressure Recorder

Labor rate $/hr, L = 30.00

1	2	3	4	5	6	7	8	9	10	11	
ID Number	number of times operation is repeated	four digit tool acquisition code	tool acquisition time (sec)	four digit item removal or operation code	item removal or operation time (sec)	four digit item set-aside code	item set-aside time (sec)	operation time, sec $[(8) + (6)] \times (2) + (4)]$	operation cost, cents $[(9) \times L/36]$	number of service items, cover parts or functional connections	Service Task Performed
1	3	5700	4.2	1710	11.3	5500	1.4	42.3	35.3	0	Remove Screws
2	1	—	—	5800	4.5	—	—	4.5	3.8	0	Reorientation
3	1	5700	4.2	4100	8.0	—	—	12.2	10.2	0	Loosen Set Screw
4	1	—	—	1500	2.4	5500	1.4	3.8	3.2	0	Remove Knob
5	1	—	—	1500	2.4	5500	1.4	3.8	3.2	0	Remove Cover
6	1	—	—	5800	4.5	—	—	4.5	3.8	0	Reorientation
7	1	—	—	4401	6.4	—	—	6.4	5.3	0	Unplug Sensor
8	2	5700	4.2	1700	8.0	5500	1.4	23	19.2	0	Remove Screws
9	1	—	—	1500	2.4	5500	1.4	3.8	3.2	1	Remove PCB
								104.3	86.9	1	Efficiency Calculation:
								T_d	C_d	N_m	

$T_s = T_d + T_r$

For Service: Efficiency, $\eta_B = (9 \times N_m)/T_s \times 100\%$

Figure 14.3 Disassembly worksheet.

Assembly Name: Pressure Recorder Labor Rate $/hr, L = 30.00

1 ID Number	2 number of times operation is repeated	3 four digit tool acquisition code	4 tool acquisition time (sec)	5 four digit item acquisition code	6 item acquisition time (sec)	7 four digit item insertion or operation code	8 item insertion or operation time (sec)	9 operation time, sec $[(4)+(7)]\times[(6)+(8)]$	10 operation cost, cents $[(9)\times L]/36$	Service Task Performed
1	1	—	—	5601	3.4	0001	4.9	8.3	7.0	Add PCB
2	2	5700	4.2	5600	1.4	0401	13.8	34.6	28.8	Screw Fasten
3	1	—	—	—	—	3000	4.4	4.4	3.7	Plug in Sensor
4	1	—	—	—	—	5800	4.5	4.5	3.8	Reorientation
5	1	—	—	5600	1.4	0001	4.9	6.3	5.3	Plastic Cover
6	1	—	—	5600	1.4	0001	4.9	6.3	5.3	Knob
7	1	5700	4.2	—	—	2700	8.0	12.2	10.2	Fasten Set Screw
8	1	—	—	—	—	5800	4.5	4.5	3.8	Reorientation
9	3	5700	4.2	5600	1.4	0401	13.8	49.8	41.5	Screw on Cover
								130.9	109.4	Efficiency Calculation:
								T_r	C_r	

$T_s = T_d + T_r$ For Service: Efficiency, $\eta_s = (9 \times N_m) / T_s \times 100\%$

Figure 14.4 Reassembly worksheet.

The serviceability efficiency of the design is determined by considering each disassembly operation and item removal and judging whether they are necessary according to three categories given below:

- The part/subassembly removed is or contains the service item(s) or is the service operation itself.
- The part/subassembly removed is a cover part which must fully enclose the service item or protect the end-user from the service item.
- The part/subassembly must be removed to isolate the service item or the subassembly containing the service item.

Only the components of an assembly that fall into one of these categories are considered to be justified for removal or unfastening in the service task. A functional cover part is defined as a part that must fully enclose the service item from the surrounding environment to keep out dust or keep in fluids and so on. The plastic cover in the Pressure Recorder example does not enclose the PCB board, thus it can not be justified as a functional cover. Examples of items which must be removed to isolate the service item may be the blades on a fan motor, the valve on a pressure vessel and so on. In general these are items which can not be mounted or attached elsewhere and which must be removed for replacement of the master item. In essence, the removal of an item to isolate the service item must be done because the item in question must be connected to the service item for functional purposes.

If a part or operation does not fall into any of these categories then it is not considered to be a theoretically necessary part of the service procedure and a value of "0" is entered onto the worksheet. When items or operations fall into one of the categories then a value less than or equal to the total number of items or operations performed in that step is entered into the worksheet. In the Pressure Recorder example, only the removal of the printed circuit board itself is justified. The sum of the numbers in Column 11 gives an estimate of the theoretical minimum number of item removals and operations, N_m which are justified as necessary for performance of the service task.

The next step in the process is to complete the corresponding Reassembly Worksheet. The worksheet is almost identical in format to that of the Disassembly Worksheet and completion of it simply requires reference to the appropriate database charts for item insertion and securing operations. A completed Reassembly Worksheet for the pressure recorder example is given in Figure 14.4.

When both worksheets have been completed the service and/or repair efficiency of the design can be calculated for replacing the printed circuit board. The total service time, T_s, is first obtained by adding the disassembly time, T_d, with the reassembly time, T_r.

$$T_s = T_d + T_r \tag{1}$$

Thus for the present example,

$$T_s = 104.3 + 130.9 = 235.2 \text{ seconds}$$

The ideal service time for a particular task is based on the minimum amount of time required to perform item removal, item set-aside, item acquisition, and item insertion operations. The following assumptions were made to determine the ideal time for a service operation:

1. All parts necessary for the service task are placed within easy reach of the service area and no tools are required for the 'ideal' service task.
2. In the DFA methodology (Boothroyd and Dewhurst, 1990), for the purpose of estimating the minimum assembly time, it is assumed that in an ideal design approximately one third of the parts are secured immediately on insertion by an efficient self-securing method. Similarly, for ideal service conditions it will be assumed that one in every three parts will need to be unsecured and later resecured by efficient methods. For the current work, snap fit fastening will be assumed as the ideal securing method. It will also be assumed that these snap fits have snap-release features that allow for easy disassembly in which the snaps are released simultaneously.

Using these assumptions, the ideal service time for parts that do not need additional assistance can be given by:

$$t_{min} = \frac{2 \times T_{Rem.Un} + T_{Rem.Snap}}{3} + \frac{2 \times T_{Ins.Un} + T_{Ins.Snap}}{3} + T_{Acq} + T_{Setaside} \qquad (2)$$

where

$T_{Rem.Un}$	= Item removal time for unsecured items
$T_{Rem.Snap}$	= Item removal time for snap fit items
$T_{Ins.Un}$	= Item insertion time for unsecured items
$T_{Ins.Snap}$	= Item insertion time for snap fit items
T_{Acq}	= Item setaside time
$T_{Setaside}$	= Item acquisition time

Substituting the values from the databases developed at URI (Whyland, 1993; Subramani and Dewhurst, 1994; Abbatiello, 1995), Eq. (2) becomes:

$$t_{min} = \frac{2 \times 2.4 + 3.6}{3} + \frac{2 \times 3.8 + 2.2}{3} + 1.4 + 1.4 = 8.87 \approx 9.0 \qquad (3)$$

Thus, the time based efficiency measure can be given as:

$$\eta_{time} = \frac{t_{min} \times N_m}{T_s} \times 100\% \qquad (4)$$

or

$$\eta_{time} = \frac{9 \times N_m}{T_s} \times 100\% \qquad (5)$$

where

N_m = The theoretical minimum number of part removals or operations that can be justified for performance of the service task
T_s = Estimated time to perform service task (seconds)

Hence, for the PCB replacement in the Pressure Recorder, $N_m = 1$, and substitution in Eq. (5) gives

$$\eta_{time} = \frac{9 \times 1}{235.2} \times 100\% = 3.8\%$$

In order to calculate the time-based service efficiency for a system comprised of several service procedures, it is proposed that the separate index values should be weighted according to the expected failure frequencies This gives an expression for system service efficiency as:

$$\eta_{total} = \frac{\eta_1 \times f_1 + \eta_2 \times f_2 + \cdots + \eta_n \times f_n}{f_1 + f_2 + \cdots + f_n} \tag{6}$$

where

$\eta_1, \eta_2, \cdots, \eta_n$ = Time based efficiency values for tasks 1, 2, ..., n respectively
$f_1, f_2, \cdots, f_n$ = The failure frequency for items 1, 2, ..., n respectively

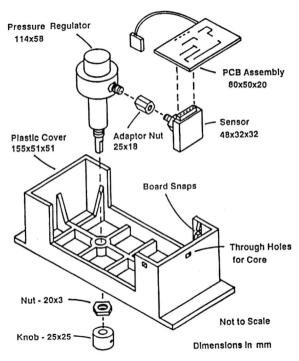

Figure 14.5 Exploded view of the redesigned pressure recorder assembly.

If the service efficiency is considered to be inadequate, then redesigns of the assembly should be considered, focusing on the goals of simplifying the assembly structure, using securing methods which are efficient to disassemble and allowing convenient access for items which are to be serviced. The simplification of the assembly structure is also the goal of DFA (Boothroyd and Dewhurst, 1990), so that DFS considerations parallel to those of DFA but with checks for item removal difficulties. For example, Figure 14.5 shows a DFA redesign of the Pressure Recorder. Considering the PCB to be the primary service item the structure has been arranged so that the board is on the outermost layer. Using the same database values, the estimated time for removal and reinsertion of the PCB is estimated to be 16.5 seconds. The efficiency measure from Eq. (5) is now

$$\eta_{time} = \frac{9 \times 1}{16.5} \times 100\% = 54.5\%$$

14.3 DESIGN FOR SERVICE OPTIMIZATION

The task of optimizing a product for ease of service is fundamentally different from that of optimizing a design for ease of initial assembly. For initial assembly the goal in design is the simply stated one of minimizing assembly time or cost. For service, on the other hand, there are inevitable conflicts between the different service tasks which must be carried out on a particular product. The perfect design for service would have all of the items to be replaced in service, and all of the service operations to be performed, immediately accessible on the outer surface of the product. Clearly, this is not generally possible and so decisions have to be made about which items are to be most easily accessible. Even when these decisions have been made, the ease with which the service tasks should be performed is still a matter for deliberation. At one extreme the product user should be alerted that a problem has occurred and corrective action should be possible without documentation or tools and with the absolute minimum of disassembly operations. At the other extreme, the efficiency of initial manufacture should not be compromised in any way to allow easier disassembly for service. For example, spot welding may be a preferred fastening technique even though spot welds may have to be drilled out occasionally to allow service tasks to be performed.

In this section, the authors describe a procedure for establishing serviceability goals for a new product. The procedure is based on knowledge of the possible failure modes of the product, the likely frequencies of these failures occurring, and the likely consequences of such failures. For these reasons, the establishment of service goals should be preceded by a complete evaluation of the possibilities for failure of a product using a systematic approach such as Failures Modes and Effects Analysis (FMEA) (Sundarajan, 1991). FMEA will be described briefly in the following section.

14.3.1 Failure Modes and Effects Analysis

Failure modes and effects analysis (FMEA) is a procedure that identifies potential component failures and assesses their effect on the system. If the criticality of the effect is also considered in the analysis, the analysis is then referred to as the failure modes, effects, and criticality analysis (FMECA). A FMEA or FMECA analysis is used to detect potential weak spots in the system design and improve them through design changes focused on increasing the reliability of the system (Sundarajan, 1991; Moss, 1985; Priest, 1988).

Questions considered during FMEA analysis vary depending on the system being analyzed as well as the scope and purpose of the analysis. However the following five questions are typically considered for every part of the system (Sundarajan, 1991).

1. How can the component fail? (There could be more than one mode of failure.)
2. How often will the component fail?
3. What will be the effects of the failure?
4. How critical are the consequences of the failure?
5. How will the failure be detected?

In completing an FMEA, all failure modes are identified, their detection documented, frequency of failure recorded, and their effects on the system as well as the potential criticality of failure are considered.

Failure modes and effects analysis attempts to do all of the following (Sundarajan, 1991):

1. Ensure that all conceivable failure modes and their effects are understood.
2. Aid in the identification of design weaknesses.
3. Provide a basis for comparing design alternatives during the early design stages.
4. Provide a basis for recommending design improvements.
5. Provide a basis for corrective action priorities.
6. Provide a basis for implementing test programs.
7. Aid in trouble shooting existing systems.

A well prepared FMEA will benefit the design team by identifying any weak spots in the system design allowing the team to improve the system reliability by focusing design efforts in these areas.

Table 14.4 Service task importance analysis worksheet

1	2	3	4	5	6	7	8	9
Failure Source	FF (per/yr)	Freq. Rank	Function of Failure Source	Potential Failure Mode	Effect of Failure	Con. Rank	Potential Cause of Failure	Import. Rank

FF - Failure frequency   = failures per year
Importance Rank = Frequency rank x Consequence rank

14.3.2 Service Task Importance Analysis

Obviously the service task efficiency of every part can not be considered. Therefore, a procedure is needed to determine where the resources available to increase serviceability of a product should best be applied. Important factors that determine the level of serviceability which should be designed into a product are the frequency with which different failures are likely to occur and the consequence(s) of the failures occurring. The more frequently that a failure occurs, the simpler the service procedure should be, and likewise the greater the consequences of a failure, the simpler should be the procedures for preventive maintenance.

The service task importance analysis proposed here is similar in format to a failure modes and effects analysis. It is intended to complement FMEA, since its focus is to increase the efficiency of the necessary service tasks after all possible reliability improvements have been made. Also in the course of an FMEA analysis, many of the inputs for the service task importance analysis will already have been determined. In order to facilitate the analysis a worksheet shown in Table 14.4 was developed (Abbatiello, 1995). It is intended that this analysis will be performed by starting with the main assembly, then proceeding through the product structure, analyzing any subassembly that is considered to be repairable.

Before attempting to complete the worksheet it is necessary to construct an assembly structure chart. It is also recommended that all information pertaining to product failures be gathered in advance using FMEA or similar procedures. Each line on the worksheet represents one potential failure source. The required column entries are as follows.

Column 1   Failure Source
Enter the name of the part or subassembly being considered in the analysis. If a subassembly is being considered, it may be necessary to conduct a separate analysis for the components of the subassembly if it is considered repairable.

Column 2   Failure Frequency (FF)
Enter the failure frequency for the failure source identified in Column 1. The failure frequency is the number of system failures that can be attributed to the failure source. For example if the system has a failure rate of 10% per year and the failure source causes 33% of those failures per year, then the failure frequency is equal to 3.3% or 0.033 system failures per year.

Table 14.5 Failure rate ranking

Rank, Fr	Likely Failure Rate
10	Failure occurs one or a few times per week
9	Failure occurs one or a few times per month
8	Failure occurs a few times per year
7	Failure frequency 0.5 to 0.2 per year
6	Failure frequency 0.2 to 0.1 per year
5	Failure frequency 0.1 to 0.05 per year
4	Failure frequency 0.05 to 0.01 per year
3	Failure frequency 0.01 to 0.005 per year
2	Failure frequency 0.005 to 0.002 per year
1	Failure very unlikely - FF < 0.002 per year

Column 3   Frequency Rank, F_r
The number for the frequency rank is placed into this column. The frequency rank is a number used to compare the likelihood of different component failures. A suggested failure ranking scheme is given in Table 14.5.

Column 4   Function of Failure Source
Enter the function of the item under analysis. If the item has more than one function with different potential modes of failure, list all the functions of the item separately.

Column 5   Potential Failure Mode
Potential failure mode is defined as the manner in which a component, subsystem or system could potentially fail to meet the design requirements. The potential failure mode may also be the cause of a potential failure mode in a higher level assembly, or be the effect of one in a lower level component. If a

component or subassembly has more than one potential failure mode list each one separately.

Column 6   Effect of Failure

Effect of failure is defined as the effects of the failure mode on the function of the system, as perceived by the customer. The effects of the failure should be described in terms of what the customer might notice or experience. The effects should always be stated in specific terms relative to the system, subsystem or component being analyzed.

Column 7   Consequence Rank, C_r

Consequence rank is an assessment of the seriousness of the effect of the potential failure mode to the next component, subassembly, system or to the customer if it occurs. A suggested consequence ranking scheme is given in Table 14.6. It is important to notice that the suggested ranking scheme does not just focus on the possible effects on the system or involved people, but also includes the economic consequences of failure to the end-user. If the end-user's business depends on the system for economic survival, the failure of the system becomes significantly more important than if the failure has little effect on business even though the system may still be inoperable.

Table 14.6 Failure consequence ranking

Rank, Cr	Criteria for Consequence Ranking
10	Catastrophic failure with no warning & a high probability of personal risk.
9	Total loss of operating capability causing substantial economic damage or posing personal risk.
8	Total loss of operating capability causing major disruption to important activity or causing major damage to other items.
7	Total loss of operating capability causing minor disruption to important activity or causing minor damage to other items.
6	Total loss of operating capability but causing only minor inconvenience.
5	Performance severely effected by failure.
4	Significant loss of performance.
3	Minor effect on performance.
2	Slight effect on performance.
1	No effect.

Column 8   Potential Cause of Failure

Potential cause of failure is defined as an indication of a design weakness, the consequence of which is the failure mode. List every conceivable cause of failure for each failure mode. Although knowing the cause of the failure has little impact on the serviceability of the system, it is important that the service engineer realizes the potential cause of the failure. Some typical causes may include, but are not limited to:

- Inadequate design life assumption
- Over-stressing
- Wear
- Poor quality in manufacturing

Table 14.7 Quality of serviceability design requirements

Criteria	Service Quality Category
$I_r > 70$	Design for obvious diagnosis and easiest possible service procedure. Service task efficiency very important due to high frequency and high consequence of failure. Necessitates the use of fasteners designed for rapid disassembly, easy access to service locations, minimization of the number of unrelated items to be removed, and so on. Preventive maintenance schedules necessary.
$50 < I_r \leq 70$	Rank necessitates efficient service procedure and some diagnostic capability. No permanent fastening methods allowable, threaded fasteners should be avoided and easy access to service location should be enabled. Preventive maintenance schedules recommended.
$30 < I_r \leq 50$	Changes in design to improve upon serviceability efficiency are recommended. Use of threaded fasteners allowable but permanent fastening methods should not be considered an option.
$15 < I_r \leq 30$	Serviceability should be considered but major changes to improve design for service improvements are not justifiable. Investigate improvements that will increase manufacturability and serviceability.
$I_r \leq 15$	Do not compromise manufacturing efficiency for increased serviceability. Investigate improvements that will increase manufacturability and serviceability. Permanent fastening methods may be acceptable.

Column 9   Importance Ranking Number

The importance rank is a number that signifies the importance of designing for quick and easy service of the particular failure source under consideration. Importance rank, I_r, is calculated by multiplying the frequency rank and the consequence rank numbers; i.e.

$$I_r = F_r \times C_r \tag{7}$$

which yields a number between 1 and 100. The higher the importance rank, the easier to perform should be the associated service task. In general, regardless of the resultant importance rank, special attention should be given to items when the frequency of failure and/or the consequence of failure are high. Suggested service quality evaluation criteria are given in Table 14.7.

Currently research is being conducted to determine if a relationship exists between the importance and service efficiency index of a given procedure (Abbatiello, 1995). Initial investigation of case studies suggests that the higher the importance rank, the higher the corresponding service efficiency should be.

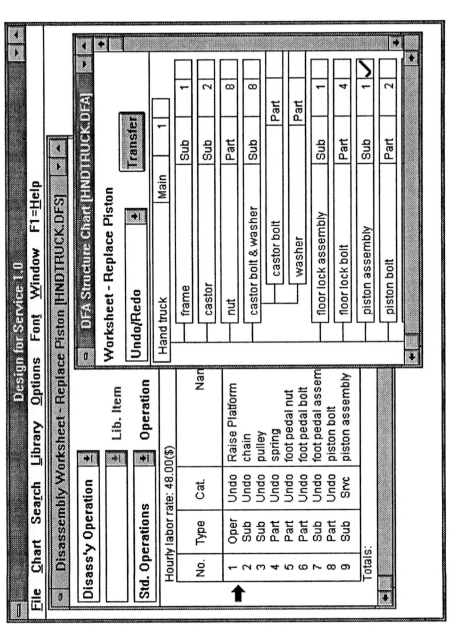

Figure 14.6 Sample screen of Design for Service program.

14.4 DFS SOFTWARE

Recently the procedure described in Section 14.2, has been released as a Windows DFS program by Boothroyd Dewhurst, Inc. (1994). The program can be used as a stand-alone analysis tool or in conjunction with DFA. Figure 14.6 shows the DFS program windows which form the link between DFS and DFA. The window in the foreground contains the product structure chart which resulted from a DFA analysis. The background window is a Disassembly Worksheet where the operations required to access the service location and to carry out the service task are listed. This list is built up effortlessly by simply clicking on items or operations on the structure chart in the order which they must be removed or undone. A drop-down list above the structure chart allows the user to choose the category of the item or operation. Most will be assigned the category "undo/redo", which means remove and set aside for later reassembly, or unsecure and later resecure. Other categories are "discard" for items such as seals or fluids which must be renewed simply because they have been disassembled or drained, and "service" for items or operations that represent the goal of the service task. For example, for the headlamp replacement shown in Figure 14.1, 31 of the 32 items are "undo/redo" and only the bulb has the category "service."

The program allows all of the service tasks for a given product to be included in a service list in the same file. The disassembly worksheets are automatically reversed to complete the service tasks. The time-standard DFS databases described in Section 14.2, are used by the program to estimate service times. In addition, a user library system in the program allows users to build up lists of their special disassembly, reassembly or miscellaneous service operations complete with equations based on their own variables. The program also allows the input of labor rates, service occurrences or failure rates, costs of service items, and special tools, to build up a complete representation of the service costs. Types of tools and difficulties of operations are transferred from DFA and can be altered or new difficulties can be entered. Final reports include service costs per occurrence, life-cycle service costs that may represent warranty costs if the warranty period is used for the calculations, and service index efficiency measures for each service task and for the entire product.

14.5 SUMMARY

From case studies of DFA, DFS and more recently from considerations of recycling, it seems that the common thread which runs through these separate requirements is simplicity of product structure. A simplified structure includes fewer items, fewer interfaces and fewer fastening items or operations. This offers increased potential for ease of disassembly provided that important disassembly sequences are assessed at the early stage of design when assembly and manufacture are being considered. This has been the principle motive for the development of a DFS analysis method which can be used at the same time as DFA studies are being carried out.

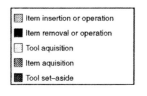

Item insertion or operation
Item removal or operation
Tool aquisition
Item aquisition
Tool set–aside

Figure 14.7 Service time breakdown.

It has been suggested that designing for initial assembly efficiency will negatively impact on the efficiency of serviceability. However, after completing numerous serviceability analyses on products ranging from a coffee maker to a refrigerator, it has been found that usually about 50% of the total service time is spent reassembling the product. Figure 14.7 shows the total time breakdown of thirty five service tasks. The mean time for reassembly in these tasks was 54% with a 95% confidence interval of 51.95 to 56. Clearly then, by considering efficient assembly in early design, improvements in the reassembly stage of service will also be made. This conclusion is, of course, independent of any positive effect on service tasks produced by reducing the number of items in the assembly through DFA analyses.

REFERENCES

Abbatiello, N. (1995) Development of Design for Service Strategy, *M.S. Thesis*, University of Rhode Island.

Boothroyd, G., Dewhurst, P. (1990) *Product Design for Assembly Handbook*, Boothroyd Dewhurst Inc, Wakefield RI.

Dewhurst, P. (1993) Efficient Design for Service Considerations, *Manufacturing Review*, **6** (1).

DFMA Case Studies Report, Boothroyd Dewhurst, Inc., Wakefield, RI, 1994

Moss, M.A. (1985) *Designing For Minimal Maintenance Expense: The Practical Application Of Reliability And Maintainability*, Marcel Dekker Inc., New York.

Priest, J.W. (1988) *Engineering Design For Producibility And Reliability*, Marcel Dekker Inc., New York.

Subramani, A., Dewhurst, P. (1994) Repair Time Estimation for Early Stages of Product Development, *Journal of Design and Manufacturing*, **4** (2).

Sundararajan, C. (1991) *Guide To Reliability Engineering: Data, Analysis, Applications, Implementation, and Management*, Van Nostrand Reinhold, New York.

Whyland, C. (1993) Development of a Design for Service Database, *M.S. Thesis*, University of Rhode Island.

Zandin, K. (1990) *MOST*$^{®}$ *Work Measurement Systems*, Second Edition, Marcel Dekker, Inc., New York.

15

EASE-OF-DISASSEMBLY EVALUATION
IN DESIGN FOR RECYCLING

Thomas A. Hanft; Ehud Kroll

This chapter presents a procedure for evaluating ease-of-disassembly for product recycling. The methodology consists of a spreadsheet-like chart and rating scheme for quantifying disassembly difficulty. Difficulty scores derived from work measurement analysis of standard disassembly tasks provide a means for identifying weaknesses in the design and comparing alternatives. To maximize feedback to the designer, the method captures the sources of difficulty in performing each task. The disassembly evaluation chart is explained and its application is demonstrated in the analysis of a computer keyboard. Derivation of task difficulty scores is described. The current method focuses on manual disassembly of business equipment. However, the same methodology may be applied to robotic disassembly processes and other products.

15.1 INTRODUCTION

Current trends in environmental protection legislation indicate that manufacturers will soon be responsible for recovering products at the end of their useful life. In Germany, the Electronic Waste Ordinance, which mandates that electronics producers "take back" and recycle used products, will become law in 1995 (Dillon, 1994). Diminishing natural resources, limited landfill space, and problems with hazardous waste disposal have increased the environmental awareness of consumers. Consequently, manufacturers are under pressure to create products that are easy to dismantle and recycle, while maintaining product quality and performance.

Design for disassembly (DFD) involves developing products that are easy to take apart and thus facilitate recycling and removal of hazardous materials. Research activity related to DFD has increased dramatically in recent years. In anticipation of "take back" legislation, manufacturers in Europe have researched ways to make products easier to disassemble since the mid-1980's (Wilder, 1990). Early investigations in this area, primarily by BMW, were limited to pilot projects but yielded general guidelines about design for recyclability (Constance, 1992). Fundamental DFD concepts, such as consolidating parts and using snap-fit joints, were also demonstrated (Bakerjian, 1992).

Products which use plastics extensively are prime candidates for DFD improvements. While the amount of plastic used in products is increasing, less than 1% of all plastics produced in the US is currently being recycled (Burke *et al.*, 1992). Furthermore, plastics suppliers, such as GE Plastics, are eager to collaborate with manufacturers on recyclability projects (Seegers, 1993). Thus, products such as automobiles, business equipment, and appliances have been the focus of corporate research. For example, researchers at IBM have published detailed discussions sharing their experience in designing computers for ease of disassembly (Kirby and Wadehra, 1993). Similarly, the evaluation method presented below targets business equipment, such as computers, printers, monitors, and keyboards.

Recently, attempts have been made to integrate basic recyclability concepts and life cycle considerations in comprehensive "design for the environment" (DFE) procedures (Navin-Chandra, 1991; Thurston and Blair, 1993). DFE is a broad approach to product development which considers the environmental impacts of a product throughout its entire life cycle (Fiskel and Wapman, 1994). In practice, DFE forces engineers to evaluate a product's fabrication, use, and disposal with respect to the environment. Since designers specify the manufacturing processes, materials, and structure of products, it is their responsibility to make choices that are ecologically sound (Eekels, 1993). However, these decisions can be daunting during the early stages of design when numerous "concurrent engineering" factors (e.g., serviceability and reliability) must be considered.

In support of designers facing new environmental obligations, some manufacturers, such as Motorola and Hewlett Packard, have developed educational programs to introduce engineers to DFE fundamentals (Eagan *et al.*, 1994; Bast, 1994). Recently, a German standard on design for easy recycling (VDI 2243) was published (Beitz, 1993). A table for selecting recycling-oriented fasteners was included in the manual. However, only qualitative ratings of fasteners in categories such as recyclability and detaching behavior are provided (VerGow and Bras, 1994). Overall, further development of evaluation tools which aid designers with complex DFE decisions is needed.

Design for disassembly is a key component of any DFE framework. Likewise, evaluation schemes which identify design weaknesses and allow alternative designs to be compared with respect to DFD are of great importance. A procedure for assessing the ease of disassembly of products is presented below. The method was developed following manual disassembly experiments on small electrical appliances and computer equipment (e.g., monitors and keyboards). It consists of a disassembly evaluation chart and corresponding catalog of task difficulty scores. The scores represent the difficulty encountered in performing the required disassembly tasks. The evaluation procedure entails manually disassembling a product or simulating the design's disassembly process, choosing difficulty scores for the tasks involved, and recording the data in the chart. Design weaknesses may then be identified through interpretation of the evaluation results. The disassembly evaluation chart and derivation of task difficulty ratings are discussed following a brief review of related literature. The procedure is then demonstrated and discussed in the evaluation of a computer keyboard.

15.2 LITERATURE REVIEW

Several approaches to product evaluation in design for disassembly and recycling have been described in the literature. These evaluation procedures vary widely in terms of the type of data measured, method of analysis, and form of information provided to the designer. However, the approaches can be categorized according to the extent of their analysis. In general, the strategies range in scope from assessment of the entire product life cycle to evaluation of a single aspect of its recyclability. While each approach has advantages and disadvantages, perhaps the best approach is a combination of all the methods.

Numerous life cycle assessment (LCA) procedures have been the focus of extensive research in recent years (Vigon and Curran, 1993; Tummala and Koenig, 1994). LCA is the systematic, comprehensive evaluation of the energy use, raw materials consumption, and waste emissions associated with a product during its entire life cycle. However, according to Veldstra and Bouws (1993), the broadness of scope and depth of evaluation of LCA are both its greatest strengths and major weaknesses. While the effects of design changes can be assessed in many areas over the life of a product, extensive data on every aspect of the product's manufacture, use, and disposal are required. As a result, widespread use of LCA as a design tool is unlikely until simpler procedures and accurate life cycle databases are developed.

Emblemsvag and Bras (1994) have proposed a similar approach which estimates the total cost of a product's life cycle. The procedure is based on activity-based costing (ABC). ABC differs from LCA in that it assesses the consumption of "activities" rather than energy or raw materials. An activity is defined as a group of actions with a logical connection. For example, all the operations required to dismantle a car are considered a single activity. Thus, a product's cost is actually the sum of the costs of all the processes performed on it during its life cycle. During analysis, matrices of alternative designs versus associated activity costs are created. These matrices may extend all the way to the design parameter level. However, the authors acknowledge that finding accurate data to create the matrices is tedious. The advantage of this method over LCA is that cost drivers can be defined in terms that are readily understood by engineers. For example, the disassembly costs for a car can be expressed in terms of dollars rather than energy expended, as in LCA.

A disadvantage of both the ABC and LCA approaches is dependence upon assumptions about future conditions, such as disposal costs and recycling techniques. Since these factors are likely to change during the product's lifetime, decisions may be based on unrealistic data. Zussman et al. (1994) have addressed this problem in their method for finding the optimum end-of-life scenario for a product. The uncertainty of future conditions is accounted for by including probabilistic distribution factors in their analysis. Probability density functions for evaluation criteria are estimated by analytical prediction methods or through the forecasts of experts. For example, future labor costs and the expected lifetimes of components are forecast. In this way, more realistic results are obtained and the effects of uncertain conditions are considered. However, this method requires extensive data and does not facilitate easy identification of design weaknesses.

Several other end-of-life approaches have been proposed. Simon (1993) has developed a method which uses a decision tree in combination with design indices; quantitative measures of design features which affect disassembly and recyclability. One index of particular relevance is disassembly cost. This metric is calculated from the time required to perform a standard disassembly task, such as unscrewing. Data for each task is determined through time and motion studies performed in a disassembly laboratory. Overall, the indices are effective in comparing designs and identifying areas for improvement (e.g., number and type of fasteners used). However, Simon notes that further development of the metrics is required.

Recently, a computer-aided design tool for recovery analysis of products was developed by Navin-Chandra (1993). Similar to Simon's approach, this disassembly planning and optimization program is based on a decision tree. The designer enters information on the product structure into a table. For example, the type of fasteners used and how each part is constrained. An algorithm then computes the most profitable disassembly scenario using a database which includes disposal costs, revenue from recycled or reused parts, and disassembly times and costs. This method is best suited for assessing the recovery process rather than identifying the effects of specific design changes.

Another computer-aided design approach has been developed by Burke et al. (1992). This method analyzes the recyclability of a product by breaking down associated costs into categories such as disassembly and material processing. Thus, the designer can identify which general areas of the design should be improved. However, the designer lacks detailed feedback on which design features are responsible for high costs. If a part incurs a high disassembly cost, the designer needs to know which aspect of its disassembly creates the added difficulty.

In terms of scope of assessment, numerous other evaluation procedures fall somewhere between LCA and end-of-life analyses. For example, design efficiency indices developed by Dewhurst (1993) evaluate ease-of-disassembly with respect to both service and recycling. These metrics are based on estimates of disassembly time and evolved from Dewhurst's previous work with Boothroyd (1987) in design for assembly. Although product repair and recycling both entail disassembly, the practices are currently very different. Products are carefully disassembled and reassembled during service, while recycling involves the rapid, sometimes destructive, separation of valuable or hazardous components. Dewhurst's indices aid engineers in the design of products that are likely to be maintained for longer lifetimes and easily dismantled when disposed of.

Each type of evaluation approach provides valuable information to designers. Life cycle approaches, such as LCA and ABC, are necessary for balancing tradeoffs in design. For example, specification of a recyclable plastic may increase the amount of material required and thus decrease the overall efficiency of a product. End-of-life and recovery analysis methods help determine the optimum disassembly scenario for a given design. Dewhurst's approach emphasizes the value of products that are easy to service and recycle. However, all the previously described approaches fail to provide detailed feedback on which particular aspect of a design is responsible for disassembly difficulty and why. The evaluation chart presented below was developed to aid designers in tracing, classifying, and avoiding specific sources of disassembly difficulty.

15.3 DISASSEMBLY EVALUATION CHART

Our method is centered around the *disassembly evaluation chart* illustrated in Figure 15.1. Each row on the chart corresponds to a separate disassembly task. Tasks are sequentially recorded and assessed during the disassembly process. Proposed designs are evaluated in the same way, by visualizing the disassembly process. Several rows may correspond to the disassembly of a single part if multiple operations are required to remove it. Each column contains data pertaining to different aspects of the disassembly evaluation. Entries are described below.

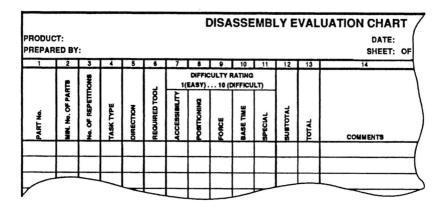

Figure 15.1 The structure of the disassembly evaluation chart.

- **Column 1: Part Number.** During disassembly, each part is assigned a number for identification or part numbers from the bill of materials may be used. Identical parts removed at the same time and under similar conditions may be assigned the same number. For example, three screws fastening the same part may be given a single part number. A subassembly is considered a single part with a separate disassembly sequence. A group of connected parts is not considered a subassembly if it is disassembled immediately after removal. The suffix "*sub*" is assigned to subassembly part numbers.
- **Column 2: Theoretical Minimum Number of Parts.** Each part is evaluated to determine whether it is theoretically required to exist as a separate component. The purpose of this assessment is to identify opportunities for eliminating or consolidating parts. A separate part is required if it satisfies any of the following three criteria stated in the design for assembly work by Boothroyd and Dewhurst (1987):

1. During operation of the product, does the part move relative to all other parts still assembled? Only gross motions that cannot be accommodated by elastic hinges, for example, are sufficient for a positive answer.
2. Must the part be made of a different material or isolated from all other parts still assembled? Only fundamental reasons concerned with material properties are accepted.
3. Must the part be separate from all other parts still assembled because otherwise assembly or disassembly of other parts would be impossible.

The total number of physical components theoretically required is entered in the chart. For example, if a part consists of three identical screws, only one may be required. In this case, a "1" would be entered in the chart. If no screws were judged necessary, the entry would be "0." The score for subassemblies depends on subsequent handling. If the subassembly is eventually disassembled after removal, it is theoretically not required as a separate component and automatically receives a score of zero. If the subassembly is not disassembled further, then it is considered a single part and subjected to the three criteria above.

- **Column 3: Number of Repetitions.** The number of times a given disassembly task is performed is entered in column 3. This entry accounts for identical tasks performed in succession. For example, removal of three similar screws will require three repetitions of the same unscrewing operation.
- **Column 4: Task Type.** The task type is recorded since some tasks are more difficult to perform than others. For example, prying a glued joint apart is more elaborate than simply picking up a loose part. Fifteen standard disassembly tasks were identified from initial disassembly experiments. Each task has a standard definition. For example, *unscrew* covers the removal of all types of threaded fasteners including screws, nuts, and bolts. The tasks are listed with their letter codes in Table 15.1.
- **Column 5: Direction.** The direction in which the tool or hand accesses the assembly with respect to a fixed Cartesian coordinate system is recorded. An XYZ system is attached to the disassembly workbench so that the Z-axis is vertical with its positive sense directed upward. The coordinates do not change if the product is reoriented (e.g., flipped on its side) during disassembly. Direction information is valuable mainly for future robotic disassembly considerations.
- **Column 6: Required Tool.** The tooling used to perform each task is recorded. In this way, tool manipulations and special requirements are noted. Twenty-eight standard robotic and manual disassembly tools have been identified and are listed with their two-letter codes in Table 15.2. A provision is made for any "special" tools not included in the standard list. If only the hands are used, the column is left blank.

   A tool manipulation occurs each time a tool is picked up or put down and is implied each time different tool codes appear on successive rows of the chart.
- **Columns 7-11: Difficulty Ratings.** Each task is assigned quantitative difficulty scores for five aspects of task performance. The scores are based on scale of "1" (easy) to "10" (difficult) and are obtained from charts which include ratings for each task performed under various conditions. Since the difficulty ratings are based on task performance time, scores higher than ten are possible. The derivation of the scores is discussed in the next section of the chapter. The five categories of task performance are:

   - *Accessibility:* A measure of the ease with which a part can be accessed by the tool or hand. This is an indication of whether or not adequate clearance exists.
   - *Positioning:* The degree of precision required to position the tool or hand. For example, a higher degree of precision is required to engage a screw with a screwdriver than to simply grasp a loose part.
   - *Force:* A measure of the amount of force required to perform a task. For example, less force is required to remove a loose part than to free a part glued to the assembly.
   - *Base Time:* The time required to perform the basic task movements without difficulty. This category excludes any time spent positioning the tool or overcoming resistance. The basic ease of performing a task is indicated by this score. For example, an *unscrew* operation will have a higher *base time* score than a simple *flip* operation.
   - *Special:* This category covers special circumstances not considered in the standard task model. For example, if the standard model includes removal of screws with only five to eight threads and a screw with twelve threads is encountered, then a score greater than "1" would appear in the *special* category.

- **Column 12: Subtotal.** The sum of the individual difficulty ratings from columns 7 through 11. This represents the difficulty in performing a single repetition of the task.
- **Column 13: Total.** The product of columns 12 and 3 is entered here to account for multiple repetitions of a task.
- **Column 14: Comments.** This space is provided for explanations of special tools or tasks required and notes about high scores in any of the performance categories. This information draws the designer's attention to obvious design weaknesses.

Table 15.1 The standard tasks with their letter codes to be entered in column 4 of the evaluation chart

Pu	– Push/Pull	Fl	– Flip	Ha	– Hammer
Un	– Unscrew	De	– Deform	Cl	– Clean
We	– Wedge/Pry	Ho	– Hold/Grip	In	– Inspect
Cu	– Cut	Pe	– Peel	Gr	– Grind
Re	– Remove	Dr	– Drill	Sa	– Saw

Once completed, the disassembly chart may be used to identify weaknesses in a design and to assess its overall efficiency. The design efficiency is defined by:

$$\text{Design Efficiency} = \frac{5 \times \sum \text{Column 2}}{\sum \text{Column 13}} \times 100\% \qquad (1)$$

Equation (1) compares the current design with an "ideal" design of the same product. The reference design consists of the theoretical minimum number of parts which are disassembled with minimal difficulty. Design efficiency may be used to discern the effects of design changes on overall ease of disassembly. Furthermore, it allows different versions of the same product to be compared quantitatively. Optimization may be achieved through several design iterations.

Possible areas for design improvement may be identified by reviewing a summary of the evaluation results. A summary usually contains the following items: the actual number of parts, the number of theoretically unnecessary parts, the total number of disassembly tasks, the number of tasks which do not result in the direct removal of a part ("non-value-added" tasks), the number of tools used, the number of tool manipulations, the total difficulty score (S column 13), and the overall efficiency rating from Eq. (1).

Opportunities to consolidate or eliminate unnecessary parts are revealed by comparing the actual number of parts to the number of theoretically not required parts. Tasks that do not directly contribute to the progress of disassembly are identified when several rows of the chart correspond to a single part. These extraneous tasks should be eliminated. The use of many tools indicates time wasted reaching for tools while a large number of tool manipulations may imply an inefficient disassembly sequence.

The most detailed feedback to the designer is given by the categorized difficulty ratings in the chart. High ratings in column 13 indicate opportunities for improvement through substitution of less difficult tasks or design modifications to simplify specific aspects of task performance which scored high in columns 7-11. For example, a difficult *wedge/pry* task may be replaced by a simple *push/pull* operation by a adding a hole to the assembly through which a tool has access to push the part out.

Table 15.2 The standard robotic and manual tools with their two-letter codes to be entered in column 6 of the evaluation chart

Unscrewing:

PS	–	Phillips Screwdriver
FS	–	Flathead Screwdriver
ND	–	Nut Driver
FW	–	Fixed-end Wrench
AW	–	Adjustable Wrench
SR	–	Socket with Ratchet
AK	–	Allen Key
PW	–	Power Wrench

Gripping & Fixturing:

VS	–	Vise
PL	–	Pliers
SG	–	Standard Gripper
NG	–	Long-nose Gripper
EG	–	Expanding Gripper
LG	–	Large (>3") Gripper

Cutting & Breaking:

KN	–	Knife
WC	–	Wire Cutter
HS	–	Handheld Shears
DR	–	Drill
PG	–	Handheld Power Grinder
GW	–	Grinding Wheel
HS	–	Hacksaw
SS	–	Power Saber Saw
BS	–	Power Band Saw

Cleaning:

BR	–	Brush
RG	–	Rag

Others:

PB	–	Pry Bar
HM	–	Hammer
CH	–	Chisel
ST	–	Special Tool

15.4 DERIVING TASK DIFFICULTY SCORES

Task difficulty scores for all the pre-defined disassembly tasks were derived from an estimation of task performance time using the Maynard Operation Sequence Technique (MOST) work measurement system (Zandin, 1980). For example, the chart of difficulty scores for the standard *unscrew* task is shown in Figure 15.2. The MOST system is a predetermined time system which provides standard time data for the performance of precisely defined motions. If a disassembly task is broken down into elementary movements, such a system can predict the time required for an average skilled worker to perform the task at an average pace (Karger and Bahra, 1987). The term "average" is used in the sense that the standard time data represent mean values determined from motion-time studies of many workers of varying skill and effort, working under various conditions, in different industries.

Manual Unscrew				Accessibility	Positioning	Force	Base Time
Single Screw or Nut	Clear		Light Resistance	1	2	3	8
			Heavy Resistance	1	2	10	8
	Obstructed		Light Resistance	2	2	3	8
			Heavy Resistance	2	2	10	8
Single Bolt with Nut	Clear Screw	Clear Nut	Light Resistance	1	3	3	8
			Heavy Resistance	1	3	10	8
		Obstructed Nut	Light Resistance	2	3	3	8
			Heavy Resistance	2	3	10	8
	Obstructed Screw	Clear Nut	Light Resistance	2	3	3	8
			Heavy Resistance	2	3	10	8
		Obstructed Nut	Light Resistance	3	3	3	8
			Heavy Resistance	3	3	10	8

Figure 15.2 The chart of difficulty scores for a "standard" unscrew task.

The MOST work measurement system is based on sequences of basic motions, or subactivities, that are consistently repeated in the movement of objects and use of tools. The subactivities which make up each sequence model are represented by letter sequence parameters. A task or "activity" is analyzed as a series of standard sequence models. For example, the task model for the Clear, Light Resistance classification of *unscrew* (see first row of Figure 15.2) includes the follow sequences:

$$A_1 B_0 G_1 A_1 B_0 P_3 L_{6+16} A_1 B_0 P_1 A_1$$

$$A_1 B_0 G_1 A_1 B_0 P_1 A_1$$

The sequence on the first line indicates that a screwdriver is picked up, placed on the fastener, twisted to loosen the screw, and then placed aside. The second sequence describes grasping and removing the loose fastener. Each parameter represents a basic motion involved in performing the task. For example, "A_1" denotes the movement of the hand to an object within reach of an outstretched arm. The numerical index assigned to each letter defines the performance time of the subactivity in TMU (time measurement units, 1 TMU = 0.036 second). Numerical indices are listed in tables for each parameter according to factors which affect performance time. Once the appropriate sequence models and corresponding parameter indices are determined for an activity, the work measurement analysis is complete. The performance time in TMU for each sequence is calculated by summing the indices of all the parameters in the model and multiplying by ten. The overall operation time is the sum of all the sequence times.

The first step in the development of the difficulty ratings was to identify and define standard sequence models for each disassembly task. Task models were determined through observation of numerous manual disassembly experiments. Key steps in the performance of each task were noted and analyzed with the MOST system. Often, slight variations in the performance of the same basic task were observed. For example, different methods for turning a screwdriver were noted. The quickest and most efficient method for performing a task, as determined from the MOST analysis, was designated the standard task model.

Once the standard task models were defined, the effects of various disassembly conditions were investigated. Factors such as obstructions, handling difficulties, and heavy resistance were considered. The effects of these conditions on performance time were assessed by assigning appropriate MOST parameter indices. For example, inadequate clearance for the placement of a screwdriver (denoted by the "P" parameter in the first sequence of the *unscrew* model above) required a "P_6" rather than the usual "P_3." Thus, factors which complicated the disassembly process received higher parameter indices and increased overall performance time.

The parameter indices were reevaluated each time new conditions were imposed for a task. Also, the sequence parameters in the task model were categorized according to the aspect of task performance they measured. These categories were previously defined as: *accessibility*, *positioning*, *force*, *base time*, and *special*. For example, the "P" parameters related to the placement of a tool were grouped in the *positioning* category. The parameter indices in each category were summed to obtain the component of the total task time consumed by that aspect of task performance. For example, the indices of all the parameters related to *force* were summed to determine the amount of task time devoted to overcoming resistance. If no parameters were assigned to a category, the time component was zero. This process was repeated until a database containing component times for each standard task classification was developed.

Finally, the component times were converted to difficulty scores on a scale of 1 to 10. A time block of 260 TMU, which corresponds to the *force* component of a common *unscrew* operation, was assigned a difficulty score of 10. The *unscrew* operation was chosen as the reference point because *unscrew* operations are well defined and occur frequently in manual disassembly. Since time components for some tasks may be greater than 260 TMU, scores greater than ten are possible. It should be noted that 260 TMU equals approximately nine seconds. When the component times are linearly transformed to a difficulty scale of 1 to 10, each unit of difficulty corresponds to roughly one second of performance time. All task component times were converted to difficulty scores using the following relationship:

Difficulty Score = 1+ [9 x (Component Time in TMU)/ 260]   (2)

Equation (2) normalizes the component times with respect to the assigned reference time and shifts the values so that the minimum score is "1." Since all scores are defined on the same scale, difficulty ratings for different tasks may be compared directly. For example, the *force* score for an *unscrew* operation may be compared to the *base time* score for a *wedge/pry* operation. The difficulty scores for each task are classified in charts similar to Figure 15.2 according to relevant performance conditions. Classification definitions and guidelines for handling situations not covered in the standard task models are also provided.

The task difficulty scores were based on the fundamental assumption that performance time is a valid indicator of disassembly effort. This assumption is supported by the basic principles of work measurement. For example, predetermined time systems are based on motion-time studies of experienced laborers working at an average rate and under average conditions. In effect, predetermined time systems measure manual work in terms of time. The fact that time is related to cost is a further incentive to measure difficulty in this way. Other measures of work, such as energy expended, are difficult to obtain and comprehend.

General assumptions about the disassembly area and disassembler were required to perform the MOST analyses. It was assumed that a "knowledgeable" disassembler performs each task. This means that the disassembler has been specifically trained to dismantle certain products and is completely familiar with the disassembly process. Therefore, no time is wasted searching for parts or deciding which task is to be performed next. It was assumed that all hand tools are placed "within reach" of the disassembler and that disassembly is performed on a workbench in front of the worker. Bins are provided around the disassembly area so that parts may be "tossed" aside as they are removed. Other equipment, such as vises, band saws, and grinding wheels, are assumed to be positioned within "one or two steps" of the disassembler.

15.5 EXAMPLE: EVALUATION OF A COMPUTER KEYBOARD

The computer keyboard shown in Figure 15.3 was disassembled and evaluated with recycling in mind. Note that the Z-axis in Figure 15.3 is vertical with its positive sense pointing downward. This indicates that the keyboard was actually disassembled with the keys facing the bench. The disassembly procedure focused on separating and preparing plastic parts for recycling. For example, labels that could contaminate the plastic during recycling were removed. A parts list with generic material descriptions is included in Table 15.3.

The disassembly evaluation chart for the keyboard is found in Figure 15.4. A summary of evaluation results obtained from the chart is shown in Table 15.4. Although some design modifications are suggested in the discussion below, they are included for illustrative purposes only. The example is not intended to present a redesigned keyboard. The economical and functional implications of implementing the design changes were not considered.

The example keyboard exhibited a fairly efficient design in relation to other keyboards we have evaluated. An overall design efficiency of 38% was calculated from the evaluation results in Table 15.4. For comparison, the recent evaluation of another keyboard yielded an overall efficiency of 25%. From Table 15.4, it is seen that the example keyboard consisted of 214 parts, of which 118 were judged not to be required as separate components. Screws (parts 1, 2, 3 and 9) accounted for 27 of the 118 unnecessary parts. Four labels (parts 22, 23, 25 and 26), three key plates (parts 12, 15 and 18), and 84 key retainers (parts 14, 17 and 20) were also not required. If the keyboard were under consideration for redesign, the designer could attempt to eliminate as many of these unnecessary parts as possible. For example, a few snap-fits may be chosen to replace some or all of the unnecessary screws.

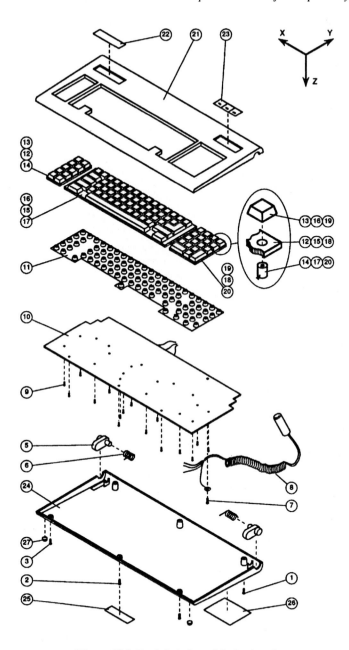

Figure 15.3 Exploded view of the keyboard.

Table 15.3 Part list of the keyboard

Part No.	Quantity	Part Name	Material
1	3	Rear screw	steel
2	1	Middle front screw	steel
3	2	Corner front screw	steel
4	–	Subassembly of parts 24-27	–
5	2	Retractable leg	plastic
6	2	Spring	steel
7	1	Ground screw	steel
8	1	Cable	mixed
9	21	PCB screw	steel
10	1	PCB	mixed
11	1	Spring mat	rubber
12	1	Function key plate	plastic
13	10	Function key	plastic
14	10	Function key retainer	plastic
15	1	Letter key plate	plastic
16	56	Letter key	plastic
17	56	Letter key retainer	plastic
18	1	Number key plate	plastic
19	18	Number key	plastic
20	18	Number key retainer	plastic
21	1	Cover	plastic
22	1	Manufacturer's label	metallic foil
23	1	LED window label	plastic
24	1	Base	plastic
25	1	Serial no. label	paper
26	1	Model no. label	paper
27	2	Foot	rubber

Table 15.4 Summary of the keyboard evaluation results

Number of parts	214
Number of theoretically not required parts	118
Number of tasks	140
Number of non-value-added tasks	5
Number of tools	8
Number of tool manipulations	22
Total difficulty score	1,255
Overall efficiency	38%

DISASSEMBLY EVALUATION CHART

PRODUCT: PC'S LIMITED AT, model no. 2188002XX keyboard
PREPARED BY: Thomas Hanft
DATE: 1/11/95
SHEET: 1 OF 1

PART No.	MIN. No. OF PARTS	No. OF REPETITIONS	TASK TYPE	DIRECTION	REQUIRED TOOL	DIFFICULTY RATING 1(EASY)...10(DIFFICULT)					SUBTOTAL	TOTAL	COMMENTS
						ACCESSIBILITY	POSITIONING	FORCE	BASE TIME	SPECIAL			
1	0	3	Un	-z	PS1	2	2	3	8	1	16	48	Screwdriver with magnetic tip used to pick up screws
2	0	1	Cu	-z	KN	1	1	2	2	1	7	7	Cut hole in label (part 25) to access screw
		1	Un	-z	PS1	1	2	3	8	1	15	15	
3	0	2	Un	-z	PS1	1	2	3	8	1	15	30	
4(sub)	0	1	Fl	-z		1	1	1	1	1	5	5	Lift base subassembly and put aside
5	2	2	Re	-z		1	1	1	1	1	5	10	
6	2	2	Re	-z		1	1	1	1	1	5	10	
7	1	1	Un	-z	PS1	1	2	3	8	1	15	15	
8	1	1	Cu	-z	WC	1	1	2	2	1	7	7	
9	0	21	Un	-z	PS2	1	2	3	8	1	15	315	
10	1	1	Re	-z		1	1	1	1	1	5	5	
11	1	1	Re	-z		1	1	1	1	1	5	5	
12	0	1	Ho	-z		1	1	1	2	1	6	6	Grip plate until keys and retainers removed
13	10	10	Pu	z		2	1	2	1	1	7	70	Keys dropped in bin
14	0	1	Fl	z		1	1	1	1	1	5	5	All retainers fall freely into bin
15	0	3	Ho	-z		1	1	1	2	1	6	18	Reposition grip 3 times as keys and retainers removed
16	56	56	Pu	z		2	1	2	1	1	7	392	Keys dropped in bin
17	0	1	Fl	z		1	1	1	1	1	5	5	All retainers fall freely into bin
18	0	1	Ho	-z		1	1	1	2	1	6	6	Grip until keys and retainers removed
19	18	18	Pu	z		2	1	2	1	1	7	126	Keys dropped in bin
20	0	1	Fl	z		1	1	1	1	1	5	5	All retainers fall freely into bin
21	1	1	Fl	-z		1	1	1	1	1	5	5	
		1	Ho	-z		1	1	2	1	1	6	6	Grip cover until labels removed
22	0	1	We	-z	PB	2	1	3	2	3	11	11	Heavy pressure applied during 3 wedging motions
		1	Pe	-z		1	2	1	2	1	7	7	
23	0	1	We	-z	PB	2	1	3	2	2	10	10	Heavy pressure applied during 2 wedging motions
		1	Pe	-z		1	2	1	2	1	7	7	
Disassembly of Subassembly 4													
24	1	1	Ho	-z	VS	1	1	1	8	1	12	12	
25	0	1	Gr	X,-X	PG	1	1	1	1	11	15	15	Grind label off in 10 seconds
26	0	1	Gr	X,-X	PG	1	1	1	1	31	35	35	Grind label off in 30 seconds
27	2	2	Dr	Y	DR	1	2	1	1	16	21	42	Drilling takes 15 seconds for each foot

Figure 15.4 Disassembly evaluation chart for the keyboard.

A total of 140 disassembly tasks were performed, of which five did not directly result in the removal of a part. This reveals that little time was wasted on non-value-added tasks, such as cutting a hole in the label over the middle front screw (part 2) and flipping the assembly to access the other labels (part 22, 23 and 26). All the non-value-added tasks, and several others, could be eliminated by replacing the labels with molded insignias.

It should be noted that nearly all the directions listed in column 5 of Table 15.4 are along the negative Z-axis. Thus, for most operations, the assembly was accessed from vertically above. This is usually the case for small products which are easily flipped and reoriented to facilitate vertical disassembly. In general, it is easier and more natural to disassemble a product from above. However, for larger, heavier products, movement in the X-Y plane is often necessary since it is more difficult to reorient the assembly than to access it horizontally.

Eight different tools were used during the disassembly and 22 tool manipulations were required. A large number of tool changes may indicate an inefficient design or disassembly sequence. In this case, the keyboard design is at fault. At least 16 tool manipulations were necessary since each of the eight tools had to be picked up and put down. The variety of tools needed to dismantle the keyboard should be minimized so that time is not wasted reaching for tools.

The most detailed feedback is obtained from the actual disassembly chart (Figure 15.4). Obvious design weaknesses may be identified through the composite difficulty scores for each task, found in column 13. For example, a quick glance at the chart reveals a score of 315 in column 13. This score corresponds to the *unscrew* operation for part no. 9. However, the individual difficulty ratings for this operation (columns 7 through 11) appear to be relatively low. Thus, the disassembly task itself was not the primary source of difficulty. A look at column 3 then reveals that the task was repeated 21 times. Although the task was simple, disassembly efficiency was compromised by many repetitions. In this way, the designer is led to the specific cause of disassembly difficulty. A final look at column 2 indicates that the 21 screws that comprise part no. 9 were theoretically not required. Thus, the screws may be eliminated or reduced in number, as described earlier. Interestingly, if all 21 screws were eliminated, the overall design efficiency would jump from 38% to 51%.

Further inspection of column 13 identifies scores of 70, 392 and 126. These scores all pertain to the removal of the keys (parts 13, 16 and 19, respectively). As in the previous case, the primary source of difficulty for these operations is not the task performed, but the large number of repetitions involved (a total of 84). The keys are theoretically required since they must move relative to each other. Thus, reducing the number of separate keys is difficult. However, it may be possible to interconnect the keys with thin strips of flexible plastic which permit relative motion between keys and allow the keys to be lifted from the assembly as a single unit. In this way, 84 pieces may be removed in a single operation. A similar concept is demonstrated in the elegant design of the keyboard's spring system. Instead of individual metal springs beneath each key, as in many other keyboards, the example keyboard employs a rubber mat (part no. 11) with small bulges under the keys. The bulges act independently as springs but are all part of the same mat. Thus, a single *remove* operation extracts all 84 springs.

The most difficult single operation in the disassembly process is identified by reviewing the task difficulty subtotals in column 12. The *grind* operation required to remove a specification label (part no. 26) was by far the most difficult task with a subtotal of 35. An inspection of the individual difficulty ratings in columns 7 through 11 reveals a *special* score of 31. As described in the comments for this task, 30 seconds of grinding were required. Thus, the designer should attempt to eliminate the label or minimize the process time.

15.6 SUMMARY

The structure of the disassembly evaluation chart provides a systematic method for analyzing design weaknesses. Task difficulty scores facilitate understanding and quantifying sources of disassembly difficulty. Since evaluation results are sensitive to the disassembly sequence followed, alternative disassembly sequences may be compared for the same product. Furthermore, the chart allows evaluation of all levels of disassembly. If the purpose of dismantling the keyboard had been to recover only the circuit board, the disassembly procedure could have been terminated at an appropriate stage. The total difficulty would then represent the effort applied in achieving that state of disassembly.

The current method focuses on manual disassembly of business equipment. However, the same methodology may be applied to robotic disassembly processes and other products. Initial investigations indicate that automated disassembly is very complicated. Worn or broken products, the variety of disassembly processes, and flexible parts, such as wires, present considerable obstacles to robots. Further development of "intelligent" robots that can handle uncertain conditions is required before automated disassembly becomes a viable option.

In the overall picture of design for disassembly, the evaluation chart represents the first step towards improving the design: analysis. Future work will focus on the redesign phase of product development. A collection of guidelines for overcoming specific design weaknesses will be created. Suggested improve ments will aid engineers in implementing design for disassembly.

ACKNOWLEDGMENT

Support for this work was provided by the Office of the Vice President for Research and Associate Provost for Graduate Studies, through the Center for Energy and Mineral Resources, Texas A&M University, and the American Plastics Council, Washington, DC.

REFERENCES

Bakerjian, R. (1992) Environmentally Responsible Product Design, *Design for Manufacturability*, Vol. **6**, in the Tool and Manufacturing Engineers Handbook Series, 4th ed., SME, Dearborn, Michigan, 10.61-10.64.

Bast, C. (1994) Hewlett-Packard's Approach to Creating a Life Cycle (Product Stewardship) Program, *Proceedings of IEEE International Symposium on Electronics and the Environment*, 31-40.

Beitz, W. (1993) Designing for Ease of Recycling, *Journal of Engineering Design*, **4**, 11-23.

Boothroyd, G., Dewhurst, P. (1987) *Product Design for Assembly*, Boothroyd Dewhurst, Inc., Wakefield, Rhode Island.

Burke, D.S., Beiter, K., Ishii, K. (1992) Life-Cycle Design for Recyclability, *Proceedings of 1992 ASME Design Theory and Methodology Conference*, ASME DE-Vol. **42**, 325-332.

Constance, J. (1992) Can Durable Goods Be Designed for Disposability?, *Mechanical Engineering*, **114**, 60-62.

Dewhurst, P. (1993) Product Design for Manufacture: Design for Disassembly, *Industrial Engineering*, **25**, 26-28.

Dillon, P.S. (1994) Mandated Electronic Equipment Recycling in Europe: Implications for Companies and US Public Policy, *Proceedings of IEEE International Symposium on Electronics and the Environment*, 15-20.

Eagan, P., Koning, J. Jr., Hoffman, W. III (1994) Developing an Environmental Education Program Case Study: Motorola, *Proceedings of IEEE International Symposium on Electronics and the Environment*, 41-44.

Eekels, J. (1993) The Engineer as Designer and as a Morally Responsible Individual, *Proceedings of 9th International Conference on Engineering Design*, 2, 755-764.

Emblemsvag, J., Bras, B. (1994) Activity-Based Costing in Design for Product Retirement, *Proceedings of 20th Design Automation Conference*, ASME DE-Vol. 69, 351-361.

Fiskel, J., Wapman, K. (1994) How to Design for Environment and Minimize Life Cycle Cost, *Proceedings of IEEE International Symposium on Electronics and the Environment*, 75-80.

Karger, D.W., Bayha, F.H. (1987) *Engineered Work Measurement*, 4th ed., Industrial Press, New York.

Kirby, J. R., Wadehra, I. (1993) Designing Business Machines for Disassembly and Recycling, *Proceedings of IEEE International Symposium on Electronics and the Environment*, 32-36.

Navin-Chandra, D. (1991) Design for Environmentability, *Proceedings of Design Theory and Methodology Conference*, ASME DE, 31, 119-125.

Navin-Chandra, D. (1993) ReStar: A Design Tool for Environmental Recovery Analysis, *Proceedings of 9th International Conference on Engineering Design*, 2, 780-787.

Seegers, H.J.M. (1993) Automotive Design for Recycling in GE Plastics, *Proceedings of 9th International Conference on Engineering Design*, 2, 812-819.

Simon, M. (1993) Objective Assessment of Designs for Recycling, *Proceedings of 9th International Conference on Engineering Design*, 2, 832-835.

Thurston, D.L., Blair, A. (1993) A Method for Integrating Environmental Impacts into Product Design, *Proceedings of 9th International Conference on Engineering Design*, 2, 765-772.

Tummala, R.L., Koenig, B.E. (1994) Models for Life Cycle Assessment of Manufactured Products, *Proceedings of IEEE International Symposium on Electronics and the Environment*, 94-99.

Veldstra, M., Bouws, T. (1993) Environmentally Friendly Design in Plastics, *Proceedings of 9th International Conference on Engineering Design*, 2, 820-826.

VerGow, Z., Bras, B. (1994) Recycling Oriented Fasteners: A Critical Evaluation of VDI 2234's Selection Table, *Proceedings of 20th Design Automation Conference*, ASME DE- 69, 341-349.

Vigon, B.W., Curran, M.A. (1993) Life-cycle Improvements Analysis: Procedure Development and Demonstration, *Proceedings of IEEE International Symposium on Electronics and the Environment*, 151-156.

Wilder, J. (1990) Designing for Disassembly; Durable-goods Makers Build in Recyclability, *Modern Plastics*, 67, 16-17.

Zandin, K.B. (1980) *MOST Work Measurement Systems*, Marcel Dekker, Inc., New York.

Zussman, E., Kriwet, A., Seliger, G. (1994) Disassembly-Oriented Assessment Methodology to Support Design for Recycling, *Annals of the CIRP*, 43, 9-14.

16

DESIGN FOR QUALITY

Gian Francesco Biggioggero; Edoardo Rovida

This chapter presents a technique of Design for Quality. It is based on what is usually called concept generation and selection. As many solutions as possible to a given function are first generated and the best is then chosen according to a set of evaluation criteria. Criteria closely related to Quality are especially important in the context of Design for Quality. A software package has been developed to support this activity using Computervision's Personal Designer. A case study is given to illustrate the technique and the use of the software.

Quality is generally defined as "compliance with requirements", that is the degree to which the specific range of characteristics of a machine conform to the requirements. If they match well, then the quality is high; otherwise quality is considered poor. This point of view is also maintained by Hubka (1989): "Quality ... concerns statements about the «what» and «how» of an object or process," and "... a combination of conditions ... for one product may not be applicable to another product", and also is considered more appropriate the expression «vive quality» instead of the common (and abused) expression «high quality». Koudate (1991) declares that "switable quality" is achieved with definite limitation of time and cost.

16.1 DESIGN AND QUALITY MANAGEMENT

The application of rational methods is essential for a complete design and its correct management. In this section, quality systems and the role of design management are reviewed, and a method of setting quality objectives for effective design is discussed.

16.1.1 Machine Design Process

The engineering design is a primary datum-point to obtain the product quality. The explicit development of the quality exigency confirm once more the validity of the propositions of the methodological design that is very suitable for the Quality Systems applications. Figure 16.1 shows a general flowchart of methodical design "philosophy" (Biggioggero and Rovida 1990). Such a systematic procedure of design process is an essential starting point for correct design for quality as with each rational step in a design project.

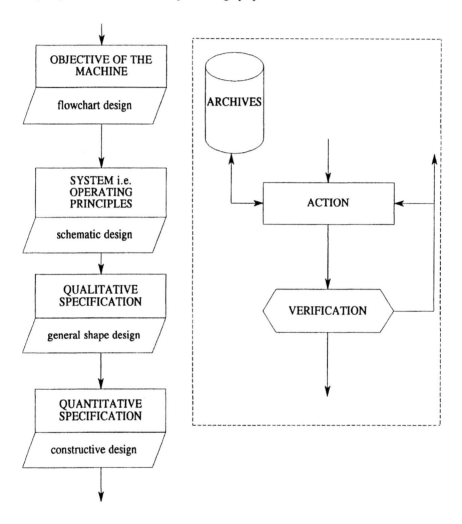

Figure 16.1 Flowchart of design process of archives, controls and revisions.

The systematic machine design procedure in the figure includes the following four stages:

a   objective of the machine being designed, expressed in terms of:

       a.1   function (or set of functions)

       a.2   qualitative and quantitative characteristics

b   system, i.e. operating principle and scheme (frequently drafting scheme) with which the designer intends to achieve his objective. This system can be identified by means of the two following steps:

       b.1 known systems, i.e. part of the technical heritage, taken from:

           b.1.1   history

           b.1.2   scientific-technical literature

           b.1.3   technical-commercial literature

           b.1.4   standards

           b.1.5   patents

       b.2 "new" systems that enlarge upon the heritage of technical knowledge, stemming from considerations and combinations of:

           b.2.1   new physical phenomenon used

           b.2.2   new geometry of functional surfaces

           b.2.3   new kinematics of functional surfaces

           b.2.4   new statics of functional surfaces

           b.2.5   new generation of forces

           b.2.6   new operating systems of operating surfaces.

c   Qualitative specification, i.e. definition of general shapes and families of materials pertaining to the configuration of the machine and corresponding to a specific system;

d   Quantitative specification, corresponding to the definition of all construction information (shapes, dimensions, materials, admissible errors) of the machine.

In addition, the flowchart of the design process shown in Figure 16.1 introduces control and revision as an efficient means of implementation and conformity with the objectives specified by the quality policy. The introduction of actions, verifications and references to archives not only avoid possible process errors but also keep an account of continuous evolution.

16.1.2 Quality Systems

Quality systems started with the use of control charts proposed by W.A. Shewhart to analyse the inspection data (Lamprecht 1992). This has developed further into a school of statistical quality and reliability control techniques. Today the growth of markets and trade together with the continuous increase in competition has resulted in a need for not only transparency but also documentation of the quality of products and services. Quality considerations has now become an explicit and dominant factor in corporate, commercial and service procedures, rather than implicit in every good design process.

Between 1987 and 1994, this requirement resulted in the publication of international standards of the ISO 9000 series enforced by Italian Standard Associations. These standards have been by now widely diffused and expounded upon (Lamprecht 1992; Jensen 1992). Criteria are established that should be followed in order to obtain "certification" of corporate quality systems by the requisite authorizing bodies. "Certification" is now essential in order to work either in Europe or abroad.

16.1.3 Quality objectives

Figure 16.2 shows several important points. First, management expresses their aims and objectives through the corporate policy. They are then translated into quality objectives. The second point indicated in the figure is that goals and targets must satisfy a number of requirements. For example, they should be identifiable qualitatively and quantitatively, pertinent and easy to understand but also achievable, and defined and measurable. Obviously, the risks and advantages involved need to be properly estimated. Finally, quality objectives do not remain static but dynamic, following evolutionary trends. This is achieved by considering technology development and customer expectations. The quality of the products and services is kept under constant verification and approval.

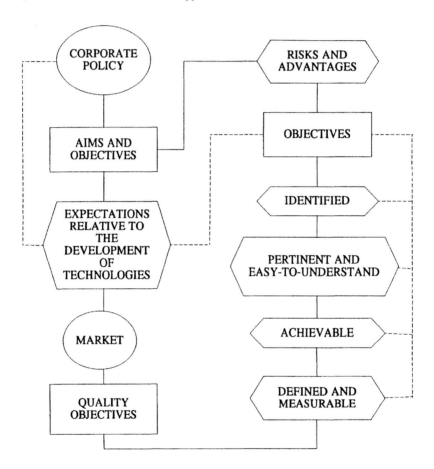

Figure 16.2 Flowchart of conversion of general objectives in "quality objectives".

16.1.4 Design management

ISO 9000 is comprehensive, generic, and wide-ranging. Figure 16.3 shows the structure and attributes of design and quality management. As indicated by the bi-directional arrows in the figure, it can be seen that design plays an important role in setting and achieving the quality objectives. Both quality and design management are bracketed together and quality function is well structured in the same way as methodical design.

In the context of ISO 9000, the term design is interpreted as the design of products and services. A "system of quality " is developed within the company through two design stages: a specific action based on traditional design, and planning based on the design procedures inherent in each phase of the product life.

Design and Quality management has four attributes. The first attribute is standardization and coordination. Design activity is present in every phase of the "product life" where quality should be assured. Therefore, ISO 9000 standards require that procedures should be followed in every phase and methodologies and archives are standardized and coordinated throughout the product life. This can be achieved by drawing up "control lists". The use of such checklists prevents some of the requisites that should always be defined and documented from being excluded. Table 16.1 shows several general quality categories under which specific quality items should be specified, synthesized, and analysed along the total product life cycles.

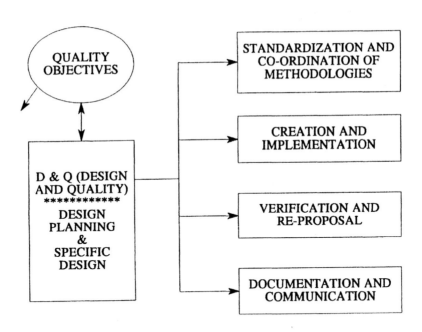

Figure 16.3 Structure and attributes of Design and Quality management.

Table 16.1 Specific quality items for each phase of the life of the product

ITEMS ╲ LIFE CYCLE STEPS	PURCHASE	MANUFACTURING	ASSEMBLY	FINAL TESTING	ASSISTANCE	USE
PERFORMANCE AND LIFE CYCLE						
SHAPS AND DIMENSIONS						
MATERIALS AND TREATMENTNS						
SAFETY AND RELIABILITY						
MAINTENANCE POSSIBILITIES						
HANDLING PACKAGING STORAGE PRESERVATION						
DISMISSION						

INITIAL INSTRUCTIONS
- CRITICAL PARAMETERS
- ACCEPTANCE CRITERIA
- SATISFACTION CONTROL

PLANNING REVISION
- INTEGRATION AND COORDINATION WITH OTHER PRODUCTS
- PARALLEL EXECUTION OF ACTIVITIES

DESIGN REVISION
- MULTIPLE ACCOMPLISHMENT
- SIMPLIFICATION
- LEGISLATIVE ADJUSTMENTS

The second attribute of Design and Quality is creation and implementation. Quality is first designed and then built into a product. A clear organization chart should be outlined where quality and design management are bracketed together. Quality function and design function are organized in similar ways. Suitable staff organization policies are needed to encourage involvement from different functional areas. For example, marketing handles customer expectations; Research and Development deals with technology development; Design and Quality is responsible for coordination; and general management provides instructions and directives. They all participate in the formulation and accomplishment of quality objectives and their periodic re-assessment.

The third attribute of Design and Quality is verification and re-proposal of modifications. This is particularly applicable to design planning procedures. Table 16.1 again plays an important role in verification and revision. The cell entries to the table include three categories of information. The first category includes initial instructions about critical parameters, acceptance criteria, satisfaction control. The second category is for planning revision, including integration and coordination with other products, and parallel execution of activities. The third category is for design revision, including multiple accomplishments, simplification and legislative adjustments.

The fourth attribute of Design and Quality is documentation and communication. Well structured documents provide media for effective communication. The scope of documentation is not limited to that of design specifications, design drawings, operation instructions, etc. All elements indispensable to quality must be documented, including identification of critical parameters which should always be measurable and quantifiable, definition of acceptability criteria, indication of compliance controls, specifications and instructions of corporate organization and product and service typology. This should apply to each step of the product life cycle. An account should be kept not only of integration and coordination with other products but also of the execution of other parallel and/or sequential activities.

16.2 A DESIGN FOR QUALITY METHODOLOGY

Machine design can be structured such that quality is explicitly emphasized. In addition to functional requirements, other pre-set objectives are met in methodical design. This section presents a Design for Quality technique which is basically concerned with concept generation and selection. The following major steps are involved in the method:

1. Determine functions
2. Determine archives
3. Determine characteristics
4. Evaluate solutions

16.2.1 Determine Functions

This step is mainly concered with the establishment of objectives corresponding to what the machine has to do. They can be expressed by means of functions and the quality and quantity specifications. When defining design objectives it is very important that the function to be carried out by the machine is expressed with precision. A functional analysis is carried out to establish specific functions and sub-functions the product and its components must provide.

16.2.2 Determine Archives

This step deals with the identification of all physical principles according to which the machine will be constructed. An archive contains all solutions for a specific function. Archives can be established in a number of ways. First, solutions are collected by chance when the designer is addressing a specific problem; secondly, the designer systematically and gradually collects and codifies the experience and observations; and thirdly, systematic collections of solutions can be established by specialist organizations, for example, academics, research centres, technical associations, and manufacturers who are involved in the business of providing specific functions. Biggioggero and Rovida (1990) provide more guidance on how to create archives.

Archives must be made available to the designer to produce machine design of good quality. With the help of archives, the designer can browse through all the operating principles and all the construction solutions corresponding to a specific function at hand, rather than limited to those known to him or her.

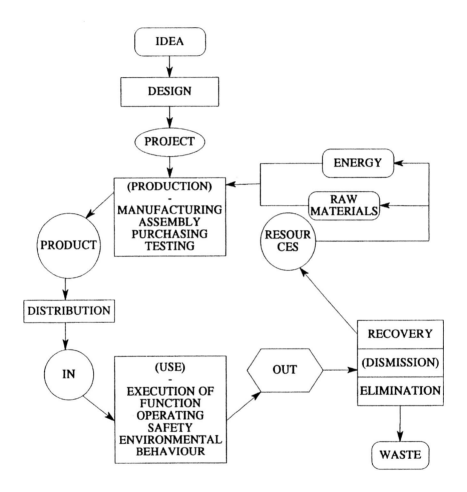

Figure 16.4 Asimow's cycle revised for the actual quality requirements.

Table 16.2. List of general characteristics that "pilot" the quality-designer

1) DISTRIBUTION	
a) packaging	(economic and safe way) its shape and dimensions must be compatible with packaging systems; in the event of this being necessary, any moving parts must be able to be blocked; it must be able to be packaged in different lots;
b) handling	(safely lifted) equipped with devices (hooks, handles, rings) which are sufficiently strong and securely fixed. These should be suited to standard lifting devices and capable of being disassembled for lifting purposes (in this case, the different machine elements must comply with the aforementioned requirements);
c) storage	easy, safe way, without incurring any damage, for prefixed periods of time;
d) preservation	correctly and definitively maintained;
e) transport	(easy and safe) its shape and dimensions should conform to standard transportation methods; it should be equipped with a blockage device (present on the machine or in the packaging) designed to keep the moving parts in place during transport; it should also be able to be transported in different lots (the above must apply to each lot).
f) delivery	in order to ensure that the client has no problems in terms of receipt, storage, unpacking or use.
2) OPERATING	
a) resistance to stress	corresponding to the normal and emergency manoeuvres established by the manufacturer;
b) visible and identifiable	without the possibility of any ambiguity;
c) consistent with start-up motion	capable of guaranteeing rapid and safe movement;
d) rapid and safe	capable of ensuring consistency between the movement of the operating mechanism and controlled movement;
e) ergonomic	compatible with operator positions and with the efforts and strokes required by same as well as in line with ergonomic principles;
f) not ambiguous	complete with clear operating instructions, in the event of operating referring to several different actions;
g) with protection	complete with instructions regarding necessary protective measures to be used by the operator, such as, gloves, helmets, glasses;
h) with "instrument" instructions	complete with instrument instructions (visible to the operator) pertaining to various actions and normal and abnormal operating conditions;
i) suitable operating station	with an operating station from which risk areas (to people) are clearly visible;
l) not involuntary	possibility of preventing involuntary start-up of actions that could put people or objects at risk;
m) with emergency block	be equipped with emergency blockage devices, to prevent the occurrence of dangerous situations.

Table 16.2. List of general characteristics that "pilot" the quality-designer (*continued*)

3) BEHAVIOUR	
a) geometric	the machine geometry must be consistent with size requirements;
b) kinematic	the kinematics of the machine must be consistent with the requirements of movement;
c) to static loads	(reaction) the machine and all components must be able to withstand the static loads present during the use foreseen by the manufacturer;
d) to deformation	the deformations present during machine operating must be compatible with the values specified during design;
e) to fatigue	the machine elements subjected to fatigue stress must be inspected with regard to material resistance, keeping account not only of the shape of the elements themselves but also of the stress characteristics; in the event of some parts requiring periodic replacement, the relevant methods should be specified by the manufacturer;
f) to vibrations	(in the presence of) in order to avoid negative phenomena, such as in the presence of dangerous frequencies or resonance, the reaction of the machine to vibrations should be the object of an accurate study;
g) to impact	the parts of the machine subject to impact stress should be scaled to conform to stresses; if periodic replacement is necessary, the relevant methods should be specified by the manufacturer;
h) to wear	the parts subject to wear should be suitably sized. Replacement should be specified by the manufacturer;
i) thermal	(reaction to heat) if, during operating, the machine, or parts of it, are subjected to high or low temperatures, design should keep account of all variations not only pertaining to size but also of the reactions of the materials to various temperatures;
l) chemical	if, during operating, the machine, or parts of it, are subjected to aggressive chemical actions, the relative materials should be chosen by keeping account of the nature of the chemical aggressor. Furthermore, design should ensure that duration and length of contact with the aggressor are consistent with the resistance characteristics of the material. Regardless of the presence of a chemical agents, the materials chosen should keep account of the possibility of corrosion caused by electro-chemical contact phenomena.

Table 16.2. List of general characteristics that "pilot" the quality-designer (*continued*)

4) SAFETY	
a) of stability	the machine and its components should not, even under extreme conditions, be subject to any risks of overturning or undesired movement. If necessary, suitable support and fixing devices should be provided. Phenomena of dynamic instability, such as resonance or critical speeds, should be non-existent;
b) reliability	(guarantee against breakage) the machine and its components should be designed in such a way to exclude the possibility of pieces or components breaking due to static, fatigue, impact and wear stresses;
c) against falling and projection of object	(danger the of falling or projection of objects) suitable protection must be provided in the case of the falling or projection of objects;
d) protection against cutting elements	(sharp corners) any machine parts that might come into contact with people should not have sharp corners or cutting edges;
e) protection from moving elements	all the mobile elements of the machine that could come into contact with people, should be suitably protected. These protections should have the following characteristics: - be fixed or mobile: the former are safer since there is no possibility of assembly "oversights", while the latter are more suitable in the event of the need for frequent inspections and maintenance; - be sufficiently robust; - not cause additional risks; - not be able to be easily rendered ineffective; - be situated at a suitable distance from dangerous areas; - not limit vision of the mobile units that they protect, or impede maintenance and inspection operations; - as long as the machine is accessible to people it should be equipped with devices aimed at preventing the start-up of moving elements ; - the removal of the mobile safety guards should only be carried out by a mechanism specially designed for this purpose and should not occur accidentally.

Table 16.2. List of general characteristics that "pilot" the quality-designer (*continued*)

5) ENVIRONMENTAL BEHAVIOUR	Refers to interaction between the machine and the environment.
a) low energy consumption	in order to ensure low energy consumption during the operating phase, the machine should be designed in conformity with the following aspects: - accurate study of energy transformations that take place within it; - tribologic study of all the parts subject to relative motion; - aero- and fluid-dynamic study of all the parts moving in an aeroform or liquid dimension; - lightness; - correct operating; requiring suitable training of operators;
b) reduction of emissions	(to a minimum) - mechanical emissions: every precaution should be taken to ensure that risks due to vibrations, impact, the machine should be suitably insulated. With regard to explosions, the manufacturer should take all the necessary precautions to ensure that these do not occur (avoid the concentration of dangerous substances, avoid the formation of sparks) and should reduce all possible consequences; - heat emissions: heat emissions (positive and negative) towards the outside should be reduced to a minimum; the presence of very hot or very cold parts can be touched should be avoided; - acoustic emissions: the risks due to noise emission should be minimum. If necessary, suitable protective measures, such as headphones, should be worn by the operator; - electric emissions: risks due to parts under tension that could be touched by people should be avoided. All the necessary measures aimed at preventing the formation of static electricity should be taken and, in any case, devices capable of discharging this, should be present; - chemical emissions: it is necessary to avoid or reduce the emission of harmful substances to a minimum. This should be done through the accurate study of the discharge devices and processes that take place within the machine. If necessary, emission containment devices should be used such as, for example, catalytic mufflers; - optical emissions: the machine should not cause optical disturbances such as dazzling, which could disturb either the operator or other people; there should be no reflections on the controls or towards the outside; it is necessary to point out that from a point of view of aesthetics, the machine should comply with the functions executed by the machine and be based upon an accurate study of shapes, size, colours and finishes; - radiation emissions: the risks related to radiation emission, of any type whatsoever, should be null.

Table 16.2. List of general characteristics that "pilot" the quality-designer (*continued*)

6) MANUFACTURING	Intended as production of the basic parts (pieces) of the machine as well as all the subsequent phases pertaining to production in general has aspects that apply specifically to quality. In this paper we will only refer to those aspects that most strongly influence design intended as actual planning for quality itself
a) processing methods	it is necessary to provide general guidelines regarding production processes, and work environment;
b) monitoring	we would like to point out the importance of this function, which in specific operations is linked to the continuos observance of working procedures;
c) controls	an indication regarding drawings and specifications, controls and tests during the working phase is necessary for geometric and technological controls as well as for those more specifically related to materials;
d) verifications	(maintenance checks) machine and equipment performance can jeopardize quality if controls to check that they are not only in-tact but also correctly operated are not carried out;
e) handling specifications	here reference is made to the necessary regulations for the handling and maintenance of the quality of objects during transfer within the factory itself.
7) ASSEMBLY	Even in this phase, the aspects already described for manufacturing are present and applicable; the processing methods are assembly methods such as for example monitoring and handling specifications. With regard to controls and verifications we can add.
a) working controls	(operating controls) both for simple couplings and complex assembly it is necessary to specify the expected results; where possible this should be done by means of statistic type controls;
b) reliability controls	not only the end quality, but, above all, the components, seen in terms of reliability tests are important.
8) PURCHASE	Every company besides the materials also purchases a number of parts (of varying complexity and importance) of the end product; these include equipment, machinery and services which must satisfy, either directly or indirectly, all the pre-established quality requisites. Therefore, it is necessary that all the rules already established for production are implemented both where the supplier and possibly the customer who supplies the products are in question in order to have a better understanding of the supplier's methods. To these the following general aspects can be added
a) evaluation	evaluation regulations apply not only to the suppliers but also to the production processes, both with regard to methods as well as production technologies; these are also applicable to products, direct or auxiliary, and should include a guideline showing specifications and drawings;
b) production control	control regulations and tests conducted during the production process as well as the extent of inspection.

Table 16.2. List of general characteristics that "pilot" the quality-designer (*continued*)

9) DISMISSING	Includes "waste" with the obvious warning that this should be able to be "recycled".
a) easy elimination	- use, as far as is possible, of biodegradable material; - use of materials with a high recovery value; - type of design that provides for the easy separation of different materials; - type of design enabling easy separation of components;
b) possible recovery	- use of materials with a high recovery value; - reduction of types of materials used; - marking of different materials (especially in the case of technopolymers) for easy identification purposes; - type of design that provides for the easy separation of different materials; - type of design enabling easy separation of components;
10) TESTING	By testing phases we mean receipt of goods and above all receipt of the final product. This implies accepting that the product complies with the quality objectives.
a) evaluation	(functional tests) to check whether the product conforms to the expected functions. These are conducted by means of specific instrument systems;
b) life cycle tests	to check whether the product is as reliable as guaranteed by the objectives;
c) safety tests	to check whether the product conforms to all relative guarantees, i.e. safety and environmental protection;
d) stress control tests	to check whether the machine can pass general tests of a mechanical, electronic, chemical and thermal, etc. nature;
e) validation	validity control of quality, elements, processes, procedures, etc. (especially if these are particularly complex) relative to controls, verifications and tests.

16.2.3 Determine Characteristics

Characteristics are "attributes" or "aspects" which define and describe functions and archives. A systematic identification of the aspects of interest for mechanical design is shown in Figure 16.4. It is based on Asimov's cycle (Asimow 1962) and revised according to the CEE machine directive. A checklist can be used to obtain better ideas of the aspects that affect the design project. Table 16.2 is a list of general items by which characteristics for functions and archives can be established. The use this checklist prevents some elements, such as compliance of the objects with maintenance procedures as stipulated by the EEC machine directives with a view to safety and safeguarding of health, from neglecting as has been in traditional design.

The machine must "go through" the necessary distribution, utilization, elimination and recovery phases. When determining characteristics, quality and safety must always be kept in mind as well as other important issues. In addition to these general items, there are more specific characteristics that should be added according to particular situations of the industrial sector and company background. Design and Quality management function can use this checklist to coordinate factors of different life cycles.

There are several uses of the characteristics when determined. One is to filter and then shortlist alternative solutions from the matched archive for a given function. In addition, these characteristics can be easily translated into criteria against which shortlisted alternatives are evaluated in detail. This is discussed next.

16.2.4 Evaluate Concepts

Based on functional requirements and other characteristics, a number of alternative solutions can be established for a given function. The objective of this step of concept evaluation is to address the question of which is the best solution or which is better. This solution is best in the sense that it achieves the optimum balance between meeting the functional requirements and other aspects.

The reaction of different solution systems in terms of different aspects can by expressed by a matrix below:

$$[V_{ij}]$$

where i indicates the i-th system evaluation behaviour (or solution construction) and j represents the j-th criterion.

The V_{ij} evaluations can be determined according to one or more of the following methods:

1. An evaluation is carried out instinctively based on the designer's professional judgement on what is best and what is worst. An element of subjectivity is inevitable in this case.
2. The designer carries out an evaluation based on qualitative measures such as excellent, good, fair and bad, or by means of graphic symbols associated with these meanings.
3. The designer carries out an evaluation based on quantitative measures such as scores between 1-10, 1-5, or 1-100, or normalized probability between 0-1 (0% - 100%).

Above methods may be formalized by academics, study centres, technical associations, professional bodies, manufacturing firms, and packaged into computer software. This would improves the productivity and accuracy of evaluation work carried out by the designer.

16.3 DESIGN FOR QUALITY SOFTWARE

Based on the collaboration between the Polytechnic of Milan, Computervision, the Association of Italian Industrial Designers, and the Association of Lombardy (AIL), a computer software package is being developed using the above Design for Quality methodology. It is capable of supporting the designer in all stages of his/her work, from the identification of the function to the assignment of the design information. The overall program has been developed using the language UPL, which Computervision's CAD personal Designer software is equipped with.

The design for quality software includes the following main functional components:

- Catalogue of mechanical functions. This is a collection of some functions often used in machine design. Functions considered so far include axial constraint, radial constraint, radial and tangential constraint, force transmission, force transformation, moment transmission, moment transformation, transformation of force in moment.
- File of functioning principles. For each of the above functions, a file is maintained to contain realization principles with various physical phenomena.
- File of constructive solutions. For each of the above functioning principles, a file is maintained to include constructive solutions.
- Applicative software for the iconic managing of files.
- Optimization of all constructive solutions, in relation to design requirements for the given characteristics.
- Evaluation of the given solutions in relation to evaluation of the optimal solution.
- Comparison between two given solutions. Design modelling.
- Modelling of all the components that make up the machine, and of all the design information pertaining to it.

16.4 CASE STUDY OF DESIGN FOR QUALITY

The proposed case study is an example of an application of an commercial software, to organize files of mechanical solutions of given functions (Biggioggero et al 1995). The solutions can be selected and compared by the expert system. The procedure described in the preceding section is used here in this case study.

Step 1   Determine Functions

The function in this case study is simple: to transmit motion between two rotating shafts. To be more specific in mechanical terms, the function is about "radial and tangential fit between shaft and hub."

Step 2   Determine Archives

Figure 16.5 shows the top-level menu of the software with reference to the given function - radial and tangential fit. It includes two general categories of alternative solution principles:

1. direct fits (diretto)
2. with interposed elements (con elementi interposti)

The first category of direct fits include force fit (forzamento) and form fit (di forma). The second category includes removable fits (smontabili) or non-removable fits (definitivi). There are three types of different solution principles with the removable fits: pins, splines, and friction. Each of these removable fits in turn includes a number of solution alternatives. For example, the removable fits by friction can be obtained with elastic elements, ringfeder, ringblock, bikon, with air and oil pressure.

Figure 16.6 shows some examples of the constructive solutions for the given principle "direct force fit." An early warning is necessary here that Figures 6-8 are in Italian because they are outputs from the software. Figure 16.8 shows archives for the Bikon solutions (the squared italian words are for the page change).

Step 3 Determine Characteristics

Solutions are selected from the archives by analyzing characteristics specified earlier in Step 2. Figure 16.7 gives verbal descriptions about each of the solutions shown in Figure 16.6. These characteristics can be used in several ways, for example, to shortlist appropriate solutions from an solution archive and to evaluate the selected concepts to be discussed in the next step.

Step 4 Evaluate Solutions

From the preceding step, a number of solutions are obtained to satisfy the desired function. At this step, the characteristics established in the preceding step are used to evaluate these shortlisted solutions in detail. Because these characteristics are primarily oriented towards quality, the evaluation naturally places a primary emphasis on quality. Such evaluation can be made by an expert system. This step is not included in this case study.

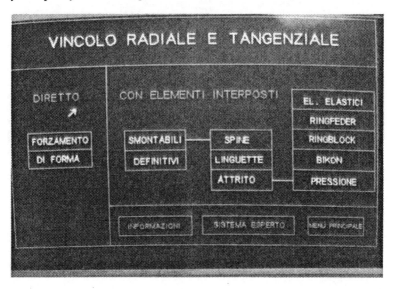

Figure 16.5 Raster image of solution principles for function: "radial and tangential fit".

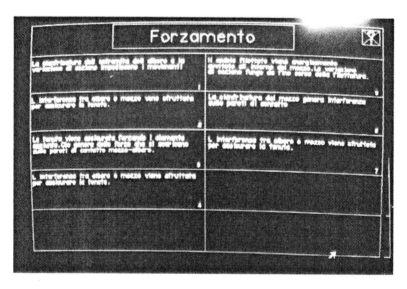

Figure 16.6 Raster image of constructive solutions of direct forced fits.

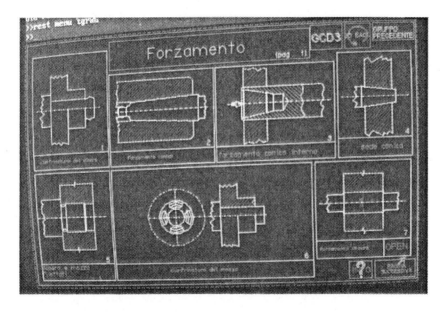

Figure 16.7 Raster image of information useful to address a first step of choice (selection).

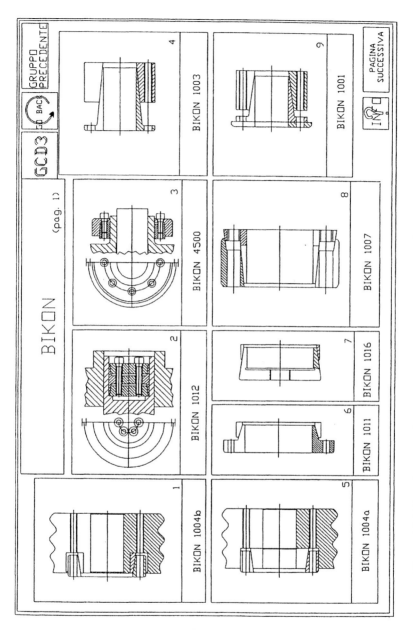

Figure 16.8 Plotted reproduction of a raster image of same representation of Bikon constructive examples.

16.5 SUMMARY

This chapter has presented a Design for Quality methodology and its associated software. It starts with a list of mechanical functions. For each of these functions, the designer can proceed to a database of physical principles that relate to that particular function. By making selections from these principles, according to the design specifications, the designer is able to make the optimum choice. Once the physical principle is chosen, the designer can access a database of configurations that will incorporate it. Again, choosing among these configurations, according to the specifications of the project, the designer is able to make the ideal choice. From the chosen configuration, the designer can proceed to the design modelling stage. Such necessary features are useful for designers, particularly in refining the development of their ideas, from initial concept to final design details of the machine. They are also of educational importance to designers, developing their creatively under the guidance of the archives provided. This will enable them to widen the archives themselves, this always identifying new principles and new configurations.

REFERENCES

Asimow, M. (1962) *Introduction to Design*, Prentice-Hall, London.

Biggioggero, G.F., Galli, P., Rovida, E. (In print) "Un software applicativo per la progettazione metodica", *Dipartimento di Meccanica Politecnico di Milano*, Milano.

Biggioggero, G.F., Rovida, E. (1990) *Disegno di Macchine: note di progettazione metodica*, CittaStudi, Milano.

Hubka, V. (1989) Design for quality, *Proceedings of ICED'89*, Institution of Mechanical Engineers, Harrogate, 1321-1333.

ISO 8402 (1994) *Quality - Management and quality assurance vocabulary.*

ISO 9000 series on quality.

Jensen, P.B. (1992) *ISO 9000 A Guide and Commentary*, APS/Hytek.

Koudate, A. (1991) *Il mangement della progettazione, ISEDI Petrini*, Torino.

Lamprecht, J.L. (1992) *ISO 9000: preparing for registration*, ASQC Quality press.

89/392/CEE   (14.06.90)

91/368/CEE   (20.06.91)

93/44/CEE   (14.06.93)

93/68/CEE   (22.07.93)

DESIGN FOR MODULARITY

Gunnar Erixon

This chapter presents a Design for Modularity technique - Modular Function Deployment. It is a systematic procedure consisting of five major steps. It starts with Quality Function Deployment (QFD) analysis to establish customer requirements and technical solutions with a special emphasis on modularity. This is followed by a systematic generation and selection of modular concepts. Module Indication Matrix (MIM) is used to identify possible modules by examining the interrelationships between module drivers and technical solutions. A questionnaire is provided for this. MIM also provides a mechanism for investigating opportunities of integrating multiple functions into single modules. A thorough evaluation can be carried out using Modularity Evaluation Charts (MEC) for each modular concept. Module Indication Matrix is used again to identify opportunities of Design for Manufacture and Assembly (DFMA) for further improvements to promising modules. While Module Indication Matrix (MIM) is applied at the level of sub-functions to establish modules, Modularity Evaluation Chart is applied at the product level as a whole. They complement each other.

The Design for Modularity method has a number of advantages. First, it structures the product development process which leads to rational product assortments. Second, it provides feedback to the synthesis phase, especially when the user becomes acquainted with the tool. Third, the tool supports a learning feedback and enhances the ability to "get it right the first time". Fourth, it enables creative thinking and encourages teamwork and therefore facilitates the implementation of concurrent engineering. Fifth, modular products are more competitive because they have grown out of a systematic procedure where every small detail has been sufficiently treated from customer requirements to finished product. Finally, the method guides the design iterations where the results of changes are measured, obsolete ideas are scrapped, promising ideas are revised and new ideas are born.

17.1 DESIGN FOR MODULARITY

Diverse customer requirements lead to wide variety of products. As a consequence production becomes complex and difficult to plan and control. Time is limited for companies to rationalise the entire product range. Instead customer needs have been met with "ad hoc" solutions, and quite often by specially built products.

It is well known that DFX (Design for X) tools are very efficient in product design. However, most DFX tools only handle product structures and components. Company studies have shown that efforts on the level of product ranges lead to further benefits as far as ease of manufacturing and assembly is concerned. Myrup Andreasen (1990) claims that the effect of the assembly principles derived on the product range, product structure and component level is in the ratio 100:10:1. This view is reflected by many researchers and practitioners (Pahl and Beitz, 1988; Stoll, 1986; Pessina and Meinhardt, 1994).

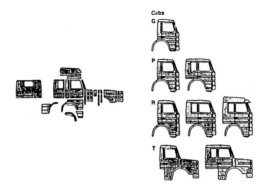

Figure 17.1 Scania´s new cab range is built up from a given set of modules.

Figure 17.1 shows one example of successful Design for Modularity at Scania AB in Oskarshamn, Sweden. All cabs are produced in one line. The cabs are built up from a standardised assortment of modules and components. Table 17.1 lists major achievements of Design for Modularity. Eight types of cab can be built, with thousands of variants within each type. The modules make it possible to produce all variants in a completely mixed flow. The changeover time for a new variant of cab is zero.

Table 17.1 Benefits of Design for Modularity of Scania´s new cab

	Before	After
Number of sheet metal parts	1400	380
Interior fitting parts	1800	600
Parts in top	7	3
Parts in front	8	3
Parts in doors	12	8
Windscreen	3	1
Sheet metal tools	1600	280

Table 17.2 Benefits of Design for Modularity

Development	Parallel design of modules, leading to reduced development time. Selecting parts of a product as modules according to future technology development. Simplified product planning. Possibility of using and creating "carry overs".
Manufacturing	Common modules lead to high volume and scale of economy advantages. Rational material handling of modules instead of products. Utilisation of investments in specialised manufacturing processes. Decreased rework by testing modules. Possibility for good work organisation.
Product variant	Possibility to adapt products to different markets by having some modules as "variants" modules.
Purchasing	Suppliers offer modules which may be cheaper to make in house. Lower logistic costs.
After sale	Possibilities of upgrading. Simplified maintenance and service. Possibility of rebuilding a product. Modules are easier to be disassembled for recycling.

A number of industrial case studies have been carried out (Erlandsson, Erixon and Östgren, 1992; Erlandsson and Yxkull, 1993). Table 17.2 summarizes typical benefits of modular design. Several general comments can be made. First, modular design is an excellent basis for continuous product renewal and concurrent development of the manufacturing system. Second, short feedback links for failure reports can be secured if modules are tested before delivery to the main line. Third, increased modularity of a product gives positive effects in the total flow of information and materials, from development and purchasing to storage and delivery. Fourth, combining a modular design with product planning will simplify the product development process and planning of corresponding manufacturing system changes. Instead of making big investments in new systems when changing the product, the manufacturing system can be developed in small steps (Erlandsson and Yxkull, 1993). Finally, the reduction of all throughput times is the most important effect of increased modularity. By an early fixation of the interfaces between modules, product development can proceed in separate parallel projects, one for each module. Short lead manufacturing times contribute to less working capital, increased quality and faster delivery (Gröndahl, 1987).

Figure 17.2 shows how the entire assortment of products can be divided into modules that are each manufactured in a so-called "module area". In the same way as modules are "Products in the Product", "Factories in the Factory" are formed as a result. The traditional final assembly line, where all the parts are assembled along the line, does not exist any more. With simple interfaces between the modules, assembly work is moved to the module assembly areas. Modules are assembled separately and supplied to the main flow, where they are attached to one another. The two-way arrow between module and module area indicates the possibility to achieve short and clear information links between development and manufacturing teams. This increases the possibility to apply real simultaneous development of products and processes. The figure shows that every module can be formed to meet its specific manufacturing requirements. At the same time successive development is made possible because a change in one module only influences a limited area without disturbing others.

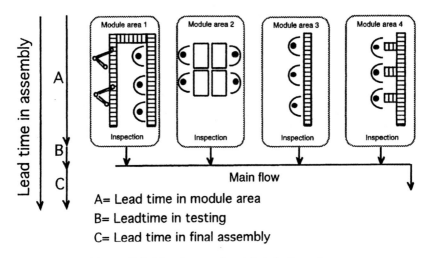

A= Lead time in module area
B= Leadtime in testing
C= Lead time in final assembly

Figure 17.2 Manufacture of modularly-built products.

17.2 OVERVIEW OF MODULE FUNCTION DEPLOYMENT (MFD)

Figure 17.3 presents the systematic procedure of a comprehensive Design for Modularity method - MFD: Modular Function Deployment. Five steps are involved. Each step of MFD is based on one formal technique. QFD, Pugh's selection matrix, and DFMA have been covered elsewhere. Module Indication Matrix (MIM) and Modularity Evaluation Chart (MEC) are explained in Sections 17.3 and 17.4 respectively.

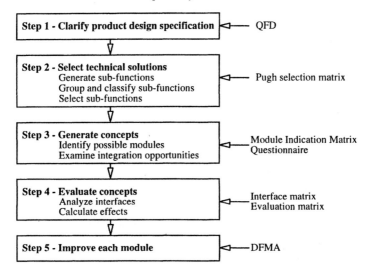

Figure 17.3 Design for Modularity - MFD flowchart.

Step 1: Clarify product design specification

The first step of Design for Modularity is to make sure that the right design requirements are derived from the customer/market needs. QFD (Quality Function Deployment) is well suited for this purpose (Akao, 1990). A multi-disciplinary team must be established to accomplish a QFD analysis. At this stage, the "modularity" requirements are of paramount importance and therefore should be specified explicitly in the QFD matrix. This creates the appropriate "mind set" for the participants right from the start. Case experience has shown that it encourages creativity and gives new dimensions to creative thinking.

Step 2: Select technical solutions

The second step of Design for Modularity is to establish technical solutions that meet product design specifications. First, when the design requirements are derived, a number of sub-functions fulfilling them are formed. Functional integration is discouraged at this stage and will be dealt with later to achieve greater modularity. Next, it is beneficial to distinguish common sub-functions required by all customers from optional sub-functions specific to customers (Pahl and Beitz, 1988). This helps establishing product variants later. Finally, several technical solutions may appear and selections have to be made. Experience shows that the Pugh's selection matrix is simple and effective for this purpose (Pugh, 1981). This selection method leads to a converging process where alternatives are sorted out and new ones come up. The process also prevents personal pushing of ideas for irrational reasons.

Step 3: Generate concepts

The aim of this step is to establish product concepts based on modules to achieve sub-functions selected at the previous step. Module Indication Matrices (MIM) is used for three activities. First, sub-functions are evaluated against so-called module-drivers. Second, every single sub-function is treated as a separate module and this results in the risk of getting a product that consists of stapled functions. The MIM works as a basis for analysing the possibility of integration. Third, a number of alternative modules may exist even after integration, to achieve a group of sub-functions. At this stage, a selection is necessary to narrow down the number. Again, the Pugh's selection matrix can be used for this purpose. In the method the selection of technical solutions is carried out only by comparing satisfactory achievement of the manufacturing goals. This can be done because all solutions suggested that have survived this far in the process, meet the design specifications.

Step 4: Evaluate concepts

Modular concepts generated at the preceding step are evaluated using the so-called universal virtues - Costs, Time, Quality, Efficiency, Flexibility, Risk and Environment (Olesen, 1992). Such evaluation is important to assess the proposed changes and to compare with the earlier situation. During a development process there are many cross-roads to pass and choices have to be made. It is also possible to use a good evaluation method in the early concept phase as an early feedback. In this case the method serves as an adaptive feedback process to the synthesis. Pahl and Beitz (1988) advocate that preliminary evaluations have to be made even in the conceptual phase as economic factors are of crucial importance in the design of modular systems. To minimise the costs of a modular system, not only the modules themselves but also their interaction must be taken into account .

Step 5: Improve each module

The design for modularity method described here should not be considered as a replacement of DFMA for design improvements on the part level. It is important to emphasise the necessity of such work within every single module in order to secure the final result. MIM is also a pointer to what is important for each module respectively, i.e. a module that is chosen mainly for service and maintenance reasons should be designed to facilitate disassembly.

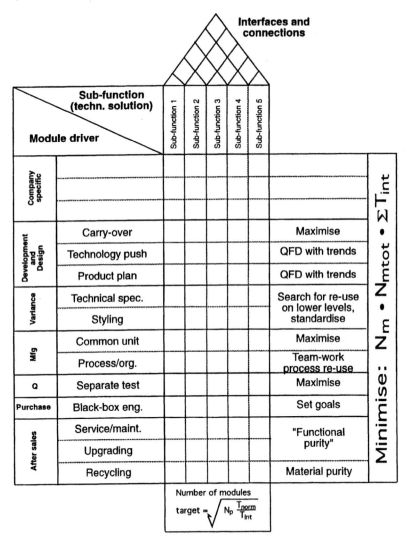

Figure 17.4 MIM - Module Indication Matrix.

Table 17.3 List of module drivers

Product development	
Carry-over	A carry over is a part or a sub-system of a product that can be re-used (carried over) from an earlier generation of a product to a new generation or from one product family to an other. This should be examined throughout the company.
Technological evolution	Technology push means that a part or a sub-system is likely to go through a technology shift during its life cycle because customer demands will change radically. The technology itself will evolve i.e. from mechanical to mechatronic, new material, resulting in changes and development of competitive products.
Product planning	When a part is a carrier of specific features the product planning might indicate a change at a specific time.
Variance	
Technical specification	Variations in technical specification should be accommodated in one or a few parts to prevent variations spreading throughout the entire product. It is also advantageous to adapt variations as late as possible in the manufacturing chain. Parametrization in one or a few of the modules may be one way of doing this.
Styling	Some products are strongly influenced by trends and fashion while others have part(s) that are strongly connected to a brand or trade mark. It is advantageous to create what can be called a styling module that can be altered more freely without causing disruptions in the whole product.
Manufacturing	
Common unit	It is possible to find parts and functions that can be common units used throughout the entire assortment of products. Common units can be found by comparing assembly drawings of several product types or by checking existing sub-assemblies. Parts that contain the basic function(s) are possible candidates.
Process / organisation re-use	Effective workcentres can be formed for modules. Work content, responsibility and authority, technical level etc. can be adjusted to suit a development opportunity, leading to increased work satisfaction. This also improves the possibility for automation since similar types of operations can be placed in the same workcentre. Modular product structures results in a high degree of similarity between different departments and makes it possible to use the same computer system for product specification (sales), purchasing and spare parts.
Quality	
Separate testing of functions	If each module can be tested before it is supplied to the main flow, immediate feedback on quality can be supplied to the operators, resulting in increased quality mainly due to prompt feedback.
Purchasing	
Supplier offers black box	Modularity makes it possible to purchase complete and standard modules (blackbox engineering) instead of individual parts. This reduces the amount of purchasing work. Modularity also reduces material cost because fewer parts are needed to build up the assortment. This means less material to ship and lower logistic costs. Dealing with one big supplier instead of many small ones also makes the administration part of the logistic cost lower.
After sales	
Service and maintenance	Quick service and maintenance are an important factor for many companies. Modules play an important role in fast service. One damaged module can be replaced by a new one and the damaged one can be repaired at a service centre.
Upgrading	Modular products provide better possibility to upgrade and rebuild. Several companies have exploited modular product structures for sales persons to give a detailed price quotations quicker.
Recycling	For better recyclability, the number of different materials should be minimized in a module and environmentally hostile material should be grouped in a module.

Table 17.4 MIM questionnaire

Carry over		
Are there	[ ] strong [ ] medium [ ] any	reasons why this part should be a separate module since it can be carried over from earlier to new product generation?
Technology push		
Is it	[ ] very possible [ ] possible [ ] some poss.	that this part will go through a technology shift during the product life cycle?
Product planning		
Are there	[ ] strong [ ] medium [ ] some	reasons why this part should be a separate module since it is the carrier of changing attributes?
Technical specification		
Is it possible to contain	[ ] all [ ] the main [ ] some	variants of the technical specification in this part?
Styling		
Is this part	[ ] strongly [ ] fairly [ ] to some extent	influenced by trends and fashion in such a way that form and/or colour has to be altered?
Common unit		
Can this part be the same in	[ ] all [ ] the most [ ] some	of the product variants?
Process/Organisation		
Are there	[ ] strong [ ] medium [ ] some	reasons why this part should be a separate module because: - it will be an ergonomic part to handle? - it has a suitable work content for a group? - the production accessories can be re-used? - it fits to our special know-how? - a pedagogical assembly can be formed? - the lead time differs extraordinary?
Separate testing		
Are there	[ ] strong [ ] medium [ ] some	reasons why this part should be a separate module because it can be tested separately?
Purchase		
Are there	[ ] strong [ ] medium [ ] some	reasons that this part should be a separate module because: - there are specialists that can deliver part as a black box? - the logistics cost can be reduced? - the capacity can be balanced?
Service/maintenance		
Is it possible to locate	[ ] all [ ] most [ ] some	of the service repair to this part?
Upgrading		
Is it possible to do	[ ] all [ ] most [ ] some	of the upgrading by changing this part only?
Recycling		
Is it possible to keep	[ ] all [ ] most [ ] some	of the highly polluting material in this part?

17.3 MIM - MODULE INDICATION MATRIX

This section explains the structure of MIM (Modular Indication Matrix) and its use. It is a QFD-like way of giving an indication of which sub-functional group(s) can form a module. With MIM, sub-functions are tested one by one against modularity criteria or module driver. An overall picture can be obtained by examining MIM to determine the reasons for forming a module and sub-functions it performs. To go through all the sub-functions of a product and test them against all module drivers has proved to be a powerful tool in the development process. To see a product as a number of sub-functions is a way of getting away from the existing form of the product and of encouraging creative and free thinking to find new forms (Tuttle, 1991). Figure 17.4 shows an example of MIM. The following components are included: (1) Module drivers, (2) Sub-functions or technical solutions, (3) Relationships between module drivers and sub-functions, (4) Goal settings, (5) Interfaces and connections (between sub-functions), and (6) Ideal number of modules.

There are three basic components in the MIM. A set of module drivers need to be first established. Table 17.3 shows some of the typical examples of module drivers. Sub-functions entered into the MIM are outcomes from the sub-function selection matrix. The questionnaire shown Table 17.4 can be used to assist establishing the relationships between module drivers and sub-functions. The three optional features are often used in combination with the MEC.

17.4 MEC - MODULARITY EVALUATION CHARTS

Figure 17.5 shows a sample Modularity Evaluation Chart. It serves as a checklist for examining product characteristics that influence a good modular design. The first column includes ten major items, listed in Table 17.5 under three main life cycles of product development, production assembly and sales and after sales, for evaluating product modularity. The second column is for specifying the goals - ideal or optimum targets or rules. They can be calculated using the equations given in the first column or assigned based on experiences. They must reflect the ambition of improvements. Following is a list of a few rules of thumb to support goal settings in the search for the best modular concept:

1. Aim at a modular concept where the number of modules per product is equal to the square root of the expected number of parts in one product.
2. The expected number of parts in a re-design project may be set to 70% of the number in an existing product.
3. Make sure that every new variant of a module can be used in several product variants.
4. Minimise the value: [number of modules per product (N_m) * total number of modules for all product variants (N_{mtot}) * total final assembly time(ΣT_{int})].
5. Refine the interfaces between the modules to minimise final assembly time, aiming for 10 seconds per interface.
6. Maximise the share of separate tested modules.
7. maximise the share of carry-over modules and purchased modules.
8. Limit the number of different material in a module (material purity).
9. Do not divide a function in two or more modules (functional purity).

Guide			
General Number of parts in average product (Np). A relevant objective for a new concept is 70%. (New Np = 0.7*Old Np).	$N_p = \dots\dots\dots\dots$		
Estimate the average assy. time relation between part assy. op. and interface assy. op. Common average part assy. op. is 10 seconds (Tnorm = 10). 10 sec. op. time is an easy interface and 50 sec. a fairly difficult one (Tnorm ≤ Tint ≥ 5 Tnorm). T_{int} = Average assembly time for interfaces. T_{norm} = Average assembly time for one part (10 sec).	assembly time relation: $T_{int} / T_{norm} = \dots\dots\dots\dots$		
Lead time in assembly	**Ideal, Optimum or Goal**	**Actual**	**Yield**
$L = \dfrac{N_p\, T_{norm}}{N_m} + (N_m - 1)\, T_{int}$ Where: N_m = Number of modules in one product.	$20\sqrt{N_p} - 10 = \dots\dots\dots$ <small>Ideal when assembly time relation = 1.</small>		%
System cost Share of purchased modules following the rules.	Goal = ...		
Product cost $C = \sqrt{N_m\, N_{mtot}\, \dfrac{\Sigma T_{int}}{3?}}$ Where: N_{mtot} = Total number of modules.	$1{,}5\sqrt{N_p} = \dots\dots\dots$		
Quality Estimate the expected average defects in figure "Expected ....". (PPM).	Ideal value (all separately tested) from figure, upper curve, 100% = ...		
Lead time in development Int. Compl. = $\dfrac{\sum\limits_{i=1}^{N_m - 1} T_{BDI_i}}{3}$ Where: T_{BDI_i} = Assembly time for inteface, i. (DFA–analysis)	$\dfrac{(N_m - 1)\,10}{3} = \dots\dots\dots$ <small>Observe Nm=the actual value for the concept evaluated.</small>		
Development cost Estimate the share of "carry overs" following the rules.	Goal = ...		
Development capacity Share of purchased modules as above.	Goal = ...		
Sales / After Sales Product variants as: $\quad E_{var} = \dfrac{N_{var}}{N_{mtot}}$ Where: N_{var} = Number of variants that can be built. N_{mtot} = Total numbers of modules needed.	"maximize"	....	....
Service / Upgrading, check the MFD for functional "purity".	No functional connections between modules.	....	....
Recycling, see separate Pareto chart.	The 80/20 rule	....	....

Figure 17.5 MEC - Modularity evaluation chart.

Table 17.5 Good modular design

Effects (life phases)	Product characteristics	Value/rule
Development		
1. Lead time in development	Interface complexity	Value
2. Development costs	Share of carry over	Rule
3. Development capacity	Share of purchased modules	Rule
Assembly		
4. Product costs	Assortment complexity	Value
5. System costs	Share of purchased modules	Rule
6. Lead time	Number of modules in product	Value
7. Quality	Share separately tested modules	Value
Sales/After sales		
8. Variant flexibility	multi-use	Value
9. Service/Upgrading	Functional purity in modules	Rule
10. Recyclability	Material purity in modules	Rule

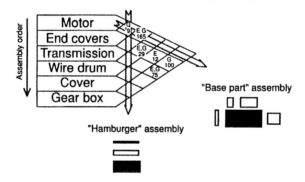

Figure 17.6. Evaluation matrix for Interface Complexity.

1. Development lead time

Development lead time depends on the Interface Complexity. The lead time in development will decrease when there is a possibility to work in parallel. This would give the smallest possible information flow between design groups and shorten the development time. Simple interfaces between the modules, without causing increased complexity within the modules are preferable. The specification of an interface is made by giving form, fixation principles, number of contact surfaces and attachments, number of energy connection points.

An interface might be fixed, moving or media transmitting. Fixed interfaces only connects the modules in a product and transmit forces. Moving interfaces transmit energy in the form of rotating, alternating forces etc. Media can be fluids electricity etc. One example of interface revisions is the questions concerning electrical connections revised by Ong (1991):

1.   Can modules be located next to each other in order to eliminate connections?

2.   Can wires be connected between two assemblies rather than using an intermediate part?
3.   Can a simpler method of wire attachment be used?
4.   Can several separate connectors be replaced by a single connector?
5.   Can parts used for wire securing be integrated with the chassis?

Figure 17.6 shows an evaluation matrix for Interface Complexity. (E) stands for moving (energy transmitting) and media transmitting force, inertia, electricity etc. and (G) for solely geometrical specification in the connection. The assembly operation times should also be entered to complete the picture. Interface principles preferred have been marked with arrows. All markings outside the areas marked show not wanted connections and should be avoided and/or be made the subject of improvements. The matrix serves as a pointer for the interfaces that have to be observed and eventually improved.

The assemblability metrics of Boothroyd-Dewhurst DFA can be used to calculate the interface complexity and measure the possibility of parallel product development. An ideal value of the interface complexity is reached when we succeed in getting ideal values for all the interfaces in the product, that is only one contact surface between each module and 10 seconds for the interface assembly operation.

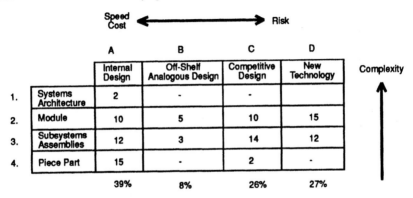

		A Internal Design	B Off-Shelf Analogous Design	C Competitive Design	D New Technology
1.	Systems Architecture	2	-	-	
2.	Module	10	5	10	15
3.	Subsystems Assemblies	12	3	14	12
4.	Piece Part	15	-	2	-
		39%	8%	26%	27%

1. Specifications are rated

2. Definitions
   - Internal Design   - Existing Xerox Product
   - Analogous Design - Similar Product Outside of Xerox
   - Competitive Design - Competitive Product Providing Same Function
   - New Technology   - No design Available Inside or Outside

3. Items 1 through 4 represent hardware or software groupings as they
   relate to an appropriate product positioning model.
   - The number 10 in the 2A element means that 10% of the specifications
   can be met by using internal design at the module level.

Figure 17.7 Xerox Re-usability matrix.

2. Development cost

Development cost depends on the share of carry-overs. The number of carry-overs has great influence on the development cost. A carry over is a part of a product (a module) that is carried over to the next generation of the product without any changes. The possibility to use carry-over modules should be discussed with the considerations of customer requirement on the new product generation, the company image, and the new modular structure. The share of carry over modules directly influences the development costs and can be used to evaluate different concepts when the development costs are vital.

Holmes (1993) describes how the Xerox Corporation focuses on re-use in product development with the help of their so-called re-usability matrix as shown in Figure 17.7. This matrix is used together with QFD matrices, in which the customer needs and wishes are evaluated against existing products, modules, sub-systems and piece parts.

3. Development capacity

Development capacity depends on the share of purchased modules. The demand for development capacity can be moderated with the help of "blackbox" engineering. By "blackbox" engineering it means that the supplier is responsible for the development of a complete module. Development capacity is closely related to system costs to be discussed in Item 5.

4. Product costs

Direct material and labour are major components in the manufacturing costs for many products. Experience from earlier research shows that the detailed design of each module has a great influence on the product cost. It is therefore important to pursue proper DFMA work for the design of each separate module. One could expect that the direct material costs would increase for modular product because of the eventual need of extra interfaces. Our case studies however showed that this is not the case. The companies had succeeded in controlling this and the measured effect on the material costs lay between an increase of 3% and a decrease of 10%, with a median of 6% decrease.

The product cost also includes the module specific capital costs, tools, fixtures etc. The size of these costs depends mainly on the number of articles, number of modules, and the complexity of the module assortment. The possibilities to control these costs on the assortment level is mainly due to greatest possible re-use of modules in the assortment and/or re-use of processes.

An assortment with the smallest possible variation of modules and interfaces that satisfies the customer requirements should be the goal. In other words, the complexity of the assortment should be minimised. Pugh (1990) has used a measure for the complexity of a product calculated on component level. If we look upon a modular product assortment in the same way, it is possible to calculate the complexity for the entire assortment. The complexity in a modular product assortment increases with the number of modules in each product variant, the number of modules totally needed to build all product variants and the number of contact surfaces in the interfaces. The number of contact surfaces might be difficult to estimate or calculate. Our experience shows that the total assembly operation time (seconds) for the interfaces can be used as basis for an approximation. A modular concept that has the lowest value for C will have the lowest product costs summarised over the entire assortment. This is a measure that is impossible to get from a standard cost calculation.

5. System costs

Modules constituting products can be purchased from outside vendors or manufactured in-house. The latter results in system costs which include the following items:

type of costs	influenced by
assembly system purchase	number of vendors
production planning	complexity, number of parts
quality control	complexity, number of parts
production engineering	number of modules, complexity
logistics	number of vendors, number of parts

System costs are very much dependent on the share of purchased modules. The higher the share the lower the system costs. There are two extremes. One is that all modules are manufactured in-house and none is purchased from vendors. This results in highest possible system costs. The other is that all modules are purchased from vendors and none manufactured in-house. This leads to lowest possible system costs. In reality, companies operate somewhere in between. A "make or buy" analysis must be carried out to determine which modules to be sourced from outside. Several question must be addressed:

1. Are there any strategic reasons for keeping production of the module in house?
2. According to the product plan, do we today and in the future have the capacity to produce this module?
3. Are there any qualified companies for producing the module?
4. Are there strategic reasons for keeping this module in house?
5. Do we have capacity today and in the future to develop and manufacture?
6. Can this module be purchased as standard from supplier?

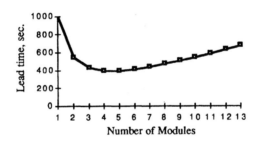

Figure 17.8 Assembly lead time as a function of the number of modules in a product.

6. Assembly lead time

The assembly lead time is largely determined by the number of modules in the product. Based on the assumption that: each module is concurrently assembled with the others to be delivered to the main assembly line where complete modules are assembled to each other an ideal value for the lead time can be calculated. One example of a lead time calculation is plotted and shown earlier in figure 17.8. It is clear that an optimum exists for an ideal number of modules. In figure 17.9 these ideal/minimum values are plotted as the number of parts in the products and the relation between interface assembly and part assembly times. This figure can be used to estimate the ideal number of modules for the division into modules. Theoretically it will be possible to shorten the lead time further by dividing each module into sub-modules. There is, however, a lowest possible limit for this kind of sub dividing when the work content in a module gets too small.

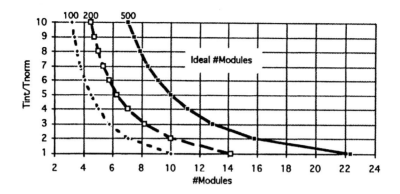

Figure 17.9 Estimation of ideal number of modules.

7. Quality

Quality depends on the share of separately tested modules. The quality improvements that can be reached by separate testing of some or all modules can now be calculated. The result of such a calculation is shown in Figure 17.10, showed earlier. Quality in the assembly system will be improved when modules are designed to admit separate functional testing. That is, only perfect modules are delivered to the main flow. The quality increase is then due to the shorter feedback time of fault reports within the module area (team work area) that is achieved. The best possible quality, according to these criteria, will come when all modules are separately tested. A modular product with the highest share of separately tested modules is the best in this sense. It should be noted that this is valid only for the quality effect of the assembly. The piece part design will of course also affect the total quality.

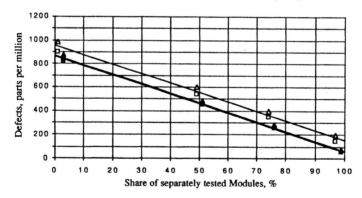

Figure 17.10 Expected average defect rate with separate testing of modules.

Nevins & Whitney (1989) gives four circumstances that encourage testing on module level: Such tests might address faults which occur very often.

1. They might address faults which are substantially cheaper to diagnose or repair at the module level than later.
2. No later tests for a fault may be available, because the test points are no longer accessible, for instance.
3. No later specific test may be available; that is, later tests may reveal the fault in question only in combination with other faults, requiring additional tests or diagnoses.

8. Variant flexibility

Variant flexibility depends on multiple uses of modules. When judging how a modular concept supports the creation of variants the re-use of modules, processes and organisations is important. A simple measure E_{var} of the variant flexibility can be obtained from the number of variants and the total number of modules. A high value indicates high similarity between product variants, resulting in many advantages such as fewer set-ups, fewer tools, simpler order planning.

It is important to avoid unnecessary variants. A German study by Schuh and Becker (1989) showed that 60% of the variants in the products studied had already existed in earlier generations of the products. The following questions can be asked to examine the significance of creating variants and a possibility to control the variant explosion:

1. Will this really be a variant?
2. How significant is it to the customer?
3. Will it be clear to the customer that it is a variant?
4. Where in the manufacturing chain, will the variant appear?
5. Which variants are necessary?
6. In which part of the product can variants be allowed?

9. Service/Upgrading

Service/Upgrading depends on the functional purity in modules. By functional purity it means that no single sub-function should be divided between two or more modules. This is already secured by the use of MIM. MEC provides a double check here.

10. Recyclability

Recyclability depends on the material purity in modules. In order to ensure a high degree of recyclability the number of different materials should be kept as low as possible within each module. A simple Pareto chart may give a picture of how this requirement is met, and the 80/20-rule can be used to set goals.

17.4 CASE STUDY: SEPSON MOBILE WINCHES

Westlings Industri AB is a medium sized company in the north of Sweden. They produce, amongst other products, 2,000 to 3,000 SEPSON mobile winches every year. Winches are for adaptation on vehicles such as tractors, rescue cars, trucks, etc. Most winches are hydraulic but some are electrical. The products are well-known for good quality and the level of know-how in the company is high. Product development, sales, manufacturing and assembly are all carried out in house. Figure 17.10 shows sample variants of SEPSON winches.

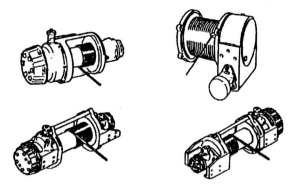

Figure 17.11 Variants of SEPSON mobile winches.

Over the years the number of variants of winches in the assortment had grown out of control. In 1992 there were 24 main families of winches with more than 10 variants in each family. Every variant had its own identification number and was uniquely identified in the MRP (Materials Requirements Planning) system. The inventory consisted of nearly 10,000 parts and a value of 5 million crowns. At the same time the competition from foreign volume producers had increased rapidly.

To deal with the situation the company decided to reengineer the winch assortment with the objective of lowering manufacturing costs and increasing inventory turnover. The goals were set at a 50% reduction in manufacturing costs and a shortening of lead time from 8 weeks to 2 weeks. It was decided to develop a modular design for the winches that should make it possible to build 80% of the product variants out of a standardised range of modules. The project team consists of six people: company manager, product manager, design manager, manufacturing engineer, one assembler and one external consultant. The method described previously was used in this project.

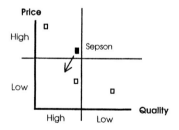

Figure 17.12 Price/Quality diagram.

Step 1: Clarify product design specification

Design specifications are re-examined for the product assortment. Some purging actions become already evident at this preparatory stage. It can be quickly determined which winch family was already fading or dying out. Three activities were conducted at the preparatory stage. They were customer survey, competition analysis and specification brainstorming.

A survey was first carried out with the help of opinion polls at five customer companies and they were also interviewed personally. The objective was to emphasize quality over quantity. The questionnaires included weights of customers needs and the recognition of company in the market place. A number of existing customer specifications were analyzed.

The customer survey was followed by a competition analysis to establish the positioning of the products. The current level of the SEPSON product was established through the customer survey. The relative position with the competitions was evaluated to set out a goal for improvement. Figure 17.12 shows that the product cost should be reduced to lower half and quality improved further.

Based on the customer survey and competition analysis, a brain storming session was conducted to conclude this preparatory stage. Customer requirements are grouped into easy to purchase, good styling, high reliability, high security, easy to adapt, easy to use, high relation performance/price. Figure 17.13 shows a QFD analysis of customer requirements in relation to design requirements. Some of the wishes that appeared could be characterised as non-technical and should be investigated for separate QFD analysis.

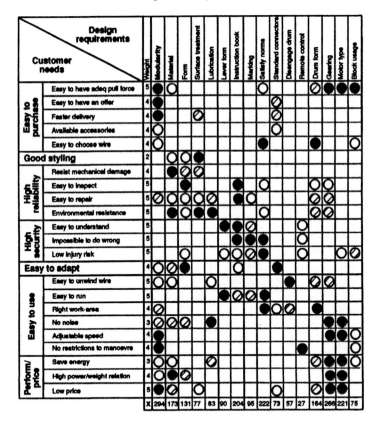

Figure 17.13 QFD for modular product planning.

Technical Concepts and Solutions	Lead time	Storage	Direct Labor	Direct Material	Purchase	Quality	Jobsatisfaction	Balancing	Sales Offers	Logistical Costs	Company specific		
Exchang. gearbox	=	=	+	-	+	=	-	=	=	-			2+	3-
Reconstr. gearbox	+	+	-	=	+	=	+	=	=	+			5+	1-
Cooper. gearbox	-	-	-	-	-	=	-	=	=	-				7-
Existing solution	D		A		T		U		M				

(Column heading above: Manufacturing Objectives)

Figure 17.14 Pugh's selection matrix.

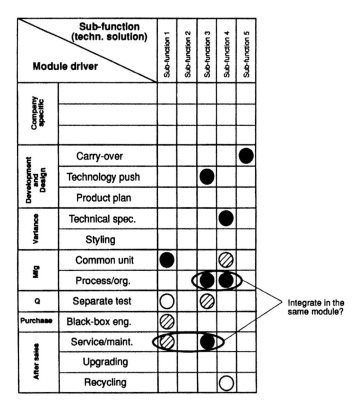

Figure 17.15 Module-Indication-Matrix for integration.

Step 2: Select technical solutions

The most important design requirements were transferred on in the analysis and the functional structure formed created the basis for the development of technical alternative solutions of sub-functions needed in the product. In order to determine which alternatives to choose an evaluation as described earlier was made. The different solutions were evaluated against manufacturing criteria as shown Figure 17.14.

Step 3: Generate concepts

How the modules should be shaped was given through the MIM in Figure 17.15. Different types of integration between sub-functions were tested and evaluated. Many combinations were removed early because of technical difficulties in doing the integration. It was clear that the technical solution for sub-function had to exist in order to test the integration technically.

Step 4: Evaluate concept

With the help of the Modularity Evaluation Chart, this new design concept was evaluated. The result is shown in Figure 17.16. It can be seen from the evaluation chart that the new modular design produces better results in shortening the assembly lead time and lowering the product costs, measured across the entire assortment. This leads to further reduction in development lead times and improvements in product quality.

Step 5: Improve modules

Two principally different alternative concepts were proposed for the modular winch design. One was based on a single gearbox concept where the gearing is varied by interchanging the internal gears. The other concept was based on smaller gear-box modules where the gearing is altered by choosing the number of gear modules. The final choice was based on the criteria specific to this company. The company had the manufacturing capability for producing gearboxes for the second alternative. On the other hand, extra investment was needed for the company to produce gearboxes for the first alternative. The product manager was also of the opinion that the second alternative (modular gearbox) would be more flexible. He saw that it would be possible to give the customer even greater options than planned from the beginning. Although the first alternative consisted of much fewer parts and would be cheaper to produce, the second alternative was favoured to avoid further investment and exploit extra flexibility.

As a result, a typical winch contains an average of seven modules and twenty-eight variants have been generated so far, as shown in Figure 17.17. A new concept of modular winches was generated. Figure 17.18 shows some assortments of modularly built winches. The new product range consists of six modules. Three modules are variant and the other three modules are common units. A typical winch consists of seven modules, including two gear-box modules. Twenty eight variants of winches are built up through various combinations of the six modules. Table 17.8 summarizes the difference between the old and modular winches.

Evaluation Chart

Guide	Ideal, Optimum or Goal	Actual	Yield
General Number of parts in average product (Np). A relevant objective for a new concept is 70%. (New Np = 0.7*Old Np).	$N_p = \underline{84}$		
Estimate the average assy. time relation between part assy. op. and interface assy. op. Common average part assy. op. is 10 seconds (Tnorm = 10). 10 sec. op. time is an easy interface and 50 sec. a fairly difficult one (Tnorm ≤ Tint ≥ 5 Tnorm). T_{int} = Average assembly time for interfaces. T_{norm} = Average assembly time for one part (10 sec).	assembly time relation: $T_{int} / T_{norm} = \underline{4,5}$		
Lead time in assembly $L = \dfrac{N_p\, T_{norm}}{N_m} + (N_m - 1)\, T_{int}$ Where: N_m = Number of modules in one product.	$20\sqrt{N_p} - 10 = \underline{170}$ Ideal when assembly time relation = 1.	503	% 34
System cost Share of purchased modules following the rules.	Goal = ...	/	
Product cost $C = \sqrt{N_m\, N_{mtot}\, \dfrac{\Sigma T_{int}}{3}}$ Where: N_{mtot} = Total number of modules.	$1,5\sqrt{N_p} = \underline{13}$	21	62
Quality Estimate the expected average defects in figure "Expected ....". (PPM).	Ideal value (all separately tested) from figure, upper curve, 100% $= \underline{150}$	900	17
Lead time in development Int. Compl. = $\dfrac{\overset{N_m - 1}{\underset{i=1}{\Sigma}} T_{BDI_i}}{3}$ Where: T_{BDI_i} = Assembly time for inteface, i. (DFA–analysis)	$\dfrac{(N_m - 1)\,10}{3} = \underline{20}$ Observe Nm=the actual value for the concept evaluated.	132	15
Development cost Estimate the share of "carry overs" following the rules.	Goal = ...	/	
Development capacity Share of purchased modules as above.	Goal = ...	/	
Sales / After Sales Product variants as: $E_{var} = \dfrac{N_{var}}{N_{mtot}}$ Where: N_{var} = Number of variants that can be built. N_{mtot} = Total numbers of modules needed.	"maximize"	2,8	....
Service / Upgrading, check the MIM for functional "purity".	No functional connections between modules.	V	OK ....
Recycling, see separate Pareto chart.	The 80/20 rule	/	....

Figure 17.16 Modularity evaluation chart.

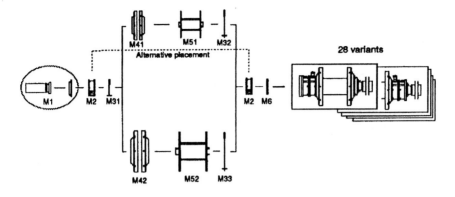

Figure 17.17 Modular winch assortment.

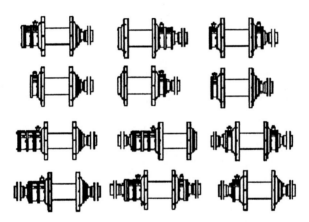

Figure 17.18 Some of the variants of modularly built winches.

Table 17.8 Comparison between the new and the old concept

	Old	Modular
Number of modules per product, N_m	-	7
Number of modules in assortment, N_{mtot}	-	10
Number of parts in product, N_p	119	84
Number of different parts in product	51	28
Number of different parts in assortment	107	32
Assembly time per product, sec	1200	751
Theoretical possible lead time, sec		503
Final assembly of modules, T_{int} sec		398
Number of drawings in A4 equivalents	91	45

17.5 SUMMARY

This chapter has presented a Design for modularity tool which is applicable across the entire product range. Through the use of the method, it becomes easier to plan products and predict their performance, as well as to control and manage the development. The use of module drivers makes the manufacturing aspects clearer and gives a direct link between customer needs and the design requirements. This facilitates achieving concurrent development of product and processes increases.

With the help of the guidelines and ideal/optimum values given in the evaluation chart, it is possible to set the adequate targets for improvement projects. Besides they also give an early feed back at the beginning of a project and make it possible to judge the economic results at an early stage.

REFERENCES

Akao, Y. (1990) *QFD - Integrating Customer Requirements into Product Design,* Productivity Press, 1990.

Barkan, P. (1992) *Benefits and Limitations of Structured Methodologies in Product Design,* ME 217 A 92/93.

Boothroyd, G., Dewhurst, P. (1987) *Product Design For Assembly Handbook,* Boothroyd Dewhurst Inc.

Brannan, B. (1991) Six Sigma Quality and DFA - DFMA Case Study/Motorola Inc., *Boothroyd & Dewhurst DFMA Insight,* **2**, Winter.

Erlandsson, A., Erixon, G., Östgren, B. (1992) Product Modules-The Link Between QFD and DFA?, *International Forum on DFMA,* Newport (RI).

Erlandsson, A., von Yxkull, A. (1993) Product plans and their interaction with the development of the manufacturing system, *Proceedings of ICED'93,* The Hague, Holland.

Fujimoto, T., Sheriff, A. (1989) Consistent patterns in Automotive Product Strategy, Product Development, and manufacturing Performance - Road Map for the 1990´s, *IMVP'89.*

Gröndahl, P. (1987) *Ledtidsförkortning genom förändring av tillverkningssystemet - effekter av flexibel automatisk montering,* Ph.D. Thesis, The Royal Institute of Technology, Department of Manufacturing Systems, Stockholm, Sweden.

Holmes, B. (1993) Competing on delivery, *Manufacturing Breakthrough,* January/February.

Andreasen, M.M., Olesen, J. (1990) The Concept of Dispositions, *Journal of Engineering Design,* **1** (1).

Nevins, J.L. Whitney, D.E. (1989) *Concurrent Design of Products & Processes,* McGraw-Hill Publishing Company.

Olesen, J. (1992) *Concurrent Development in Manufacturing - based on dispositional mechanisms,* Ph.D. Thesis, Institute of Engineering Design, Technical University of Denmark.

Ong, N.S. *et al.* (1991) Assembly times for electrical connections and wire harnesses, *International Journal of Advanced Manufacturing Technology,* **6** (2), 155-179.

Pahl, G., Beitz, W. (1988) *Engineering Design - a Systematic approach,* Springer-Verlag.

Pessina, M., Mainhardt, B. (1994) Customized Manufacturing Is Ultimate in Meeting Customer Demands, *Total Quality Newsletter,* **5** (11).

Pugh, S. (1981) Concept selection - A method that works, *Proceedings of ICED'81,* Rome, Italy.

Pugh, S. (1990) *Total Design*, Addison-Wesley Publishing Company.

Schuh, G., Becker, T. (1989) Variant Mode and Effect Analysis A new Approach for Reducing the Number of Product Variants, *International Forum on DFMA*, Newport (RI).

Stoll, H.W. (1986) Design for Manufacture: An Overview, *Applied Mechanics Review,* **39** (9).

Tuttle, B. L. (1991) Design for Function: A Cornerstone for DFMA, *International Forum on DFMA*, Newport (RI).

Östgren, B. (1994) *Modulindelad produkt ger effekter i hela tillverkningssystemet*, Licentiate Thesis, The Royal Institute of Technology, Department of Manufacturing Systems, Stockholm, Sweden, TRITA-TSM R-94-2.

18

DESIGN FOR OPTIMAL ENVIRONMENTAL IMPACT

Leigh Holloway; Ian Tranter; David W. Clegg

This chapter presents a design for environment (DFE) methodology to assist designers to address environmental problems of their product designs. It considers the complete product life cycle from "cradle to grave". Environmental impacts of design decisions are generally discussed in Section 18.1. The DFE method is detailed in Section 18.2. A case study is given in Section 18.3 to demonstrate how the method works.

In the past environmental problems were seen, and dealt with, as specific problems affecting certain areas, such as waste disposal sites containing hazardous materials or certain stretches of river and waterways being polluted. Traditionally manufacturing and environmental problems were treated very much independently and little or no concern was given to the environment during the course of product development. As our understanding and awareness of these problems develops it is becoming apparent that design and manufacturing have a very immediate effect on the environment and can, to a large extent, dictate the effects which products and their related systems have on the eco-systems around us. If the environmental problem is to be addressed it appears that design practices will have to change. Design activities can dictate up to 70% of the total manufacturing cost of a product, so it would be reasonable to conceive that a large proportion of the environmental cost of a product can also be dictated at the design stage. The complexity of the product design process necessitates approaches such as concurrent engineering which utilises a number of methodologies and tools to assist designers and keep product development times low. The inclusion of further concerns, such as environmental, threaten to complicate design even further and as such the development of an environmental concurrent design methodology, Design for the Environment, is required.

18.1 ENVIRONMENTALLY CONSCIOUS DESIGN

Terms such as Green Design, Design for Environment, and Environmentally Conscious Design are used alternatively. Design for Environment (DFE) will be used in this chapter. DFE is design carried out within current product development frameworks, that addresses all the environmental impacts associated with a product or system throughout its complete life cycle, with a view to reducing these impacts to a minimum but without compromising other criteria such as function, quality, cost and appearance.

18.1.1 Drives for Green Design

The need to take environmental considerations into account has been driven by a number of factors, each of which may facilitate a different approach to address a particular problem:

- **Financial.** Resource depletion and pollution are costly. Many organisations are beginning to realise that environmental performance is no longer a moral dilemma but is becoming a commercial imperative.
- **Legislative.** Legislation is one of the major "pushing" factors behind the shift to consideration of the environment within industrial practices. The severity and type of legislation varies depending on the country involved but in all cases the measure is set to increase. With an increased awareness of the environment public consumer pressure is also a major consideration in design.
- **Market Pressure.** In the past products offering better environmental performance differed considerably from their more mainstream counterparts in terms of functional performance and appearance. For most people a compromise in performance allied to a higher price was not acceptable. However with the advent of new technologies and adaptation of existing design methods these problems may be overcome and truly 'environmentally friendly' products marketed.
- **Environmental concerns.** It may be difficult to visualise the global effects of what may seem very minor design decisions. Design has the power to influence the environment in many ways. In order to fully understand how design and designers can help us achieve a sustainable future we need to look at the various problems which they will have to address. It helps to classify these problems into definite areas of concern as shown in Table 18.1 Guineé *et al.* (1993).

18.1.2 Environmental Effects of Design Decisions

All of the environmental problems described above are a direct affect of modern industrial practices. As designers occupy a central position in the product development process they have the power to positively influence environmental effects arising from their decisions and actions. When considering the concept of Design for the Environment it becomes apparent, as in many concurrent engineering imperatives, that there are many ways in which engineers may achieve their goals. By studying the different approaches to environmental or 'green design' it will become apparent that each particular method will have its pros and cons but also that there is a main core of objectives and actions which may be compiled and presented as generic goals in design for optimal environmental impact.

Design decisions can have a very profound and complex effect on the overall environmental impact of a product or system. In order that these decisions can be considered easily and quickly during the product development process, there is a need for methods and guidelines that help designers to include these concerns in their work. By choosing the material content and composition as well as processing routes, component arrangement, efficiency during use and the scope for maintenance or easy recycling the design team has fixed the main parameters of environmental effect.

Table 18.1 Environmental issues

Depletion	Pollution	Disturbances
Abiotic resources	Ozone depletion	Desiccation
Biotic resources	Global warming	Physical ecosystem degradation
	Photochemical oxidant	Landscape degradation
	formation	Direct human victims
	Acidification	
	Human toxicity	
	Ecotoxicity	
	Nitrification	
	Radiation	
	Dispersion of heat	
	Noise	
	Smell	
	Occupational health	

Materials Selection. Material choice has been informed by the trade off of functional performance with availability, processability and cost. Required mechanical strength and stiffness, electrical properties, UV and corrosion degradation are balanced with the amount of material needed and its ability to be transformed into the desired component. But material choice also brings a large number of complex environmental effects. The 'winning' of the material will deplete resources (be they renewable or not), cause pollution and use energy. The pre-processing of the material, for example refining bauxite into aluminium or chemical modification of oil products into thermoplastic, will add to the burden. Post-use activities such as recycling or disposal will also result in pollution and waste. The disposal phase of products however may be the most complicated to assess in environmental terms. Additives and material alloying may prejudice straightforward recycling while incineration and energy recovery may result in emissions which are unacceptable. All these effects can be weighed up during material choice, they should be clearly stated and wherever possible be quantified.

Processing Routes. Manufacturing processes can dictate the amount of waste material generated during transformation into a component. Near net shape processes such as casting, forging, injection moulding and blow moulding are inherently more efficient than subtractive processes such as machining and etching. The material removed by such subtractive processes may be reused as raw material but in most cases it will require some form of reprocessing thus increasing energy usage and pollution as a result. For those applications which require simple uniform shapes the use of pre-formed stock such as extrusions offers environmental advantages in terms of material wastage and energy usage. Paradoxically some near net shape processes, such as casting, are energy intensive in themselves but this is outweighed in environmental terms by the overall saving on material and secondary process energy.

Functions and Form & Size. The focus of design for efficiency during use changes with the nature of the product. Non-energy using products such as packaging have what may be termed as passive energy consumption during transport and as such weight reduction is a key issue. Reducing the weight of goods which are transported and their packages will have a direct effect on the fuel consumption of the transporting vehicle thus reducing energy usage and pollution. For products which actively use energy, efficiency is of paramount importance. Again we come across an ambiguity in that the definition of efficiency, or those factors which effect efficiency, will depend upon the nature of operation of the energy consuming product. For instance vehicles create most of their life time environmental impact during use through fuel consumption. In such a case efficiency may be looked at from different viewpoints. Efficiency of the engine within the vehicle or efficient choice and use of materials to reduce the weight of the vehicle and in turn reduce the fuel consumption. In a simpler product such as a fan heater energy efficiency of the motor and heating elements will obviously reap the largest environmental gains. In an intermediate category of product, energy conversion is not the product aim but life long use is significant. In the case of personal computers, addressing energy consumption at the design stage can yield 40% savings, with more available as peripheral technologies change, such as the development of energy efficient displays and drives. Mobile telephones, through a technology change to digital transmission require less power, smaller batteries and exploit miniaturisation of parts and cases. All this yields environmental benefits in production and use compared to their analogue predecessors. In this class of products a balanced view of environmental effects of production and use is essential.

Fits & Fixings. Detailed design decisions have a major impact on the possibilities for service, re-manufacture and recycling. Statistical approaches to reliability assessment and failure mode and effects analysis can provide a rational approach to providing serviceability in a product. Inseparable mixing of materials in components by using such processing routes as co-moulding, bonding or other permanent fixing prevents cost effective recycling and in some cases prevents any form of recycling as materials become chemically bonded together. Application of films, labelling & printing in incompatible materials can have the same effect.

Table 18.2 List of green design guidelines

• Consider every stage of the products life cycle in environmental terms
• Increase efficiency in the use of materials energy and any other resources
• Use recycled, renewable and biodegradable materials
• Choose materials that will minimise other environmental damage or pollution
• Ensure that the life expectancy of the product is appropriate, try to extend this as much as possible
• Consider the actual use of the product with a view to minimising the long term environmental effects.
• Design for ease of recycling, reuse or re-manufacture

18.1.3 Green Design Guidelines

Having looked at the environmental problems apparent and the way in which design decisions can affect these problems a main core of environmental or 'green' objectives for designers may be drawn up. In recent years there have been an increasing number of green design methods developed to address this main core of environmental design objectives. Table 18.2 lists some of these green design methods. As well as these very general guidelines there are also more specific considerations which it may be necessary to take into account when designing a product or system (Holloway and Tranter, 1995), for example:

• Environmental problems particular to the field you are working in
• Current DFE practices apparent in the field in which you are working

- Size of the organisations operations
- Geographical position of the operations and any related environmental problems

A key discipline, when considering the disposal of a product, is design for disassembly which pays attention to the conectivity of components, ensuring easy segregation for re-use or recycling and embodies recommendations for co-locating high value recyclables accessibly within a product.

18.2 A METHODOLOGY OF DESIGN FOR ENVIRONMENT

In order that a methodology for DFE may be developed it is important that we attempt to fully define what DFE is. As with other concurrent engineering imperatives DFE is explained very well by its name. Just as Design for Manufacture attempts to address all the possible problems and opportunities which may arise when manufacturing a product and Design for Assembly looks at the process of assembling a product with a view to improving that process by design, DFE looks at all the possible environmental problems and opportunities which arise during the manufacture of a product and addresses them as fully as is feasible by the use of appropriate design strategies.

Like some other concurrent DFX disciplines such as Design for Quality and Design for Cost, DFE is extremely wide ranging and uses a cradle-to-grave approach. The whole life cycle of a product is considered from winning of raw materials to ultimate disposal. DFE cannot be carried out independently for each separate stage of a products life as there are too many trade-offs to be made. For instance if DFE was carried out at the material selection stage of design with no reference made to other later life stages it could cause problems. A material which has the minimum environmental impact may facilitate processing which is very environmentally damaging. Overall one of the alternatives allied to it's particular processing route may well be more environmentally friendly. This type of trade-off needs to be considered for each stage of the product's life and concurrent engineering is the best approach to support this.

18.2.1 Overview of DFE

Inputs. Most of the environmental effects related to a design will arise from material choice, processing, usage and final disposal. Therefore to create a realistic picture of the environmental effects of a product or system all these stages must be considered. Material usage must be defined in terms of the actual raw materials used and the amounts of each.

Each of the materials will be subject to some amount of processing which will contribute to the environmental effect. Each processing route will result in energy usage and waste which is directly related to the amount of material processed. Therefore for each of the materials used in the design full processing routes must be defined. The usage statistics of a product or system may be the most difficult stage to interpret. In the case of DFE the projected amount of consumption of energy or fuel must be defined at the input stage. Separate studies may have to be carried out to ascertain these figures. Finally information concerning the predicted disposal of the product must be defined. It should be made clear what will happen to each of the materials used, will they be recycled, incinerated or sent to landfill?

Data Processing. All of the information given in the previous stage will give rise to large amount of environmentally relevant data. This data will be a more detailed breakdown of the inputs. For example, having defined the use of 0.3kg of high density polyethylene at the input stage, the data processing stage will break this down into the total amount of raw materials and energy used and emissions and waste generated. This procedure is carried out for material usage, processing routes, projected in-use data and disposal practices. Once the data has been processed for each of these stages the data is collated into one large overall 'environmental profile' of the product or system in question.

Outputs. The output of DFE exercises will be a breakdown of all the inputs and outputs of the product or system presented as emissions to air, emissions to water, energy usage and waste produced, known as the environmental profile. The environmental profile calculated may be very complex and therefore difficult to present to designers in a way which is useful to them. The outputs of DFE must be presented in a number of different ways. Comprehensive tables of data should be presented to the designer to allow full environmental analysis to be carried out. However in many cases the amount of data will be so large that an aggregation and summing system may be utilised to allow quick comparison of different designs. Graphical representation of results in discreet or aggregated form may also be of advantage when comparing more than one design in environmental terms.

Data Analysis. To allow this each emission will have to be ranked in relation to all the others. A problem now arises in how do we decide which pollutant is more serious that another and if we establish this how much more serious is it? Work is being carried out in this area but currently there is no accepted methodology in use. One way of doing this which is used in the case study later in this section is the use of MAC and O.v.D values. These values are government specified levels of pollutants which render air or water unsuitable for use by humans, and therefore considered polluted. Each atmospheric emission has an associated MAC value in mg/m^3 and each waterborne emission has an associated O.v.D value again in mg/m^3. In the case study there are environmental indices, used to compare alternative designs, called Air Pollution Indices (API) and Water Pollution Indices (WPI). These indices are calculated by taking the value of each emission from a design and dividing it by the appropriate MAC or O.v.D value respectively. For example chlorine has an O.v.D value, in water, of 200,000 mg/m^3 so a release of 200mg of chorine into water would result in the following WPI:

$$\text{WPI} = \frac{200}{200,000} = 0.001$$

The units are strictly in m^3 but in this work they are used as comparative indices only and as such the units are ignored. Doing this calculation for each waterborne emission and summing all the WPI values together gives an overall WPI which is used to compare designs on an environmental basis. The API is calculated in the same manner. It is useful to use these API and WPI values for comparison but it must be noted that all the discreet emissions data should be available to the designer as by using only single figures to quantify air or water pollution vital detail may be lost. Legislation is a very important issue in data analysis. All DFE exercises must make sure that the proposed product or system meets and if possible exceeds any legislation. It is also very important to keep abreast of up and coming legislatory developments and attempt to meet these before they come into force. If there are a large amount of environmental opportunities present and only a few have been addressed the reasons for this must be explored. The table or checklist will differ slightly for different product sectors.

Design Refinement. When attempting to refine a design in environmental terms it may be useful to compare it to a checklist as shown in Table 18.3 This allows the designer a quick and simple assessment of which environmental goals have been achieved. If some of the goals have not been achieved then the designer may look for ways in which they can be by refining the design. This may be done in an iterative process by analysing data, comparing the design to the checklist and if necessary going back to the input stage of DFE and defining different materials, processing routes etc., analysis data and so on. In many cases not all of the environmental goals may be achieved and so an attempting to address the most important ones is a priority. Once a final decision has been made on the design all it's environmental effects, and advantages over other designs, should be fully documented.

Table 18.3 Environmental Checklist

Life Cycle Stage	Design Goals	Addressed?	Method of Achievement	Level of Achievement
Materials	Min energy content	Yes/No		
	Min air pollution	Yes/No		
	Max. use of recyclate	Yes/No		
	Min use of material	Yes/No		
Processing	Minimise waste	Yes/No		
	Minimise energy usage	Yes/No		
Disposal	Max. recyclable material	Yes/No		
	Max. biodegradable material	Yes/No		
	Energy recovery possible	Yes/No		
Legislation	Relevant legislation met/exceeded	Yes/No		

18.2.2 Components of DFE

Most DFE exercises will follow the same pattern and are made up of a core of main components: materials, processing, use and disposal. Depending on the disposal route chosen the relationship of these components will differ slightly, shown in Figures 18.1 to 18.3. In all cases the production and processing of raw materials will result in emissions of substances to both air and water and waste products. The use of a particular product will have specific effects on the environment. For example the use of a car will result in emissions from the burning of fuel while the use of a milk bottle will only really result in environmental damage from the washing of the bottle each time it is re-used. The disposal routes will also differ depending upon the product type, materials used and available methods of disposal for that particular product or material.

Recycling. When recycling is specified as the intended disposal route for materials or products it affects the environmental profile in two ways. Extra energy and raw material inputs are required to recycle materials and the overall material requirement of the product system is reduced. When carrying out DFE it is important to establish that the environmental damage resulting from recycling a material is less than that resulting from using virgin material. in the case of non renewable resources recycling is usually the most environmentally sensible option.

Incineration. Using incineration as a disposal route has a number of advantages and disadvantages. While it recovers energy and reduces the amount of waste needing to go to landfill it adds to the overall airborne emissions of a product or system. Figure 18.2 shows how incineration components fit into the overall life cycle of a product or system in DFE.

Landfill. When landfill is specified as a disposal route there are less components to DFE. although the case of landfill is a very complex one. At this time emissions resulting form landfill are unknown. Contamination of land is possible to quantify but emissions from waste degrades over time is very difficult to assess and as a result of this currently models for landfill calculations consider landfill as contributing to waste levels only.

In a real DFE exercise all these different components will be used together in differing degrees. It is very unlikely that materials from a product or system will all be recycled some incineration or landfill will be likely. Indeed in many cases 100% recycling is not the best option. In the case of recycling paper the optimum level is about 60% recyclate.

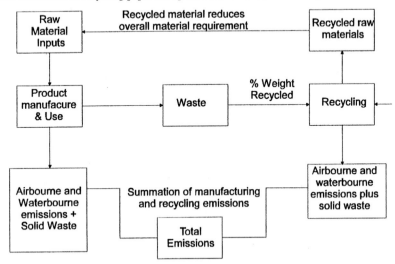

Figure 18.1 Recycling Calculation Model.

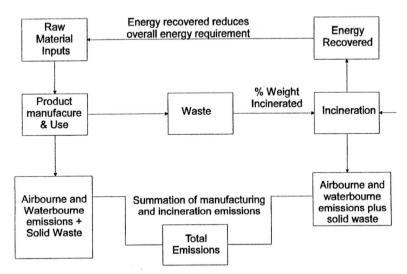

Figure 18.2 Incineration Calculation Model.

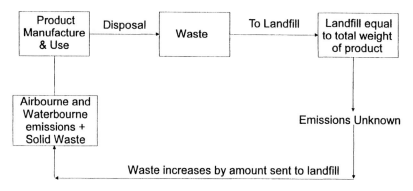

Figure 18.3 Landfill Calculation Model.

The other main component of DFE is assessment and refinement. This is an iterative process and is represented in Figure 18.4. The environmental profile of a design is assessed and the main problem areas highlighted. By studying these problems the designer may come up with possible solutions. These solutions must then be checked to ensure that there are no other problems apparent and when this is ascertained a final design proposed. In proposing the final design the environmental advantages over the original must be documented.

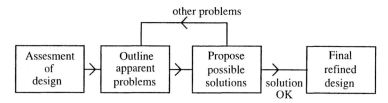

Figure 18.4 Design Refinement.

18.2.3 Procedure of DFE

Figure 18.5 shows the flowchart of our DFE methodology, made up of the following steps:

1. Firstly state all the functional and environmental objectives of the exercise. For example the aim of the exercise may be to produce a container which will hold 1 litre of liquid while keeping energy usage and waste to a minimum. At this stage only the objectives should be considered and no possible solutions put forward. The output of this first stage is a checklist similar to the one shown in the previous section. This checklist should be a comprehensive as possible and cover all the stages of the products life.

2. The next stage in DFE is to forward proposals of materials, processing and disposal routes for the product or system in question. Both mechanical, environmental and in some cases aesthetic data should be collected at this point. At this stage the designers need to have an idea about the actual amount of materials to be used. The environmental profile to be calculated in the next stage is dependent on this information.

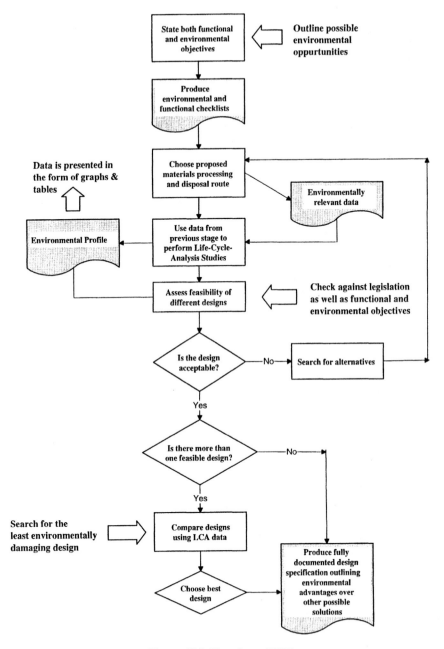

Figure 18.5 Flowchart of DFE.

3. Using the data gathered in the previous stage a Life Cycle Assessment should be carried out for each of the design alternatives. These assessment will result in and 'environmental profile' for each of the proposed designs. The results of LCA can be very complex and should be presented in a number of forms, both tabular and graphical. These tables and graphs will show all the environmentally relevant inputs and outputs throughout the life of the product.

4. By studying the environmental profiles of each proposed design checks may be carried out against both environmental legislation and the objectives set out earlier. At this stage it may become apparent that some or all of the proposed designs are unacceptable for a number of different reasons, be they mechanical or environmental. If no designs are acceptable then the designers must return to the earlier stages of choosing materials and processing etc. In extreme cases it may become apparent that the objectives laid down in the first stage cannot be met. If this happens the designers will have to rethink and redefine the objectives .

18.2.4 Computer Implementation of DFE

As an aid to the design team the development of computerised support tools may provide a powerful system to support concurrent engineering. For use in the area of design for the environmental the tool should supply the designer with up to date environmentally relevant information in a readily usable form. The tools should also operate in such a manner as to emulate the design team by including the facility to suggest changes which will improve the performance of the design from an environmental life-cycle perspective. Bowden and O'Grady (1989) have outlined the principle requirements of such a system:

*   It should be flexible enough to allow the design problem to be approached from a variety of viewpoints
*   It should allow the designer to design despite the absence of complete information
*   It should handle the large volume, variety and interdependence of life-cycle information
*   It should exhibit high performance in terms of speed and reliability
*   It should readily interface to database management and CAD systems
*   It should have a good user interface and be able to explain itself in a manner comprehensible to humans
*   It should support design (and in this case environmental) audits and be easily updateable as new information becomes available.

In order that the system fits into developing design practices its operation should follow the main steps of DFE as shown in Figure 18.5. The initial stage of defining objectives is obviously done by the designer before using the computer system but once the aims and scope of the design exercise have been decided then the computer may be utilised. The computer program goes through the following steps:

1. **Material Selection.** The computer contains a database of a large number of materials, all of which have extensive environmental data associated with them. All of the materials to be used in the design are chosen from this list. The computer then goes through this list of materials and asks the user to input the weight of each to be used.

2. **Processing.** Once the data on materials selection has been specified the computer then goes through the list of materials again and offers the user a number of processing options available for that material. For example if the material is steel the computer will

offer a list of processes such as machining, forging, casting, cutting etc. while of the material is a polymer the computer may offer a list including injection moulding, blow moulding, vacuum forming, etc.

3. **Use.** The use of a product is the most difficult stage to assess environmentally. In pre design assessments only predictions as to usage statistics can be made. Because of this and the vastly differing nature of product usage across different product sectors the environmental effect of usage is mainly decided by the user of the computer. The designer has to tell the computer the amount of energy/type of energy the product or system will use over its life cycle.

4. **Disposal.** The disposal routes are then specified for the materials chosen. Problems such as disassembly and separation of materials are not taken into account at present. The designer specifies which materials of those chosen will be recycled and which will be sent for disposal. Of those sent to disposal the computer then asks for percentage fractions of incineration and landfill. In most waste collection infrastructures the recyclate is removed and then the rest in either landfilled or incinerated or a combination of both.

At this stage the computer has all the information it needs to build up an environmental profile of the product or system life-cycle. All the emissions and waste generated as a result of the life-cycle, as defined by the designer, is added together and is presented as a list of inputs, emissions, waste and energy or material recovery. A sample output screen from the computer program is shown in Figure 18.6.

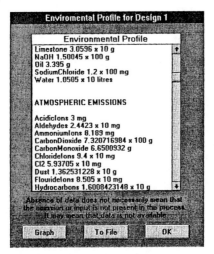

Figure 18.6 Part of a Tabular Environmental Profile calculated by Computer Program.

The computer program may then be used to refine the design, working along the lines of the procedure shown in Figure 18.4. The user of the computer specifies what the problem areas are. For example the amount of NOx emissions may be too high in legislatory terms and a way of reducing these must be found. The computer will then search for alternative combinations of materials, processing and disposal routes which will address these problems.

The materials used must meet certain requirements such as mechanical performance etc. and the computer takes this into account. Having searched for alternatives and checked them for suitability the computer offers suggestions to the user on material substitution etc. Figure 18.7 shows a simple computer output for optimising NOx emissions through material choice.

By using such a computer based system DFE exercises may be structured and accelerated thus becoming more readily accepted by designers.

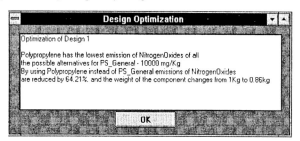

Figure 18.7 Computer calculated Design Refinement.

18.3 CASE STUDY

18.3.1 Introduction

In order to illustrate the Design for the Environment methodology described in Section 18.2, a simple case study will be presented. The selection of materials for the manufacture of bottles will be considered. Clearly a large number of issues must be considered. Packaging poses some particularly difficult problems. It usually has a very short life-cycle, being discarded after the product it is protecting and/or containing is used. The answers to these problems appears to include the choice of low environmental impact materials, minimisation of waste and the promotion of reuse and recycling.

A number of assumptions and simplifications will be made in this study for the sake of brevity. It will be assumed that the bottles are of equal capacity i.e. one litre and that they will be used to contain soft drinks, although not necessarily carbonated. The latter requirement would place additional constraints on the choice of materials in terms of permeability and strength which add a further complicating dimension to the study, and so this aspect will be avoided in order to allow the consideration of a slightly wider range of materials. The study is somewhat constrained by the limited availability of environmental data for some materials.

18.3.2 Preliminary Selection

The selection process begins with the use of software such as Plascams. The criteria for selection include easy processing, transparency, adequate strength, good toughness and food compatibility. An initial sifting of materials on the basis of adequate transparency and mouldability using standard blow moulding techniques is carried out and a short list for further consideration produced. This includes a number of plastics including PET, PVC and LDPE. These together with glass are taken forward for a more detailed analysis when the other properties are considered. The most promising materials can then be evaluated for environmental impact by the use of LCA. Clearly there is a danger in proceeding in this way because environmental impact is relegated to second place in that a particularly environmentally acceptable material could be eliminated on the grounds of a marginal inadequacy in terms of mechanical or other properties. Care must clearly be exercised to avoid this. For this reason it is preferable to consider all functional and environmental requirements together as proposed in Section 2. However, at the present time this may well not be possible because of limitations in the availability of environmental data.

18.3.3 Environmental Considerations

It is instructive to consider the most important environmental factors in relation to DFE. They are summarised below:

1. Energy usage in manufacture and transportation
2. Material inputs
3. Atmospheric emissions in manufacture
4. Waterborne emissions in manufacture
5. Solid wastes
6. Recovered(recycled) materials
7. Reused materials

The Environmental Checklist in Table 18.3 should be utilised. We can return to this later to monitor the environmental effectiveness of our design and materials selection

It should be noted that in the check-list above reused materials are those that are used again without reprocessing e.g. a traditional milk bottle. Recycled or recovered materials are those that are fully reprocessed e.g. remelted and fabricated again.

Let us consider our short list of PET, LDPE and glass. For these materials comprehensive environmental data is available. The data is available in the following categories:

1. Inputs (energy, fuel for electricity generation and materials)
2. Atmospheric emissions
3. Waterborne emissions
4. Solid wastes
5. Proportion of achievable recycled material.

Clearly an environmental profile can be produced for each material. The design using a particular material specifies a weight of material to be used and this is used to produce the profile. The profile can then be compared for the alternative materials. Given that a particular processing route is to be used for each material then the profile should clearly include both the production of the material and its processing into a bottle. We can then extend this to include recycling. Table 18.3 shows input, emission and waste data corresponding to the three materials for materials production, processing and recycling. However, the glass used in bottle manufacture is assumed to be 100% recycled whereas the polymers are assumed to be virgin. Other proportions of recycled/new glass would yield different data.

The emission figures used here are European averages. They may vary for other geographical regions. While energy usage in itself can be compared it is instructive to know the emissions produced in the generation of 1MJ of energy. This is given below:

Emissions produced in the generation of 1MJ of energy:

CO_2 100381mg
CO 88mg
DUST 35mg
C_xH_y2.6mg
NO_x 330mg
SO_x 495mg
API 188.1

These emissions are in fact included in the atmospheric emission data. (The WPI is not calculated for energy production as no emissions to water are recorded)

Table 18.5 compares the environmental profiles for one litre bottles manufactured out of PET, LDPE and glass, and including recycling.

Table 18.4 Environmental assessment of chosen design

Life Cycle Stage	Design Goals	Addressed?	Method of Achievement	Level of Achievement
Materials	Min energy content	Yes/No	Material/process choice	High
	Min air pollution	Yes/No	"	High
	Max. use of recyclate	Yes/No	Glass uses most recyclate	Zero
	Min use of material	Yes/No	Low density material	High
Processing	Minimise waste	Yes/No	PET has lower waste	Low
	Minimise energy usage	Yes/No	Glass bottle has lower process energy	Low
Disposal	Max. recyclable material	Yes/No	Glass is more readily recyclable	Low
	Max. biodegradable material	Yes/No	Non of the materials are biodegradable	Zero
	Energy recovery possible	Yes/No	Use of polymer	Medium

18.3.4 Summary

It is not necessarily straightforward to compare the data in Table 18.4. The large amount of data makes simple comparisons difficult. Hence API, WPI and energy requirement values are included. These represent a reasonable summary of the data. Problems also exist with the poor integrity of some data.

Inspection of the API, WPI and energy figures clearly reveal that the production of LDPE bottles imposes the least environmental impact in all of these categories. Glass bottles are the most problematical in spite of the fact that the use of 100% recycled glass is assumed. If the proportion of recycled glass were to fall to 56%, then the API figure rises to 2006.4 which is a 40 % increase, over twice as great as that for PET and over 10x greater than for LDPE. Glass is penalised because of the high weight of bottles. This no doubt contributes to the high values for energy, API and WPI. It also implies high transport costs and energy requirements in this respect.

Table 18.5 Environmental Profiles of Bottles

1 litre LDPE bottle (25g)		1 litre PET bottle (35g)		1 litre glass bottle (650g)	
Inputs		**Inputs**		**Inputs**	
Bauxite	7.5 mg	Bauxite	10.9 mg	Energy	7.15 MJ
Clay	0.5 mg	Clay	0.035 mg	Fuels for Elec.	0.34 MJ
Energy	2.5 MJ	Energy	3.3 MJ	Glass	696 g
Ferromanganese	0.023mg	Ferromanganese	0.035 mg	Soda	8.15 g
IronOre	5 mg	Iron Ore	19.2 mg	Sundries	1.94g
Limestone	3.75 mg	Limestone	9.45 mg		
Sodium Chloride	200 mg	Manganese	1.75 mg	**Atmospheric Emissions**	
Water	0.6 litres	Metallurgical Coal	8.05 mg	Aldehydes	6.47 mg
		Phosphate Rock	1.05 mg	Ammonium Ions	1.3 mg
Atmospheric Emissions		Sand	0.7 mg	Carbon Dioxide	638 g
Acidic Ions	1.5 mg	Sodium Chloride	172 mg	Carbon Monoxide	76 mg
Ammonium Ions	0.125 mg	Water	0.595 litres	Dust	555 mg
Carbon Dioxide	53 g			Hydrocarbons	1.45 g
Carbon Monoxide	41.9 mg	**Atmospheric Emissions**		Hydrogen Flouride	31 mg
Chloride Ions	3.25 mg	Carbon Dioxide	118 g	Lead	21 mg
Dust	8.3 mg	Carbon Monoxide	662 mg	NH3	1.29 mg
Hydrocarbons	526 mg	Dust	146 mg	Nitrogen Oxides	2.08 g
Hydrogen Chloride	1.75 mg	Hydrocarbons	1.4 g	Other Organics	4.5 mg
Hydrogen Flouride	0.125 mg	Hydrogen Chloride	3.85 mg	Sulphur Oxides	3.45 g
Metals	373 mg	Metals	0.35 mg		
Nitrogen Oxides	372 mg	Nitrogen Oxides	826 mg	**Waterborne Emissions**	
Other Organics	0.025 mg	Other Organics	329 mg	BOD	1.3 mg
Sulphur Oxides	334 mg	Sulphur Oxides	1.05 g	COD	1.9 mg
				Diss. Solids	1.99 g
Waterborne Emissions		**Waterborne Emissions**		Oil	25.87 mg
BOD	5 mg	Acidic Ions	6.3 mg	Sus. Solids	1.29 mg
COD	37.5 mg	BOD	35 mg		
Diss. Organics	0.5 mg	COD	116 mg	**Solid Wastes**	
Diss. Solids	7.5 mg	Chlorines	24.9 mg	Waste	37.9 mg
Hydrocarbons	2.5 mg	Diss. Organics	455 mg		
Metals	6.25 mg	Hydrocarbons	14 mg	**Recovery**	
Nitrates	0.125 mg	Metals	4.2 mg	Recovered Glass	644 g
Oil	5 mg	Na	52.5 mg		
Other Nitrogen	0.25 mg	Oil	0.7 mg		
Phosphates	0.125 mg	Other Nitrogen	0.035 mg		
Sus. Solids	12.5 mg	Phenols	0.035 mg		
		Phosphate	0.35 mg		
Solid Wastes		Sulphate	1.4 mg		
Industrial Waste	87.5 mg	Sus. Solids	21 mg		
Mineral Waste	650 mg				
Non Tox. Chems	20 mg	**Solid Wastes**			
Processing Waste	505 mg	Chemical Waste	4.55 mg		
Slag & Ash	225 mg	Industrial Waste	122.5 mg		
Toxic Chems.	2.5 mg	Inert Chem. Waste	66.5 mg		
		Processing Waste	1.4 g		
		Slag & Ash	336 mg		
Recovery		**Recovery**			
Recovered LDPE	23.8 g	Recovered PET	33.3 g		
Energy Requirement	**2.5 MJ**	**Energy Requirement**	**3.3 MJ**	**Energy Requirement**	**7.2 MJ**
API	**173.7**	**API**	**790.4**	**API**	**1429.9**
WPI	**37.8**	**WPI**	**73.7**	**WPI**	**129.4**

Unfortunately LDPE is unsuitable for carbonated drinks because of problems with permeability to CO_2. It is highly satisfactory for milk and other non pressurised applications. One of its other drawbacks is its relatively limited transparency.

In conclusion we can look at the checklist as shown earlier in this section and assess the final choice of design.

Of the nine main environmental design aims only four have been achieved. It must be noted though that the four that have been achieved are the most important in this list. Glass bottles use the most recyclate and are more readily recycled (mainly because of the infrastructure that is in place) and PET bottles produce the least amount of waste. In the overall life cycle when transportation is also taken into account the lightweight nature of the LDPE bottle will reduce emissions and thus make it the best choice in this field as well.

18.4 SUMMARY

The influence that design and designers can have on the environmental impact of products or systems has been recognised for some time. However, traditionally design has taught with little or no reference to the environment and now considerable responsibility can be placed on designers as they are at the centre of a holistic process. Conversely the position held by designers is the perfect stage for them to demonstrate the importance of environmental issues. Designers should now be readying themselves to deal with environmental issues by developing the following skills: Holloway *et al.* (1994):

- The ability to thoroughly research the environmentally relevant issues before undertaking a design exercise
- A general broad knowledge of environmental issues along with a more detailed understanding of those environmental issues particular to their field of work
- Access to environmental knowledge about materials, processes, technologies and legislation, relevant to the proposed design.

There are many strategies which may be adopted when carrying out DFE. If singular strategies are adopted then DFE is not fully addressed, all the opportunities available need to be explored. As with other DFX disciplines DFE will facilitate the use of design tools and modules but in this case 'The challenge here is to create modules which, in keeping with industrial ecology theory, are broad, comprehensive and system based yet well defined enough to be integrated into current design practices' Allenby (1994). Jakobsen (1991) concludes that 'In good design there exists a harmonic relationship between geometric shape, material and the production method use. In order to achieve this harmony it is necessary to use a procedure which considers the treatment of these elements as an integrated activity.' It is now apparent that the environment is an element which should now be included in this harmony.

The method of DFE presented in this chapter considers the environment as an integrated concern in design activity and by following very closely the course of other design methods should be easily integrated into concurrent engineering practices. The use of computer tools such as that described in this chapter will both structure and accelerate environmental impact assessment and as with many other DFX disciplines become an integral part of DFE.

By employing an holistic approach and developing DFE by adapting current design methods to take environmental considerations into account, it may become a common feature in future product development programmes.

REFERENCES

Mackenzie, D. (1991) *Green Design - Design for the Environment*, Laurence King Ltd.

Burall, P. (1991) *Green Design*, The Design Council.

Guineé, J.B. *et al.* (1993) Quantitative Life Cycle Assessment of Products-Classification, valuation and improvement analysis, *Journal of Cleaner Production*, **1** (2).

Holloway, L., Tranter, I. (1995) Environmental Design - What is best practice? *Inaugural Conference of the European Academy of Design - Design Interfaces*, University College, Salford.

Cross, N. (1991) *Design Methods*, John Wiley & Sons Ltd.

SETAC. (1991) *A Technical Framework for Life Cycle Analysis*, SETAC Foundation for Environmental Education.

ICI (1990) *Bopol Press Release*.

Henstock, M. E. (1988) *Design for Recyclability*, University of Nottingham.

Bowden, J., O'Grady, P. (1989) Characteristic of Support Tool for Life-cycle Engineering, *LISDEM Technical Report*, North Carolina State University.

Ryding, S. *et al.* (1993) The EPS System - A Life Cycle Assessment Concept for Cleaner Technology and Product Development Strategies and Design for the Environment, *EPA Workshop on Identifying a Framework for Human Health and Environmental Risk Ranking*, Washington DC.

Holloway, L. *et al.* (1995) Expert Systems for EcoDesign, *Interdisciplinary Conference on the Environment*, Boston, Massachusetts.

Holloway, L., Tranter, I. (1995) An Expert System Based Advisor for Assisting Predictive Environmental Impact Assessment, *IKIM'95*, Nanjing, China.

Holloway, L. *et al.* (1994) Integrating Environmental Concerns into the Design Process, *Materials & Design*, **15** (5), 259-267.

Allenby, B. (1994) Industrial Ecology Gets Down To Earth, *IEEE Circuits and Devices Magazine*, **10** (1), 24-28

Jakobsen, K. (1991) The Interrelation between Product Shape, Material and Production Method, *Proceedings of the ICED'89*, Heurista, Zurich.

DESIGNING FOR THE LIFE-CYCLE:
ACTIVITY-BASED COSTING AND UNCERTAINTY

Bert Bras; Jan Emblemsvåg

This chapter presents a method for developing an Activity-based Cost (ABC) model for use in life-cycle design under the presence of uncertainty. The crux in developing an ABC model is to identify the activities that will be present in the life-cycle of a product, and afterwards assign reliable cost drivers and associated consumption intensities to the activities. Uncertainty distributions are assigned to the numbers used in the calculations, representing the inherent uncertainty in the model. The effect of the uncertainty on the cost and model behavior are found by employing a numerical simulation technique - the Monte Carlo simulation technique. The additional use of detailed process action charts and sensitivity charts allows the influence of the uncertainty to be traced through the cost model to specific product and process parameters. The method is illustrated using a detailed product demanufacturing cost model.

Concurrent engineering represents a common sense approach to product realization in which all elements of the product life-cycle from conception through manufacturing to disposal are integrated into a single continuous feedback-driven design process. The primary goal of Concurrent Engineering has always been the minimization of costs over the complete life cycle of a system while maximizing its quality and performance (Winner *et al.*, 1988). The growing importance of including environmental issues in design has amplified the impetus for companies to more formally consider the entire life-cycle of a product, from cradle to grave or even to reincarnation through recycling and reuse. Demanufacture is the process opposite to manufacturing involved in recycling materials and product components after a product has been taken back by a company.

19.1 ACTIVITY-BASED COSTING AND UNCERTAINTY

The role of a cost model is to give feedback to the design department and product realization group for making cost/revenue correct design changes. For example, in the area of environmentally conscious design and manufacture, the following two questions represent key issues for which designers like to obtain feedback:

- What is the cost associated with pursuing environmentally benign products and processes?
- Which aspects of product and process design have the largest influence on these costs?

We believe that in order to provide efficient and effective decision support in life-cycle design, costing methods should have the following characteristics:

- Assess and trace costs and revenues.
- Handle both overhead and direct costs.
- Handle uncertainty.
- Provide decision support for the process of designing.

Several costing approaches have appeared in the literature in the context of designing environmentally benign products and processes. However, when it comes to assessing costs to life-cycle and ecological issues, Activity-Based Costing (ABC) is gaining ground rapidly on conventional costing systems (Cooper, 1990b; Cooper, 1990a; Brooks *et al.*, 1993; Keoleian and Menerey, 1994). A review of relevant life-cycle costing approaches can be found in (Emblemsvåg and Bras, 1994; Emblemsvåg, 1995). Based on our review, we believe that emerging Activity-Based Costing approach has the best potential for efficient and effective cost assessments in the context of designing for the life-cycle.

19.1.1 Activity-Based Costing

Activity-Based Costing (ABC) has received its name because of the focus on the activities performed in the realization of a product. Costs are traced from activities to products, based on each product's consumption of such activities. Activity-Based Costing differs from conventional costing systems in two distinct ways:

1) In conventional costing systems, the assumption is made that each unit of a given product consumes resources (e.g. energy, material and direct labor), while in ABC the assumption is made that products or services do not directly use up resources, but consume activities. Hence, in ABC, the cost of a product equals the sum of the costs of all activities that must be performed in the realization of the product (Cooper, 1990a).

2) Conventional cost systems are based on unit-level cost drivers (or allocation bases) of the product that are directly proportional to the number of units produced. These unit-level cost drivers are referred to as allocation bases in conventional cost systems. Direct labor hours, machine hours and pounds of material are examples of such "unit-level allocation bases". An ABC system, on the other hand, uses cost drivers that can be at the unit-level, batch-level, and/or product-level. Examples of batch-level cost drivers are setup hours and number of setups. Examples of product-level cost drivers are number of parts, number of times ordered, and number of engineering change orders (Turney, 1991).

Because of the assumption that a product uses activities and the allowance for batch and product level cost-drivers, it is generally agreed that ABC systems are superior in modeling and tracking costs (Cooper, 1990a; Turney, 1991). Mostly noted is ABC's capability to separate direct from indirect costs. In depth discussions of ABC can be found in (Cooper, 1990a; Cooper, 1990b; O'Guin, 1990; Raffish and Turney, 1991; Turney, 1991).

To exemplify the difference between ABC and conventional cost systems, examples of ways to reduce cost are given in Table 19.1, plus how these reductions are achieved with conventional costing schemes and ABC. All areas of difference in Table 19.1 are a result of these two differences presented. Reducing a set-up time or material handling activities are batch level cost drivers that cannot be modeled directly in a conventional cost system, but only modeled indirectly through a corresponding reduction of a unit-level characteristic such as direct labor by an equivalent amount. As another example, in design for assembly it is well known that using common components yields cost reductions, and from Table 19.1 we see that only ABC support this point of view. In (Cooper, 1990a) it is noted that "traditional cost systems systematically undercost small, low-volume products and overcost large high-volume products". This is due to the inability to trace overhead costs correctly, which in turn results from the use of only unit-level cost drivers and the focus on resource consumption.

We have chosen to use ABC because of the noted superiority in cost-tracing, separation of direct and indirect costs, higher accuracy, and its capability to blend into the Activity-Based Management (ABM) systems that more and more companies are employing (see, for example, (Turney, 1991)). A motivating example for its use in an environmental context can be found in (Brooks *et al.*, 1993) where it is described how Activity-Based Costing and environmental aspects can be combined to give companies the ability to identify more accurately those plants and products which are driving up their environmental expenditures. However, it should be noted that, although many have focused on ABC, the issues of a) how to provide efficient and effective decision support in design, and b) how to best include the uncertainty involved are still largely unaddressed. The uncertainty cannot be ignored and should be included when seeking to assess costs associated with the product life-cycle without much historical data.

Table 19.1 How conventional and ABC systems achieve different cost reductions

Examples of cost reductions	Conventional cost system	ABC system
Reduce setup time	Ignore or reduce direct labor with an equivalent amount	Reduce setup time to achieve low cost diversity
Eliminate material handling activities	Ignore or reduce direct labor with an equivalent amount	Eliminate activities to reduce the cost of handling materials
Choose an insertion process	Pick alternative with lowest unit level activities	Pick lowest cost alternative
Use common components	Using common components yields no cost savings, using non common components creates no cost penalty	Use common components wherever possible

Source: Turney, 1991

19.1.2 Including Uncertainty in a Cost Model

Although a number of researchers propose methods for assessing and reducing environmental impact (Navin-Chandra, 1993; Thurston and Blair, 1993), hardly any discussion is given to the accuracy of the data used and the sensitivity of the outcome to variations in the inputs. When dealing with ecological issues uncertainty must be included due to the predominant lack of hard data. Consider the pending legislation on automobile take-back in Europe. While car makers are already designing for recycling, the true economics and environmental impact can only be measured years from today when a car has ended its useful life.

In situations where we do not have probability information, we have to use uncertainty. Uncertainty can be modeled in a variety of ways depending on what kind of uncertainty is to be modeled. Generally speaking, we have the following possibilities:

- *We can model the uncertainty based on historical data.* This will typically involve statistical analysis along the line of Gaussian Statistics.
- *We can model the uncertainty based on experience, qualified guessing or even worse.* One way of doing this is by modeling the uncertainty as fuzzy numbers, but solving the problem numerically. This has the same advantages regarding the basis on which the uncertainty is modeled (e.g. based on experience, historical data or an educated guess) as using fuzzy theory, but the solution process is easier.

In design, and especially in original design, good historical data are often impossible or difficult to get, thus methods based on Gaussian statistics will soon become inappropriate and even impossible to apply. Our method must therefore be designed to deal with 'fuzziness'. This means that we in our model would have to guess, for example based on experience, the type of distribution to use as well as the mean, the left deviation and the right deviation. The uncertainty is therefore simply modeled by assigning distributions to every number in the model for which there exists uncertainty.

Given that we have modeled the uncertainty in a cost model, we must determine the effect of these uncertainties on the cost next. We have found it useful to use the Monte Carlo simulation technique to find the cost uncertainties resulting from our assumptions. This technique is a very simple, but powerful numerical approximation method that is simply based on performing a controlled and virtual experiment within the model. Although numerous different simulation methods exist, we have found it advantageous to employ a software called Crystal Ball[®] for this purpose. The Crystal Ball software adds into Microsoft Excel, which is a spreadsheet based software, and we are therefore talking about 'cells'. It allows the definition of 'assumption' and 'forecast' cells in a spreadsheet computer model. A forecast cell can be looked upon as a response variable, while an assumption cell can be viewed as a source variable. Consider the example in Figure 19.1 where product cost is modeled as a simple linear function of material and direct labor cost. Based upon our "assumptions" with respect to material and direct labor, we want to "forecast" the associated product cost. In each assumption cell, an uncertainty distribution is defined as one find appropriate for various reasons, associated with the particular value in that cell. In our example (see Figure 19.1);

- the 'Direct Labor' assumption cell is distributed as a triangular distribution, while
- the 'Material' assumption cell is distributed elliptically.

The Monte Carlo simulation provides random samples of numbers in the assumption cells (material and direct labor). These random numbers propagate through relationships/equations in the model and the value of the associated forecast cells (product cost, in our example) is calculated by means of the appropriated relationships/equation. In our example, the value of the forecast Product Cost is a simple summation of the random numbers for Material and Direct Labor. When all the trials have been performed, the calculated values of a forecast cell will form a new statistical distribution (see the Product Cost distribution in Figure 19.1). Due to the randomness, the numbers that have propagated through the model can be used in ordinary statistical analysis as if we were running a real experiment, e.g. to construct confidence intervals, perform T-tests, etc.

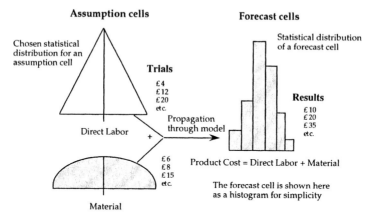

Figure 19.1 Example of the Monte Carlo Simulation.

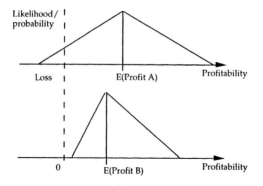

Figure 19.2 Different Resulting Cost Distributions for Design A and Design B.

After the numbers and their respective uncertainties have propagated through the cost model and the cost of each design solution is known, choosing the best design solution may still be difficult. Consider the following problem: In Figure 19.2, we see that design A is most likely to give more profit than design B, but choosing design A also include the possibility of loosing money. *The question arises what design solution should be chosen?* The answer depends on the policy of the company and the economic situation:

- A economically strong and not risk averse company would choose design A because of the expected profit is larger.
- A economically weak and risk averse company, on the other hand, would choose design B because the profit of design B is always greater than or equal to zero.

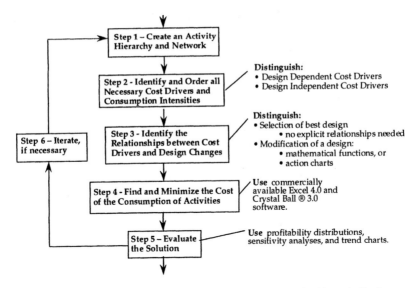

Figure 19.3 Flowchart for Development of ABC Models for Usage in Design.

Clearly, one cannot simply choose the solution with highest expected profit when uncertainty is present. We will highlight this again in Section 19.3.

19.2 DEVELOPING AN ABC MODEL WITH UNCERTAINTY FOR DESIGN DECISION SUPPORT

Our method for developing an ABC cost model that includes uncertainty for the preceding two decision support uses consists of six steps. A flow-chart of our method is given in Figure 19.3. Before discussing the details, we point out that our method has the following core components:

- *Formulation* (steps 1-3) – These steps deal with the actual formulation of the model.
- *Solution* (step 4) – The formulated model is solved (i.e. a cost assessment is obtained). Currently, we use Microsoft Excel 4.0 spreadsheet and Crystal Ball 3.0 software for this.
- *Validation* (steps 5-6) – The results from the solution process are used to verify the model and reiterate the process if necessary.

Step 1 - Create an Activity Hierarchy and Network

The purpose of this step is to break down the part (if desired) of the life cycle for which you would like to design into a hierarchy of activities. In addition, a network out of the lowest level activities occurring in a specific part of a product's life cycle needs to be established. In Figure 19.4, a general activity hierarchy is shown. The hierarchy may be broken further down into A_{ijk} activities if desired, and so on. The purpose of an activity hierarchy is to ensure that all the activities in the part of the life cycle to be studied are considered. When creating the activity hierarchy for identifying the effect of changes in design parameters on the cost, it is essential to form activities detailed enough that cost drivers can be assigned and that the lowest level of the activity hierarchy can be assigned directly to the design parameters through the cost drivers. If the model is to be used as a mathematical optimization model, then the relationships between design parameters and cost drivers must be identified.

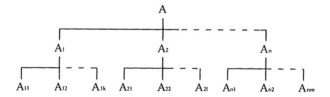

Figure 19.4 General activity hierarchy.

After identifying all the activities, a network indicating the relationships between the activities is constructed. An example activity network is shown in Figure 19.5. We use the network to identify:

- What effect a change in the design parameters will have on the consumption of activities.
- What effect a change in consumption of an activity will have on the other activities.

Furthermore, the network also provides the designer with a graphical view on how different decisions will affect the number of activities required.

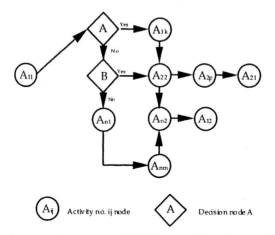

The icons are the same as used in (Greenwood and Reeve, 1992)

Figure 19.5 General activity network.

Step 2 - Identify and Order all the Necessary Cost Drivers and Consumption Intensities

The purpose of this step is to identify the cost drivers and corresponding consumption intensities that are necessary to use in order to find the cost of the consumption of activities with the desired accuracy. Before the costs of the consumption of activities can be found, proper cost drivers must be chosen. The properties of ABC depend on the cost drivers chosen. Bad cost drivers may give bad cost estimates. An example of a bad choice would be if a unit level cost driver (e.g. mass per unit) was chosen to keep track of a batch level activity (e.g. inspection). The cost of the consumption of a specific activity is then simply the cost driver(s) multiplied with the consumption intensity. Then the total costs can be found as the sum of the costs of all the activities that the design solution would impose. Based on the relationships between the design parameters and the cost drivers, we define two different types of cost drivers:

- *Design Dependent Cost Driver:* These cost drivers can be identified as a function of a set of design parameters, which imply that these cost drivers deal with the design itself. The relationship function should (ideally) be determined for each cost driver.
- *Design Independent Cost Driver:* These cost drivers are independent of the design parameters and/or the change of the design parameters. In other words, these cost drivers are not affected by design changes and in most cases a constant.

If there are uncertainties associated with specific cost drivers and consumption intensities, then uncertainty distributions should be assigned.

Step 3 - Identify the Relationships between Cost Drivers and Design Changes

The relationships between cost drivers and design parameters are the crux of a design decision support model, because they capture how much a change in one or more design parameters will affect the consumption of the activities, i.e. the cost. A key objective for using an Activity-Based Cost model in design is to identify how changes in different design parameters affect the cost and consumption of the activities. The level of detail and sophistication needed in modeling the relationships between cost drivers and design parameters depends on the purpose and usage of the cost model. In Figure 19.6, different uses and ways of modeling the effects of design changes on the cost are illustrated.

We identify two distinct usages of an Activity-Based Cost model:

1) *Evaluation of a number of discrete designs* in order to identify the economically best design, that is, the cost of a number of alternative designs is determined and a selection of a design is made based upon the result. Therefore, we do not have to model the relationships between design parameters and cost drivers explicitly. This approach works at the activity level (the top level in Figure 19.6) and selection of the most cost effective design is the primary purpose.

2) *Identification of "optimal" values for continuous design parameters*, that is, a given design is modified through, e.g. mathematical optimization in order to identify the ideal values of a number of design parameters. Rather than selection of designs, modification of an existing design is the primary purpose. This is highlighted in the gray area of Figure 19.6.

In order to improve design parameter values (second usage), it is necessary to identify the effect the design parameters have on the cost drivers. The highest amount of detail is obtained if these effects are quantified in detailed mathematical Cost Driver/Design Parameter relationships which link design changes at the property/dimension level (bottom level in Figure 19.6) to the cost drivers. This approach allows us to modify designs on a very detailed parameter values using, e.g. optimization algorithms. However, the usage of Cost Driver/Design Parameter relationships can become extremely cumbersome in the design of complicated systems where there are a) many relationships and b) many changes made over time in the relationships. In Figure 19.6, a solution is represented by introducing a level in the middle - where we do not keep track of design properties and dimensions, but rather keep track of how the design properties and dimensions affect specific actions. The aggregated effect on these actions is then transformed into an aggregated effect on the activities and the cost drivers. In Section 19.3, we exemplify this concept of using "action charts" to identify the aggregated effect of design changes on activities in the ABC model in the context of the disassembly actions needed to dismantle a telephone.

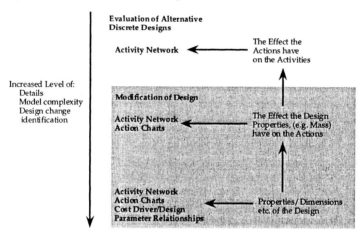

Figure 19.6 Capturing Design Changes at Three Different Levels.

Step 4 - Find and Minimize the Cost of the Consumption of Activities

The purpose of this step is to determine the most cost efficient design solution or set of design parameters by minimizing the cost of consumption of the activities. When the revenue is varying from design solution to design solution, the profitability must be the decision criterion. However, first the different uncertainty distributions associated with the cost drivers and the consumption intensities must propagate through the ABC model. As noted in Section 19.1.2, we have chosen to implement ABC models in Excel 4.0 spreadsheets in conjunction with the Crystal Ball® 3.0 software to handle the modeling and propagation of uncertainty in an easy and clear way. Given that we have this capability to handle uncertainty, various methods are available to find the most cost effective solution. For example, we could

- use an optimization algorithm, where the design parameters serve as source variables and the total cost as response variable,
- tweak the design parameters and see which set of parameters gives the best solution, or
- evaluate different designs and achieve the best solution by trial and error.

We would like to point out that it is arguably difficult to speak of "best design" or "optimum values" in light of the uncertainty present. Nevertheless, valuable trends and indications can result from such a formal assessments and optimizations. Specifically, because we do not ignore uncertainty, but explicitly model uncertainty in the form of distributions. More on finding the cost of the consumption of activities is given in Sections 19.3 where we discuss the results for a cost model for demanufacturing a telephone.

Step 5 - Evaluate the Solution

This step simply ensures that the final design solution is realizable within the given constraints. The step is to control if the found solution is a valid solution or not. If the final solution is rejected, the calculations, the formulation of the activities, the chosen cost drivers and consumption intensities must be checked. As will be discussed in Sections 19.3.2, 19.3.3, and 19.3.4, profitability distributions and sensitivity analyses are some of the tools available to support this step.

Step 6 - Iterate, if necessary

Step 4 and 5 must be reiterated until a cost effective solution is found or the project is stopped. In addition, larger iterations may be necessary because of modifications needed in the ABC model. We suggest the following approach to make the process of interpreting the results organized and structured:

(a) *Get to know the model.* The purpose of this step is to 'get to know the model' - meaning that we should see
   - if the model provides answers that are very bad - indicating logical errors in the model,
   - if the uncertainty in the model is very unequally distributed - making sensitivity charts less reliable (see Sections 19.3.3 and 19.3.4), or
   - if the number of trials performed is too low - resulting in a large mean standard error.
(b) *Find out what information in the model should be improved.* This step is especially needed if we encounter problems in the preceding step. The easiest way of identifying what information should be improved is to use sensitivity charts because we can read directly what information is most important (largest correlation coefficient). This is highlighted in Sections 19.3.3.
(c) *Use the results in design.* When steps (a) and (b) are completed we can finally use the results to spot what (product) design changes should be made to increase the profitability of the design. This can most easily be done by employing a sensitivity chart of a model in which perfect process information is assumed. This is highlighted in Section 19.3.4.

19.3 A CASE STUDY

In this section, we will illustrate our method by developing an ABC model for assessing and tracing the cost of demanufacturing products for usage in design. We will develop an ABC model and employ the model for assessing the cost (or profit) of demanufacturing a telephone. Step 6 of our method will not be addressed in this case study. More information can be found in (Emblemsvåg, 1995).

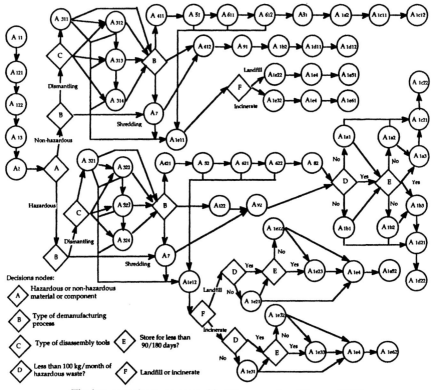

The icons are the same as used in (Greenwood and Reeve, 1992)

Figure 19.7 Demanufacturing Activity Network.

19.3.1 Model Formulation

The steps in formulating the ABC model for product demanufacture will be first discussed. We will start with Step 1, the creation of an activity network and hierarchy.

Step 1 - Create an Activity Hierarchy and Network

A demanufacture process (i.e. the process opposite to manufacture associated with recycling a product) can be broken down into activities and sub activities as in Table 19.2. This is not the only way of breaking down the recycling activities into sub-activities, but the activity table/horizontal hierarchy was created to be as generic as possible, while at the same time obtaining a desired level of accuracy. As stated in Section 19.2, the purpose of an activity hierarchy is to ensure that all the activities in the part of the life cycle to be studied are considered. As depicted in Table 19.2, three different levels of activities are present. Gray cells represent lowest level activities. The notation refers to the shaded cells. For example, activity A1 ('Collect') consists of the level 2 activities 'Buy-back', 'Transport', and 'Store' (A11, A12, and A13, respectively). Of those activities, only 'Transport' has lower level activities, namely, 'Load' and 'Move' (A121 and A122).

After identifying all the activities, a network indicating the relationships between the activities is constructed. In Figure 19.7, the activity network is shown for a demanufacture process corresponding to the activity hierarchy in Table 19.2. In the demanufacture case study we are interested in two different process scenarios:

- *Dismantling*; this scenario includes only activities related to the process of dismantling a product as much as possible/feasible in order to recover reusable components.
- *Shredding;* this scenario includes only activities associated with shredding a product. There is no dismantling.

Both process scenarios are included the network. Another approach could have been to develop separate models. It is important to note that, in general, an ABC activity network does not have a one-to-one corresponds with a process network. In an activity network, connections and relationships between process activities are given. A single activity may , however, consist of several process actions. Consider activity A311 'manual dismantling'. This activity contains all manual dismantling actions, no matter where or in what sequence they occurred in the demanufacturing process.

Step 2 – Identify Cost Drivers and Consumption Intensities

The cost drivers associated with activities in the demanufacture process are presented in Table 19.3. Please refer to Table 19.2 for the legend to the activities. The choice of cost drivers is also not unique. It is important, however, to reflect the real world situation as much as possible in order to achieve accurate cost assessments. Note that a single activity can have multiple associated cost drivers.

Having identified the cost drivers, the consumption intensities for each cost driver should be determined, because the cost of the consumption of a specific activity is the cost driver(s) multiplied with the consumption intensity. In addition, uncertainties in cost drivers and consumption intensities should be modeled at this stage. It is beyond the scope of this chapter to list all cost drivers and associated consumption intensities with uncertainty distributions used in the demanufacture cost model. Some illustrative examples are listed in Tables 19.4 and 19.5. From Table 19.2 we see that activities A411 and A412 are sorting of non-hazardous reusable components and recyclable material, respectively. The unit of the cost driver, hours per batch, indicates that the direct labor cost driver is a batch level cost driver.

As can be seen in Tables 19.4 and 19.5, normal and triangular uncertainty distributions are assigned, respectively, to both the cost drivers and consumption intensities in terms of a mean and left and right deviation. The Crystal Ball software allows twelve different distribution types. Among them the triangular and normal distributions as used in Table 19.4 and 19.5, but also other types such as the uniform, Weibull, exponential, and user-defined custom distributions.

Step 3 - Identify the Relationships between Cost Drivers and Design Changes

As stated in Section 19.2, the next step in our method is to identify the relationships between cost drivers and design changes. The level of detail and sophistication needed in modeling the relationships between cost drivers and design parameters depends on the purpose and usage of the cost model. The most detailed approach is to identify detailed mathematical Cost Driver/Design Parameter relationships, but this can become very cumbersome. For our demanufacturing case study, we propose to use a concept 'action charts' for capturing relationships between design changes and corresponding changes in cost.

Table 19.2 Demanufacturing Activities

Level 1 activity	Level 2 activity	Level 3 activity	Notation
Collect	Buy back	-	A11
(A1)	Transport	Load	A121
		Move	A122
	Store	-	A13
Pre-Clean	-	-	A2
Dismantling	Non-hazardous	Manual dismantling	A311
(A3)	dismantling	Dismantling using handtools	A312
	(destructive/ non-destructive)	Dismantling using light equipment	A313
		Dismantling using special equipment	A314
	Hazardous dismantling	Manual dismantling of haz. comp.	A321
	(destructive/ non-destructive)	Dismantling of haz. comp. using handtools	A322
		Dismantling of haz. comp. using light equipment	A323
		Dismantling of haz comp using special equipment	A324
Sort	Non-hazardous sort	Sort non-haz. reusable comp.	A411
(A4)		Sort non-haz. recyclable mat.	A412
	Hazardous sort	Sort haz. reusable comp.	A421
		Sort haz. recyclable mat.	A422
Clean reusable comp.	Clean non-haz. reusable comp.	-	A51
(A5)	Clean haz. reusable comp.	-	A52
Inspect reusable comp.	Inspect non-hazardous reusable comp.	Inspect visually non-haz reusable comp.	A611
(A6)		Test non-haz. reusable comp.	A612
	Inspect hazardous reusable comp.	Inspect visually haz. reusable comp.	A621
		Test haz. reusable comp.	A622
Shredding	-	-	A7
Collect reusable comp.	Collect non-haz. reusable comp.	-	A81
(A8)	Collect haz.reusable comp.	-	A82
Collect recyclable mat.	Collect non-haz. recyclable mat.	-	A91
(A9)	Collect haz. recyclable mat.	-	A92
Store reusable comp.	Keep records	-	A1a1
	Keep storage	-	A1a2
(A1a)	Keep max. storage	-	A1a3
Store recyclable mat.	Keep records	-	A1b1
	Keep storage	-	A1b2
(A1b)	Keep max storage	-	A1b3
Transport reusable comp.	Non-haz. transport of reusable comp.	Non-haz. loading	A1c11
		Non-haz. moving	A1c12
(A1c)	Haz. transport of reusable comp.	Haz. loading	A1c21
		Haz. moving	A1c22
Transport recyclable material	Non-haz. transport of recyclable material	Non-haz. loading	A1d11
		Non-haz. moving	A1d12
(A1d)	Haz. transport of recyclable material	Haz. loading	A1d21
		Haz. moving	A1d22

haz. = hazardous, comp. = components, mat. = materials

Table 19.2 Demanufacturing Activities *(Continued)*

Manage waste (A1e)	Collect waste from disassembly stations	Collect non-haz. waste	A1e11
		Collect haz. waste	A1e12
	Store waste for landfilling	Keep records	A1e21
		Keep storage	A1e22
		Keep max storage	A1e23
	Store waste for incineration	Keep records	A1e31
		Keep storage	A1e32
		Keep max storage	A1e33
	Transport waste to final destination		A1e4
	Landfill	Non-haz. landfill	A1e51
		Haz. landfill	A1e52
	Incinerate	Non-haz. incinerate	A1e61
		Haz. incinerate	A1e62

Table 19.3 Demanufacturing Activities and Cost drivers

Activity	Cost driver	Activity	Cost driver	Activity	Cost driver
A11	Buy back	A52	Direct labor	A1c22	Number of batches
A121	Number of batches		Tooling time		Fuel
A122	Number of batches		Number of set-ups		Number of set-ups
	Fuel	A611	Direct labor	A1d11	Number of batches
A13	Volume	A612	Direct labor	A1d12	Number of batches
A2	Direct labor		Number of tests		Fuel
	Tooling time	A621	Direct labor	A1d21	Number of batches
A311	Direct labor		Number of set-ups		Number of set-ups
A312	Direct labor	A622	Direct labor	A1d22	Number of batches
	Tooling time		Number of set-ups		Fuel
A313	Direct labor		Number of tests		Number of set-ups
	Tooling time	A71	Tooling time	A1e11	Tooling time
A314	Direct labor	A81	Tooling time	A1e12	Tooling time
	Tooling time	A82	Tooling time		Number of set-ups
A321	Direct labor		Number of set-ups	A1e21	Direct labor
	Number of set-ups of safety equipment	A91	Tooling time	A1e22	Volume
A322	Direct labor	A92	Tooling time	A1e23	Direct labor
	Tooling time		Number of set-ups		Volume
	Number of set-ups	A1a1	Direct labor	A1e31	Direct labor
A323	Direct labor	A1a2	Volume	A1e32	Volume
	Tooling time	A1a3	Direct labor	A1e33	Direct labor
	Number of set-ups		Volume		Volume
A324	Direct labor	A1b1	Direct labor	A1e4	Number of batches
	Tooling time	A1b2	Volume		Fuel
	Number of set-ups	A1b3	Direct labor	A1e51	Volume
A411	Direct labor		Volume		Mass
A412	Direct labor	A1c11	Number of batches	A1e52	Volume
A421	Direct labor	A1c12	Number of batches		Mass
	Number of set-ups		Fuel	A1e61	Volume
A422	Direct labor	A1c21	Number of batches		Mass
	Number of set-ups		Number of set-ups	A1e62	Volume
A51	Direct labor				Mass
	Tooling time				

Table 19.4 Cost Driver Examples

Activity	Cost driver				
		Distribution	Mean	Left dev.	Right dev.
A411	Direct labor	Normal [h/batch]	20.0	10.0	30.0
A412	Direct labor	Normal [h/batch]	15.0	10.0	20.0

Table 19.5 Consumption Intensity Examples

Activity	Cost driver	Consumption intensity			
		Distribution	Mean	Left dev.	Right dev.
A411	Direct labor	Triangular [$/h]	20.0	18.0	23.0
A412	Direct labor	Triangular [$/h]	20.0	18.0	23.0

What are action charts? Activities are formed by grouping actions that have a logical connection together (Cooper, 1990a; Cooper, 1990b). This is done mainly because it is impossible to achieve credible cost information for every little step in a process. It would require an enormous amount of cost drivers and consumption intensities, associated with an even larger amount of possibly uncertainty. Actions that occur in a process are grouped into activities. Forming the activities in a way that roughly describes the process is advantageous because this will make the ABC system much easier to understand and use as it coincides with our perception of the process. However, this is not necessary as the activities can be formed in any way as long as the costs are captured correctly. Remember that the purpose of the activities is simply to capture costs, not to provide a detailed description of a process.

The grouping of several process actions into a smaller number of activities opens up the possibility of designing a model with a generic set of low level activities which has design specific inputs - the actions. This is a powerful approach since any process can be described with a set of activities that will always be present, no matter what product we are dealing with. The definition of activities is a function of the desired degree of generality, accuracy and traceability. Increased generality will in general give decreased accuracy and traceability.

The usage of aggregated actions is a way of capturing how design changes affect the costs and revenues. It is an approach *in between* a) not modeling any relations at all and merely assessing cost of designs and b) modeling the relationships in detailed mathematical relationships and computing the most cost effective values of design parameters. By aggregating the actions in so-called action charts, we keep track of how the design properties and dimensions affect the actions. The aggregated effect on the actions is then transformed into an aggregated effect on the activities and the cost drivers.

In Figure 19.8, a sample from a dismantling action chart of a telephone is shown. This action chart is derived from disassembly charts outlined in (Beitz *et al.*, 1992). Note the level of details, which is typical for a good action chart. The less detailed an action chart is, the less suitable it is for design modifications.

What purpose does the dismantling action chart serve? The action chart in Figure 19.8 allows the detailed documentation of a disassembly process. Manual disassembly can be considered as a single activity in an ABC cost model and a reduction of overall disassembly time is clearly advantageous. But on which product component should a designer focus? Time is not the only cost driver in disassembly. A different material or fluctuations in material prices also affects overall revenue. This kind of product design related information is embodied in the dismantling action chart. In essence, an action chart forms an interface between detailed product information and a general demanufacturing ABC model, in our case.

In order to support the ABC model, each action must be associated with a sufficient set of information. In our opinion, the following set of action information seems to be sufficient input for each activity in a demanufacturing model:

- All actions related to the specific activity, and
- For each action:
    - the number of units,
    - the mass and material composition for each unit,
    - the time to perform an action, the tools used,
    - the process efficiencies,
    - the hazardousness of the units, and
    - the danger in performing the actions.

Uncertainty distributions can be assigned in the action chart (e.g. for specific disassembly times or material mass) and the effect of variations in the product design can be traced. In the next section, we discuss the results obtained from using a demanufacturing action chart in our demanufacturing ABC model.

Step 4 - Find and Minimize the Cost of the Consumption of Activities

Having created the ABC model and modeled the associated uncertainty, we must now proceed to find (and minimize) the cost associated with the consumption of the demanufacture activities. To support this step, we have implemented the entire model in Microsoft Excel 4.0 spreadsheet files on a Macintosh platform.

In Figure 19.9, the file structure of the demanufacturing ABC model is presented. The arrows represent the information flow, and the direction of the arrow shows which file is receiving information and which file is sending information. Clearly, there is a large amount of interaction. Most notably in Figure 19.9 is the breakdown of needed information for the cost assessment in files which can be associated with different departments in a company. As can be seen, even EPA storage requirements have been included.

There are 135 assumption cells in 10 model files and a number of assumption cells in the action chart. In other words, the number of assumptions are so large that only the most important ones will be mentioned. Some of the main assumptions made in the model are presented in Table 19.6. The most important assumption of all the assumptions in the entire model is the assumption that the plant capacity is less than market demand because "The shredder population is getting out of sync with the available auto hulks, so more and more machines are having to shred material that hasn't normally been shredded", explains Phillips, sales director of Lindemann Recycling Equipment Inc. Scott Newell, chairman of Newell Industries Inc. (San Antonio), agrees, observing that "most processors complain that they don't have enough auto bodies and white goods to run their shredders..." (Kiser, 1992). In other words; without this assumption we may end up in a situation where recycling is not feasible because of lack of items to shred. This is a market problem and these problems are not treated explicitly in our model. However, by remodeling some of our assumption cells we can simulate cases with less than 100% market demand. It should be noted that we can also model the uncertainty in the assumptions. For example, in Figure 19.10 we have given the uncertainty distributions for the telephone buy-back price and the special equipment investment cost.

As stated in Section 19.1.2, we have chosen to use a software called Crystal Ball® (Version 3.0). As mentioned before, it allows the definition of uncertainty distribution in assumption cells in a spreadsheet (the graphs of Figure 19.10 are an example of this) and finds resulting uncertainty distributions in forecast cells numerically using a Monte Carlo simulation. When all the trials have been performed Crystal Ball® creates a frequency plot for the forecast cell and computes other statistical information (like mean and variability) based on the trial values in the assumption cells and the forecast cell. In the following sections, we will discuss the results obtained in this fashion for the demanufacture cost assessment and we will perform step 5 of our method; Evaluate the solution.

Step 5 – Evaluate the Solution

Once the model is implemented in spreadsheet structure, several tools are available for evaluating the solution, results, and the effects of the assumptions and design decisions made. In the next sections, we discuss the results and illustrate two tools available, i.e. profitability distributions and sensitivity charts, which assist in evaluating the solution.

19.3.2 Profitability Distributions for Telephone Dismantling and Shredding Scenarios

We start with the 'Dismantling' scenario and the resulting profitability distribution for this scenario are presented in Figure 19.11. This distribution is generated by the Crystal Ball software using a Monte Carlo simulation, as described in Section 19.1.2. These distributions provide a good indication of what the effects of the uncertainties in the assumptions are. In our opinion, such distributions provide more valuable information for designers than merely a single number. With respect to telephone dismantling, the mean profit is estimated to be - $2.30 if we pay, on average, $1.00 for the telephones. In other words; the present telephone design is not expected to be economical feasible with respect to dismantling.

The results from the 'Shredding' scenario are presented in Figure 19.12. We see that the revenues nearly balance the costs. In fact, if there would not have been a $1.00 buyback price, then the shredding option would probably be economically feasible. Taking into account that many assumptions have been made and the fact that the upper limit of the distribution is close to a break-even, we should not rule out the possibility of a break-even in the real world.

19.3.3 Identifying Largest Cost Contributors Using Sensitivity Charts

Assuming that we want to pursue dismantling of automobiles, what process changes should we make in order to boost profitability? In Figure 19.13, a sensitivity chart of the dismantling process is presented. Such a sensitivity chart is generated for each simulation run by the Crystal Ball software. The sensitivity chart is based on the so called Spearman Rank Correlation and measures the degree to which assumptions and forecasts change together. The larger absolute value of the correlation coefficient, the stronger is the relationship. Positive coefficients indicate that an increase in the assumption cell is associated with an increase in the forecast cell.

As can be seen, the sensitivity chart allows us to pin point the factors/assumptions for the 'Dismantling' scenario that correlates most with the forecast cell. This facilitates studying the importance of the different assumption cells in the model. Such as study is important because it will tell you what cells should be updated with better, more accurate information. There is a cost associated with gathering information, and in every project it is therefore important to decide when the cost of gathering information outweighs the benefits of improved information.

No.	Name	Quantity	Type	A/C	Tool	Force	Time [sec.]	Tool for Assembly	Stnd./Special Pt.	Abrasion	Fatigue	Dirt/Corr.	Nondest-ructive	Material	Mtl. Recycl.	Mass [kg]
1	Snap off base-handset cable	1	CE	4		4	4	Snap on	Std	4	4		yes	copper, plastic	4	0.0030

Handset Disassembly

No.	Name	Quantity	Type	A/C	Tool	Force	Time [sec.]	Tool for Assembly	Stnd./Special Pt.	Abrasion	Fatigue	Dirt/Corr.	Nondest-ructive	Material	Mtl. Recycl.	Mass [kg]
2	Remove top from bottom	1	SP	3	Crowbar	2	30	Snap on	Std	4	4		No	ABS	4	0.0060
3	Remove Mass	1	SP	3	Crowbar	3	5	Glue on	Std	4	4	D	yes	Lead	4	0.0140
4	Remove mic. cables and CB	1	SA	3	Crowbar	3	15	Snap on		4	4		yes			

And so forth

No.	Name	Quantity	Type	A/C	Tool	Force	Time [sec.]	Tool for Assembly	Stnd./Special Pt.	Abrasion	Fatigue	Dirt/Corr.	Nondest-ructive	Material	Mtl. Recycl.	Mass [kg]
27	Disassemble circuit board	1	SA	3		4	5									
28	From 26; circuit board	1	SA						Std	4	4		yes	Mix	1	0.0580
29	From 26; spring	1	SP						Std	4	4		yes	Steel	4	0.0001
30	From 26; cover	1	SP						Std	4	4		yes	Thermo-set	1	0.0001

Notes:

CE: Connecting Element,
SA: Sub Assembly,
SP: Single Part,
A/C: Accessibility;
Std: Standard part,
Mtl. Recycl.: Material Recyclability,
(4 is good, 1 is bad).

Qualitative assessments:
1 => Bad,
2 => Below average,
3 => Above average,
4 => Good

Quantity, Time and Mass are assessed quantitatively.

Nondestructive:
Yes => assembly can be dismantled without destroying totally or partially any part or sub assembly in the assembly.
No => assembly cannot be dismantled without destroying totally or partially any part or sub assembly in the assembly.

Figure 19.8 A phone dismantling action chart.

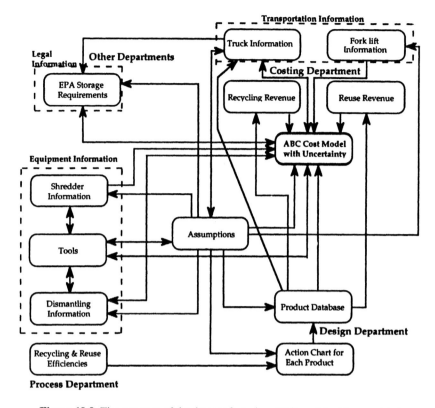

Figure 19.9 The structure of the demanufacturing ABC model with uncertainty.

Table 19.6 The Main Default Assumptions in the Demanufacturing ABC Model

Area of Assumption	Assumption	Quantification
Market	The plant capacity is lesser than market demand	
Recycling plant	Operation hours	7.00 [h/day]
	Operation days	5.00 [days/week]
	Business weeks	50.00 [weeks/year]
Investment capital cost	Truck	37,500.00 [$/year]
	Shredder	30,000.00 [$/year]
	Special equipment	10,000.00 [$/year]
Environmental overhead cost	Machine related	50,000.00 [$/year]
	Hazardous material and component related	15,000.00 [$/year]
Other overhead cost	Management related	100,000.00 [$/year]
Buy back Price*	Telephone	1.00 [$/unit]

*Assuming that we would have to pay for the phones to be demanufactured.

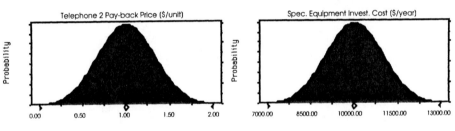

Figure 19.10 Uncertainty in the Assumptions.

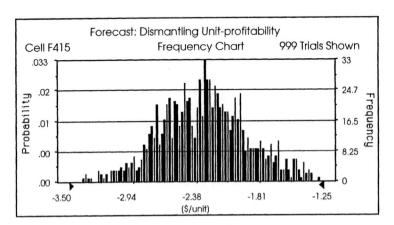

Figure 19.11 Telephone 'Dismantling' Unit-profitability Distribution.

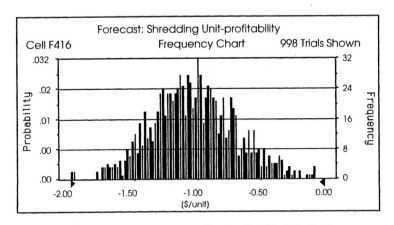

Figure 19.12 Telephone 'Shredding' Profitability Distribution.

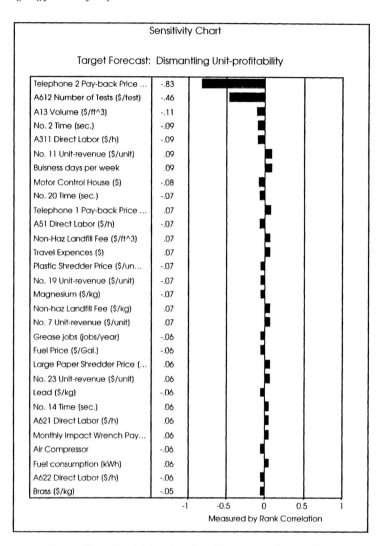

Figure 19.13 Sensitivity Chart for 'Dismantling' Profitability.

In the sensitivity chart of Figure 19.13, the 30 assumption cells that correlate strongest with the selected forecast; 'Dismantling Unit-profitability', are given. Some of the major cost and revenue triggers are:

- Telephone 2 buyback price,
- A612 Number of tests [$/test], (Number of tests to check if components are reusable or not). From the unit, [$/test], we understand that this is the consumption intensity of activity A612 and not the cost driver.

- A13 Volume [$/ft^3], (Storage cost driver for units in to the demanufacturing plant)
- No. 2 Time [sec.], (Time to perform action number 2 from the phone action chart - 'Remove top from bottom' of the handset disassembly),
- A311 Direct labor [$/h],
- No. 11 Unit-revenue [$/unit], and (Unit-revenue associated with reusable speakers)
- Business days per week.

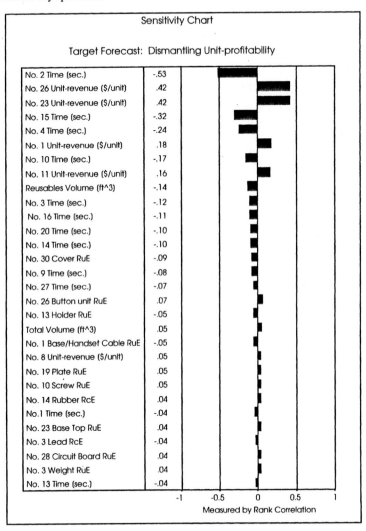

Figure 19.14 Sensitivity Chart for 'Dismantling' Profitability using Perfect Process Information.

Clearly both process and product design issues affect the profitability of dismantling telephones. It is interesting to note that variation in the buyback price has the largest strongest correlation. We have noted this as well in another case study involving the demanufacture of automobiles.

The number of trials performed has an effect on the usefulness of the sensitivity chart; the more trials - the more useful is the sensitivity chart. The reason is that the variability of the correlation coefficient estimate decreases as the number of trials increases because the number of degrees of freedom increases, and that the probability for correlation by chance decreases, see (Hines and Montgomery, 1990). With only 1000 trials performed we should be careful with relying too much on the part of the sensitivity chart where the absolute value of the correlation coefficients is lower than roughly 0.09.

19.3.4 Tracing Design Changes using Perfect Process Information

It is obvious that some design changes have to take place if the telephone company wants to recover reusable components from the telephone rather than merely recycling its materials. To spot potential areas of product design changes, the sensitivity chart from a model which has perfect process information should be employed. The reason is that only the correlation between action chart assumption cells and the forecast cells is considered in this sensitivity chart . Thus, the perfect process information sensitivity chart is not 'polluted' with process assumption cells. The sensitivity chart needed for the 'Dismantling' scenario is presented in Figure 19.14. Again we should be careful with trusting the lowest correlation coefficients, because they can be caused coincidentally when the number of trials is low.

From the sensitivity chart in Figure 19.14, we learn that action number two has the largest influence on the dismantling cost. From the action chart (see Figure 19.8) we see that action 2 - 'Remove top from bottom (of the handset disassembly) ' - was estimated to 30 seconds, and that the action included opening snap fits. As can be seen, a screw driver was used to do the job. The telephone manufacturer might argue that we should have used special tools that could handle 'their' snap fit, however, this is not a realistic option in the real world. The reason is simple; a dismantling line would have many workers that would have to deal with many different types of telephones, and if they should use special equipment for every snap fit on every telephone, the dismantling scenario would become really unprofitable. What we can learn from this is that the design of the snap fits must be changed or another fastening mechanism must be used.

The No. 26 and 23 unit-revenues listed as second and third in Figure 19.14, relate to parts made out of ABS, a relatively expensive plastic. More revenue can be obtained if larger amounts of ABS are used or if the price of ABS rises. The fourth action listed in the sensitivity chart 'No. 15 – Time' relates to the action of removing four screws from the base; always a time consuming operation. All the actions that have a correlation coefficient with an absolute value of roughly 0.09 or more should be discussed in a similar manner. The design group can, of course, make design changes that will affect actions with a correlation coefficient with an absolute value lower than 0.09, however, it is the actions with highest absolute value of the correlation coefficients that trigger most of the costs. The focus should therefore, *first* be concentrated on the actions that trigger most of the costs. This is particularly important in cases where we would have to compromise between different actions, i.e. a situation where improving the design with respect to some actions leads to worse performance for another action(s). For example, in (Scheuring *et al.*, 1994), we highlight such trade-offs between equipment utilization and product throughput.

19.4 SUMMARY

In this chapter, we discussed how an Activity-Based Cost model can be developed and used in the decision making process to obtain a cost efficient design under presence of uncertainty. We use ABC as a basis because it provides a relatively easy and clear methodology to assess *and* trace costs. Our method, as most ABC based methods, is generic in that it can be applied whenever the activities are described in sufficient detail to have cost drivers assigned.

The inherent uncertainty is handled by employing a numerical simulation technique - the Monte Carlo simulation technique - to simulate the behavior of the model when the uncertainty is modeled in terms of fuzzy numbers. The inclusion of uncertainty and usage of Monte Carlo simulation in conjunction with so-called action charts provides the capability to identify those process and product design aspects that contribute most to the cost. An action chart represents a group of associated actions which together form an activity. We used disassembly actions which were aggregated in manual dismantling activities as an example. The subsequent use of sensitivity charts enables a design group to quickly spot the most cost inefficient parts of the design which allows the group to concentrate the redesign effort and improves design efficiency.

We provided a case study regarding the demanufacture of telephones to illustrate our approach. In the context of the case study, we discussed the use of the following tools to support investigation and verification of the models and results:

- *profitability distributions* visualize the forecasted cost/profit and the associated uncertainty,
- *sensitivity charts* visualize how easy and in what direction different assumptions (like direct labor) and design aspects affect a cost forecast, and

Specifically, these tools assist in identifying:

- where we should focus our *data collection efforts*, for example, because buyback prices and direct labor have such significant influence (Figure 19.13), we should collect more accurate data in these areas.
- where we should focus our *design efforts*, for example, with respect to the telephone design, we should focus on the removal of the top from the base because it is the largest cost contributor (Figure 19.14).

It should be emphasized that especially in the early stages of design, the identification of the largest cost contributors and critical factors is more important than the actual cost.

With respect to validity, we have attempted to make our demanufacturing model as realistic as possible. For example, the EPA regulations for storing hazardous waste have been incorporated. The method presented in this paper seems to give reasonable results. In fact, the same model has been applied to assess and trace the cost of demanufacturing a car and the results were compatible with real world experiences (Emblemsvåg, 1995). In our model, we have most likely underestimated the total overhead cost for a demanufacturer, so for the 'Dismantling' and 'Shredding' scenarios we would expect that the true costs are higher than estimated. Another aspect to take into account is that much of the material revenue information is based on (Dieffenbach *et al.*, 1993) and reported to be from 1990. Most likely this information is outdated.

Our focus in future work is on the following key aspects:

- We are continuing to validate, improve, and expand our method and models. One of our objectives is to utilize the tractability of the costs and uncertainty through an activity network and identify which life-cycle activities are truly critical in a product's life-cycle. More detailed information on this is given in (Emblemsvåg, 1995).
- In the long term, we seek to utilize the ABC method not only for monetary cost assessments, but also for life-cycle assessments of environmental impact in terms of matter and energy consumption. A model which provides an environmental impact assessment in terms of energy and matter consumption and emission can be used for exploring the global (societal) environmental impact of engineering products and processes. For example, recycling products is nice, but a recycling process may cost more energy than a disposal process. An environmental impact model in terms of energy and matter is needed because it is difficult to convert all costs to the environment into a monetary value. The use of the ABC approach may overcome some of the difficulties associated with conventional Life-Cycle Assessment/Analysis tools, e.g. the cumbersome amount of work involved and the lack of common standards (Congress, 1992; EPA, 1993b; EPA, 1993a; Veldstra and Bouws, 1993). In our opinion, an Activity-Based Life Cycle Assessment is most easily done for energy where we have a single unit. It will be more difficult for materials where, for example, we would need to distinguish different grades of toxicity.

Although we have focused on environmental issues as the area of application, it should be noted that our method, as most ABC based methods, is generic in that it can be applied whenever the activities are described in sufficient detail to have cost drivers assigned.

REFERENCES

Beitz, W., Suhr, M., Rothe, A. (1992) *Recyclingorientierte Waschmaschine (recycling-oriented washing machine)*, Institut für Maschinenkonstruktion - Konstruktionstechnik, Technische Universität, Berlin.

Brooks, P.L., Davidson, L.J., Palamides, J.H. (1993) Environmental compliance: You better know your ABC's, *Occupational Hazards*, February, 41-46.

US Congress (1992) *Green Products by Design: Choices for a Cleaner Environment*, OTA-E-541, Office of Technology Assessment, Washington, D.C.

Cooper, R. (1990a). ABC: A Need, Not an Option, *Accountancy*, September, 86-88.

Cooper, R. (1990b) Five Steps to ABC System Design, *Accountancy*, November, 78-81.

Dieffenbach, J.R., Mascarin, A.E., Fisher, M.M. (1993) Modeling Costs of Plastics Recycling, *Automotive Engineering*, October, 53-57.

Emblemsvåg, J. (1995) *Activity-Based Costing in Designing for the Life-Cycle*, Master of Science Thesis, G.W. Woodruff School of Mechanical Engineering, Georgia Institute of Technology, Atlanta, Georgia.

Emblemsvåg, J., Bras, B.A. (1994) Activity-Based Costing in Design for Product Retirement, *Proceedings 1994 ASME Advances in Design Automation Conference, DE-Vol. 69-2*,

(Edited by B.J. Gilmore, D.A. Hoeltzel, D. Dutta, H.A. Eschenauer)., Minneapolis, American Society of Mechanical Engineers, 351-362.

EPA, US (1993a) *Life-Cycle Assessment: Inventory Guidelines and Principles,* EPA/600/R-92/245, US Environmental Protection Agency, Office of Research and Development, Washington DC.

EPA, US (1993b) *Life-Cycle Design Guidance Manual,* EPA/600/R-92/226, US Environmental Protection Agency, Office of Research and Development, Washington DC.

Greenwood, T.G., Reeve, J.M. (1992) Activity Based Cost Management for Continuous Improvement: A Process Design Framework, *Journal of Cost Management for the Manufacturing Industry,* **5** (4), 22-40.

Hines, W.W., Montgomery, D.C. (1990) *Probability and Statistics in Engineering and Management Science,* John Wiley & Sons, Inc.

Keoleian, G.A., Menerey, D. (1994) Sustainable Development by Design: Review of Life Cycle Design and Related Approaches, *Air & Waste,* **44**, May, 644-668.

Kiser, K. (1992) State of the Shredding Art, *Scrap Processing and Recycling,* **49** (4), 99-108.

Navin-Chandra, D. (1993) ReStar: A Design Tool for Environmental Recovery Analysis, *9th International Conference on Engineering Design,* (Edited by N.F.M. Roozenburg), The Hague, Heurista, Zurich, Switzerland, 780-787.

O'Guin, M. (1990) Focus The Factory With Activity-Based Costing, *Management Accounting,* February, 36-41.

Raffish, N., Turney, P.B.B. (1991) Glossary of Activity-Based Management, *Journal of Cost Management for the Manufacturing Industry,* **5** (3).

Scheuring, J.F., Bras, B.A., Lee, K.M. (1994) Significance of Design for Disassembly in Integrated Disassembly and Assembly Processes, *International Journal of Environmentally Conscious Design and Manufacturing,* **3** (2), 21-33.

Thurston, D.L., Blair, A. (1993) A Method for Integrating Environmental Impacts into Product Design, *9th International Conference on Engineering Design,* (Edited by N.F.M. Roozenburg), The Hague, Heurista, Zurich, Switzerland, 765-772.

Turney, P.B.B. (1991) How Activity-Based Costing Helps Reduce Cost, *Journal of Cost Management for the Manufacturing Industry,* **4** (4), 29-35.

Veldstra, M., Bouws, T. (1993) Environmentally Friendly Design in Plastics, *9th International Conference on Engineering Design,* (Edited by N.F.M. Roozenburg)., The Hague, Heurista, Zurich, Switzerland, 820-827.

Winner, R.I., Pennell, H.E., Bertrand, P.I., Slusarczuk, M.M.G.(1988) *The Role of Concurrent Engineering in Weapons System Acquisition,* IDA Report R-338, Institute for Defense Analyses, Alexandria, Virginia.

DESIGN OPTIMISATION FOR
PRODUCT LIFE CYCLE

Masataka Yoshimura

This chapter presents a optimization strategy of designing products for the life cycle. In Section 20.1, the fundamental product design decision making procedures based on clarification of the product environmental conditions are first explained. Practical methodologies corresponding to the product environment are described with applied examples in Sections 20.2, 3 and 4. In these procedures, relationships between evaluative characteristics are clarified. Information of features and behaviors of product environments is concurrently utilized for realizing the actual construction of the most satisfactory product design.

Concurrent engineering is an effective and powerful methodological philosophy for obtaining the most satisfying product design possible from an integrated and global viewpoint. The products to be manufactured are related with the product life phases of manufacturing products, selling products, using products, disposing products, recycling products and so on. In product design, the conditions relating to each issue of the product life cycle, that is, the environmental conditions of the products, should be completely comprehended and the information obtained from that knowledge concurrently utilized at the maximum level in the decision making process of product designs.

20.1 DESIGN FOR PRODUCT LIFE CYCLE

Figure 20.1 shows an optimum product design procedure including the following three general steps:

(1) Product life cycles are clarified and issues identified. Based on the clarification, characteristics to be evaluated are specified.
(2) The relationships between the evaluative characteristics are clarified.

(3)   Decision making problems are formulated and the optimum solution is obtained.

In usual product design and product optimization, step (3), the formulation of an optimization problem and solving the problem, receives the most attention. However, steps 1 and 2 are essential for obtaining the most satisfactory design solutions.

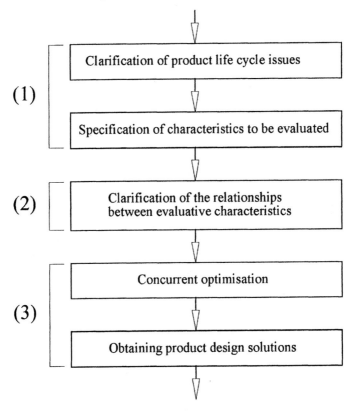

Figure 20.1 Flowchart of optimum design for product life cycle issues.

20.1.1 Product Life Phases

There are three distinct general stages in product life cycle as shown in Figure 20.2:

(i)   The environment related to "manufacturing products" where requirements from different divisions such as design and manufacturing divisions are satisfied.

(ii)   The environment related to "selling products" where market needs are to be sufficiently surveyed.

(iii) The environment related to "using products" where conditions for use of products are completely understood.

Figure 20.2 Three product life cycle phases of "manufacturing", "selling" and "using".

The product conditions determine which environment is to be regarded as important. Item (i) is an environment inside the company where the product is manufactured and so main attention is given to manufacturing. In item (ii), how products satisfy the market environment is considered so main attention is given to the market conditions. In item (iii), how suitable are products to the specific environmental conditions where the products will be used is considered so main attention is given to product users.

Clarification of the product environment is the first important step in the decision making process of product designs. Characteristics to be evaluated and the appropriate levels required for the characteristics are defined according to the assigned class.

20.1.2 Evaluation Characteristics

Working environments of machine products differ widely from one another. The formulation of product design optimization should be fundamentally different according to the specific features of each working environment. Operational accuracy, operational efficiency, operation cost (including cost for controling the operation) and manufacturing cost are fundamental characteristics considered in the evaluation of machine products. For obtaining optimum designs applicable to practical circumstances, integrated evaluation of these characteristics is essential.

For the foregoing classification of product environments, the main relationships between evaluative characteristics are as follows:

In (i), the relationships between principal characteristics of products (that is, product performance) and product manufacturing cost.

In (ii), the relationships among product attributes depending on the market needs.

In (iii), the relationships between evaluative characteristics of products (for example, operational accuracy, operating efficiency, control performance and operational cost).

For clarification of relationships between characteristics which are conflicting with each other, the formulation of the multiobjective design optimization problem where the characteristics are simultaneously evaluated on the same stage (Stadler, 1988; Eschenauer *et al.*, 1990) is effective.

20.1.3 Concurrent Design Optimization

The decision making methods are related to the methods for clarifying the relationships between evaluative characteristics used in the foregoing section. Therefore, after the relationships between evaluative characteristics are clarified, the formulation of optimum design solutions is conducted. Multi-objective optimisation methods are used to evalaute characteristics and the optimum solutions are determined from a global viewpoint.

For the design optimization of machine products, a higher operational accuracy, a smaller operational time, a smaller operation cost and a smaller manufacturing cost are preferable. So, the optimal solutions should be selected from the solutions on the Pareto optimum solution set of a multi-objective optimisation problem which has all or several of the following objectives: "maximisation of the operational accuracy", "minimisation of the operational time", "minimisaion of the operation cost" and "minimisation of the manufacturing cost".

The design solution having the maximum value of a specific satisfaction level (for example, satisfaction level of the manufacturers of the product, satisfaction level for the market needs and satisfaction level of the users of the product) is selected as the optimum one.

Table 20.1 Decision-making items in the product design and the process design

Decision-making items in product design

D_1	Mechanism
D_2	Parts constitution of product
D_3	Connected relations among parts
D_4	Purchased parts or manufactured parts
D_5	Shape of part
D_6	Shape accuracy
D_7	Surface roughness
D_8	Dimension
D_9	Tolerance
D_{10}	Material

Decision-making items in process design

P_1	Preparation process of raw material
P_2	Machining methods
P_3	Machine tools
P_4	Tools and jigs
P_5	Machining sequence
P_6	Quality of finished surface
P_7	Cutting conditions
P_8	Raw material shape
P_9	Heat treatment process

20.2 DESIGN OPTIMIZATION FOR PERFORMANCE AND COST

This section focuses on design optimization where product functional performance and manufacturing costs are of primary interest.

Step 1 - Understanding Product Environment

Generally, in product design divisions, the product performances are evaluated according to the satisfaction of the required product functions. Process designs are conducted in manufacturing divisions where the methods for practically manufacturing the designed products are determined, and the manufacturing cost evaluated. Therefore, in decision making of product design and manufacturing, the product performance and the manufacturing cost are the principal evaluative characteristics.

Table 20.1 tabulates examples of decision-making items in the product design and the process design. These decision-making items are related in a complicated manner with both the product performance and the manufacturing cost. Hence, in order to obtain the optimum design solutions from a global viewpoint of the product performance and the manufacturig cost. Hence, in order to obtain the optimum design solutions from a global viewpoint of the product performance and the manufacturing cost, decision making items in the product and process design should be concurrently and cooperatively evaluated (Yoshimura, 1993).

Step 2 - Clarifying Relationships Between Evaluative Characteristics

Product designers are principally seeking a higher product performance, while process planners are seeking a lower product manufacturing cost. Figure 20.3 shows the relation between the product performance and the product manufacturing cost (Yoshimura, 1993). The shaded part corresponds to the region feasible using present technologies, knowledge and/or theories. The designers are searching for designs along the direction of the big arrow shown in Figure 20.3. The heavy solid line PQ corresponds to the Pareto optimum design solutions (the feasible design solutions in each of which there exists no other feasible design solution that will yield an improvement in one objective without causing a degradation in at least one other objective) (Cohon, 1978) of a multiobjective optim solutions are a set of design points where both further improvement of the product performance and further reduction ofthe product manufacturing cost are impossible. The designers ultimately search for a design solution on the heavy line. Product design divisions and manufacturing divisions are fundamentally related as each arrives at a design solution on the Pareto optimum solution. When requirements from different divisions conflict with each other in this way, decision making items related with those requirements should be concurrently evaluated at the same stage (Yoshimura and Takeuchi, 1991).

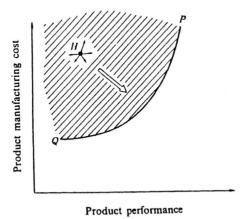

Product performance

Figure 20.3 Relationship between the product performance and the product manufacturing cost (Point H corresponds to a present design point or an initial design point).

Step 3 - Determining Optimum Solution

Evaluative performance characteristics ψ_j (j=1,2,..., k) which have conflicting relationships with other evaluative characteristics and/or the product manufacturing cost and the product manufacturing cost C are included in the objective functions Ψ as follows:

$$\Psi = [P, C], \qquad \text{where} \quad P = [\psi_1, \psi_2, ..., \psi_k] \qquad (1)$$

The integrated design optimization procedures were applied to the design of a cylindrical-coordinate robot shown in Figure 20.4 (Yoshimura, Itani and Hitomi, 1989). Characteristics selected as objective functions in this example are: the static compliance at the installation point (H) of the hand on the arm, $f_s (= \psi_2)$, the total weight of the structure, $W_T (= \psi_2)$, and the product manufacturing cost C.

The integrated optimum design of the cylindrical-coordinate robot is formulated as a three-objective optimization problem as follows:

$$\text{Minimize} \quad \psi = [f_s, W_T, C] \qquad (2)$$

subject to the constraints concerning surface roughness, tolerance and dimensions.

Main parts, 1, 2, and 3 of the robot shown in Figure 20.5, are objects of optimization design, since these parts play important roles in the product performance. Candidate materials of those parts are cast iron (FC) and low carbon steel (SC). The Pareto optimum solution sets for f_s vs. W_T and f_s vs. C are shown in Figure 20.6(a) and (b). Each combination of part candidate materials produces a Pareto optimum solution set. Point Q is a tentative solution on the Pareto optimum solution line which was intially chosen and point A corresponds to the final optimum design solution.

Here, decision making items related with divisions of product design, process design and practical manufacturing are concurrently processed.

By the methodology of concurrent design and manufacturing, not only the most preferable product design can be obtained but also a smooth flow of information from the design stage through the manufacturing stage is naturally attained.

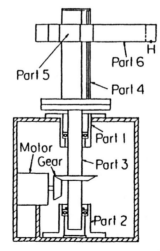

Figure 20.4 Schematic construction of a cylindrical co-ordinate robot to be designed.

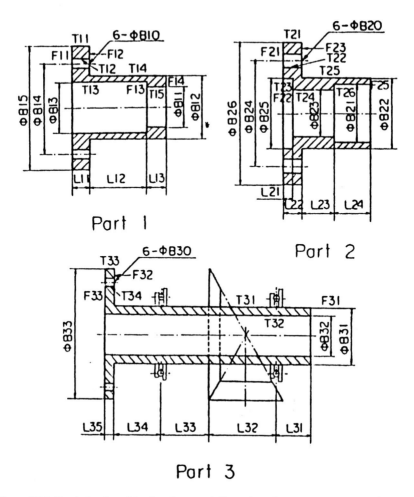

Figure 20.5 Symbols of machined surfaces and dimensions for parts to be designed (B, F, L and T are symbols added to surfaces and dimensions).

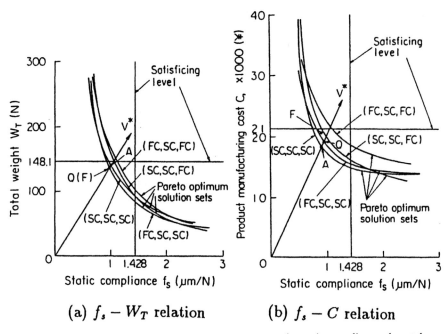

(a) $f_s - W_T$ **relation**   **(b)** $f_s - C$ **relation**

Figure 20.6 The Pareto optimum solution sets among the static compliance, the total structural weight and the product manufacturing cost (the symbols in parentheses indicate materials used for parts 1, 2 and 3 in the order): (a) Relation between the static compliance f_s and the total weight W_T; (b) Relations between the static compliance f_s and the product manufacturing cost C.

20.3 DESIGN OPTIMIZATION FOR CUSTOMER/MARKET NEEDS

Product designs which reflect market trends and needs are discussed here. The focus is placed on selling products. Here, utilization without delay of information concerning customers' needs is a key point for the decision making of product designs.

Each customer has his own requirements or desires for a product. After surveying the requirements, customers are classified based on the similarity of customers' needs. Decision making of product designs is conducted so that the satisfaction level of a target customer group becomes maximum (Yoshimura and Takeuchi, 1994).

Step 1 - Understanding Product Environment

Here, the goal of product design is to obtain design solutions which can satisfy customers' needs as much as possible. In order to quickly reflect information concerning needs to the actual manufacturing of products, the information analysis of needs and decision making of product designs are concurrently conducted.

The following survey of market needs is conducted:

(1) Check for specific product attributes (such as product performance and product cost) which can be regarded as being of major importance.
(2) Check the estimated number of purchase products and the desired value and importance level for each attribute selected in (1).

Table 20.2 One part of the completed results for the questionnaires

User No.	A	B	C	D	E x 1000	F	G
1	3	3	0.03	2.8	190	0.4, 0.2, 0.4	2
2	2	10	0.05	1.8	200	0.2, 0.3, 0.5	3
3	3	25	0.15	5.0	240	0.3, 0.5, 0.2	4
4	1	15	0.02	1.0	150	0.4, 0.2, 0.4	1
5	2	5	0.03	1.0	185	0.5, 0.1, 0.4	3
6	3	2	0.08	2.2	185	0.4, 0.2, 0.4	4
7	1	3	0.01	0.8	130	0.2, 0.2, 0.6	2
8	3	2	0.12	2.2	170	0.1, 0.5, 0.4	1
9	2	12	0.05	1.5	150	0.2, 0.1, 0.8	2
10	3	15	0.10	4.0	240	0.2, 0.5, 0.3	3
11	1	5	0.01	0.1	120	0.2, 0.2, 0.6	2
12	3	2	0.05	2.8	170	0.2, 0.4, 0.4	4
13	2	20	0.06	1.5	200	0.2, 0.2, 0.6	3
14	1	8	0.03	0.8	150	0.1, 0.4, 0.5	5

A: Type of robot (1: Cartesian coordinates robot; 2: Horizontal articulated robot; 3: Vertical articulated robot)
B: Load capacity (kgf)
C: Operational accuracy (mm)
D: Cycle time (s)
E: Product cost (202020)
F: User's importance level (Operational accuracy, Cycle time, Product cost)
G: Estomated number of purchase products

Table20.3. Grouping of potential customers by the cluster analysis

Group No.	p	q	r	s	t	u x 10000
1	28	1	5.6	0.02	0.9	139
2	20	1	12.5	0.02	1.0	150
3	40	2	3.2	0.05	1.1	168
4	38	2	8.9	0.06	1.2	204
5	25	2	12.8	0.07	1.4	224
6	35	3	2.8	0.07	2.6	169
7	22	3	9.8	0.09	4.1	205
8	27	3	26.4	0.15	5.1	283

p: Total estimated number of purchase products in the group
q: Required type of robot
r: Average value of required load capacity (kgf)
s: Average value of required operational accuracy,
t: Average value of required cycle time (s)
u: Average value of required product cost (202020)

When machine products are industrial robots, product attributes which are important for purchase decision making are types of robots, load capacity, operational accuracy (positioning accuracy and repeatability at the point of an end effector), operatio Among these product attributes, types of robots and load capacity have great influences upon the product shapes and the sizes of robots. Hence, those product attributes should be satisfied without failure in the product design and included in the constraints in the formulation of of the design optimisation. The product attributes (operational accuracy, operational efficiency and manufacturing cost) are included in the objective function. From the questionnaire survey of market needs, the results are obtained as shown in Table 20.2.

For dividing customers into groups based on similarlity of the customers' needs, the cluster analysis is conducted using data obtained by the questionnaires. Examples of results where the customers are divided into groups are shown in Table 20.3. Here, among the groups, the group which is most suitable for manufacturing in the company is selected as the product target.

Step 2 - Clarifying Relationships Between Evaluative Characteristics

When a customer desires to have a product located within the feasible design region, it is possible to design and manufacture a product satisfying the customer's requirement. When a customer desires a product which is located outside the feasible design region, it is impossible to design and manufacture a product completely satisfying the customer requirements. However, the customer may have some level of satisfaction for a compromised product outside the feasible region. Design decision making should be conducted so that the satisfaction level will be as high as possible. Here, measures for evaluating customer satisfaction levels for products are constructed to integrate demand analysis and design optimization.

Figure 20.7 shows that the utility U_{ij} of customer j for product attribute i is a function of ε_{ij}, the distance of the product attribute value z_{ij} away from the customer's desired value z_{ij}^*. Here, when the product attribute value having a minus value of eij is closer to the desried value having ε_{ij} of zero, utility U_{ij} is increased. When attribute value becomes more preferable than the desried value (that is, ε_{ij} is greater than zero), the change of increase in U_{ij} becomes smaler.

Product utility U_j, the value of the product for customer j, is expressed using utility U_{ij} and α_{ij} (which indicates the customer's importance level) of each product attribute zi ($i=1,2,...,n$). n is the total number of attributes as follows:

$$U_j = \prod_{i=1}^{n} U_{ij}^{\alpha_{ij}} \qquad \text{where} \quad \sum_{j=1}^{m} \alpha_{ij=1} \qquad (3)$$

In eq.(3), when customer's importance level α_{ij} is greater for high utility U_{ij}, product utility U_j has a greater value close to 1, but when α_{ij} is greater for low utility U_{ij}, product utility U_j has a small value close to 0. Product utility U_j always has a value close to 1 for low α_{ij}

The product utility defined in eq.(3) expresses the customer's satisfaction level for the product. When each product attribute has the level closer to or superior to the required value, the product utility has the value closer to 1. On the other hand, the lower the level is of each product attribute compared to the required value, the product utility value decreases further from 1 and of course smaller. The product utility has a greater influence on the change of the product attribute with a higher important level. Therefore, the difference among customers' preferences can also be evaluated using product utility.

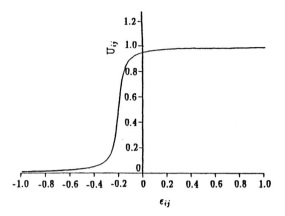

Figure 20.7 An example of the relation between a product attribute and customer utility for the product attribute.

Step 3 - Determining Optimum Solution

The satisfaction level U of each group among the groups divided by the cluster analysis is formulated using the satisfaction level U_j of customer j belonging to the group as follows:

$$U = \frac{\sum_{j=1}^{m} N_j U_j}{\sum_{j=1}^{m} N_j} \tag{4}$$

where Nj = the estimated number of customer j's products for purchase.
         m = the number of customers belonging to the group.

The satisfaction level U of the group has a value between 0 to 1. When customer j, having a large estimated number for the product for purchase, has a great satisfaction level U_j for the product, the satisfaction level U of the group is great. Designing a product having a value closer to 1 means to design the product more successfully satisfying needs of the customers in the group.

The following optimization problem, having the satisfaction level U of the group expressed in eq.(4) as the objective function, is solved   so that the design satisfies the customers' needs as much as possible.

$$\text{Maximize } U = \frac{\sum_{j=1}^{m} N_j U_j}{\sum_{j=1}^{m} N_j} \tag{5}$$

subject to $\psi_c \leq 0$ $(c = 1,2,...,M)$
             $b_v^L \leq b_v \leq b_v^U$ $(v = 1,2,...,V)$

where

b_v	:	v-th decision variable
b_v^L	:	the lower bound of b_v
b_v^U	:	the upper bound of b_v
ψ_c	:	c-th constraint function including product attributes
M	:	the number of constraints including product attributes
V	:	the number of decision variables

It can be said that design alternatives having a greater value of U more successfully meet the customers' needs. Product designs which satisfy the market needs can be obtained by the procedures described here.

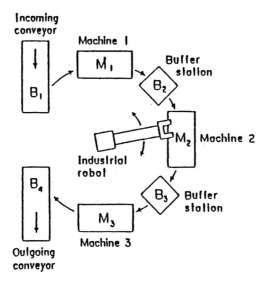

Figure 20.8 Construction of a machining cell with plural number of machine tools and an industrial robot.

20.4 DESIGN OPTIMIZATION FOR USAGE

Product designs in which the focus is placed on the use of products are discussed in this subsection. The purpose of product design of this stage is to obtain design solutions most suitable to the usage of the product (Yoshimura and Kanemaru, 1995).

Step 1 - Understanding Product Environment

Machine products are used in various environmental conditions. Product designs must consider the environmental conditions where the products are used. Therefore, a clear comprehension of the operational environment is necessary.

Industrial robots are used for many kinds of jobs. Examples of these jobs would be transporting, welding, spraying, assembling, inspecting, etc. In addition, the working environments of the robots are diversified in nature. Figure 20.8 shows an example of the construction of a machining cell with multiple machine tools and an industrial robot. The industrial robot is used for transporting workpieces. For the machine product design such as an industrial robot, integrated processing of product designs (structural design and control design) and operation planning is necessary as shown in Figure 20.9.

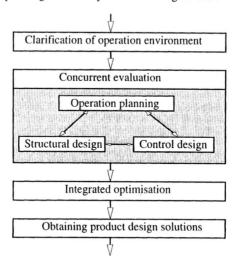

Figure 20.9 Integrated procedures of design and operation planning for industrial robots.

Step 2 - Clarifying Relationships Between Evaluative Characteristics

Relevant operational performances including cost must be evaluated in design optimization of industrial robots. They include:

(a) Evaluation of static accuracy: Static deflection y_s at the tip point of a robot arm where an end effector is installed .

(b) Evaluation of operational efficiency: The time interval t_c (that is, operational time) from the starting time of arm motion until the time when the tip point of a robot arm reaches a goal position while satisfying the dynamic accuracy constraint.

(c) Evaluation of dynamic accuracy: The maximum amplitude y_d of dynamic displacement at the tip point of a robot arm from the goal position after the planned operational time.

(d) Evaluation of operation cost: The integral of the square of feedback control input E_n (the operation torque τ of an actuator) over the operational time t_c, that is, the operational energy P (this value is here called operation cost) is used as a criterion for expressing difficulty of arm control.

Operational accuracy, both static and dynamic, operational efficiency and operation cost are related to both structural and control designs of industrial robots. To conduct the structural design and the control design simultaneously, the relationships between dynamic operational accuracy or operational efficiency and the operation cost for design variables of industrial robots must be clarified. Their relationships are important in robot design.

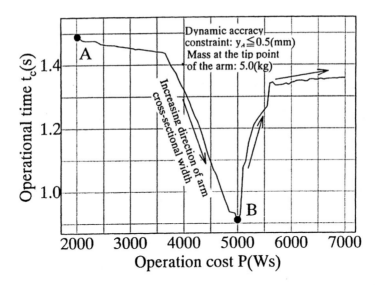

Figure 20.10 The relationship between the operational time and the operation cost under the dynamic accuracy constraint in the structural model of an articulated robot having two arms and two driving joints.

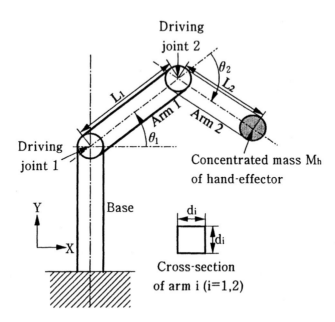

Figure 20.11 Structural model of an articulated robot having two arms and two driving joints.

Figure 20.10 shows an example of the relationship between the operational time and the operation cost under the dynamic accuracy constraint in the structural model of an articulated robot having two arms and two driving joints as shown in Figure 20.11. A smaller operational time and a smaller operation cost are preferable. Two objectives of minimization of the the operation cost and minimization of the operational time are used. The solutions on the curve from point A to point B correspond to the Pareto optimum solution set of a multiobjective optimization problem. On the Pareto optimum solution set, the operation cost and the operational time have a conflicting relationship. Solutions on the right region from point B do not include the optimum solution since the operational time also increases with the increase in the operation cost.

Step 3 - Determining Optimum Solution

Jobs in which industrial robots are used in manufacturing shops are roughly categorized into two types of jobs from the standpoints of required accuracy and operational efficiency as follows:

(1) Jobs having higher requirement levels for operational efficiency than for operational accuracy (for example, transporting jobs): Requirement levels for operational efficiency are strictly defined and higher operational accuracy brings about profits of simplification of control systems, etc.
(2) Jobs having higher requirement levels for operational accuracy than for operational efficiency (for example, assembling jobs): Requirement levels for operational accuracy are strictly defined and higher operational efficiency brings about profits by shortening of operational time, etc.

Definite evaluative requirements which must be satisfied without failure are set as the constraints in the formulation of design optimization. Evaluative requirements which bring about profits by their improvement are included in the formulation as the objective functions. Decrease of operation cost brings about profits of simplification of driving and control systems, decrease of operational power cost, etc. The operation cost is always included in the constraints.

For (1) of the foregoing categorization of jobs, the vector optimization problem of minimization of both dynamic accuracy y_d, and operation cost P under the constraint of the operational time t_c is formulated as follows:

$$\text{Minimize [dynamic accuracy } y_d \text{, operation cost P]} \qquad (6)$$

subject to $t_c \leq t_c^U$
$\quad\quad\quad\quad y_s \leq y_s^U$

For (2) of the foregoing categorization of jobs, the vector optimization problem of minimization of both operational time t_c and operation cost P under the constraint of the dynamic accuracy y_d is formulated as follows:

$$\text{Minimize [operational time } t_c, \text{, operation cost P]} \qquad (7)$$

subject to $y_d \leq y_d^U$
$\quad\quad\quad\quad y_s \leq y_s^U$

In each formulation, two objective functions exist and the constraint of the static displacement y_s is added for ensuring satisfaction of the static operational accuracy. Based on consideration for the practical working environment of industrial robots, the foregoing formulations of design decision making are practically divided into a more detailed categorization according to required levels of product perfomances. For example, the formulations of optimization problems of the articulated robot having two arms and two driving joints shown in Figure 20.11 for four assigned categorized jobs are given as follows:

Type 1. Jobs having high and definite requirement levels for operational efficiency:

Minimize [dynamic accuracy y_d, operation cost P]

subject to $t_c \leq 0.7(s)$
$y_s \leq 0.5(mm)$
$M_h = 5.0(kg)$

Type 2. Jobs having definite requirement levels for high speed transportation of heavy objects under the strict static accuracy constraint:

Minimize [dynamic accuracy y_d, operation cost P]

subject to $t_c \leq 0.8(s)$
$y_s \leq 0.5(mm)$
$M_h = 30.0(kg)$

Type 3. Jobs having definite requirement levels for operational accuracy:

Minimize   [operational time t_c,, operation cost P]

subject to $y_d \leq 0.5(mm)$
$y_s \leq 0.07(mm)$
$M_h = 5.0(kg)$

Type 4. Jobs having extremely high requirement levels for static and dynamic accuracies:

Minimize   [operational time t_c,, operation cost P]

subject to $y_d \leq 0.1(mm)$
$y_s \leq 0.03(mm)$
$M_h = 5.0(kg)$

Design variables for optimization are cross-sectional widths d_1 and d_2 of arms 1 and 2, respectively. The decision making problems having two objective functions are solved by the min-max multiobjective optimization strategy. Here, the weighting factors between the objective functions are determined interactively.

The optimized results of the design variables, the dynamic accuracy, the operational time and the operation cost for the foregoing four assigned jobs are shown in Table 20.4. It can be understood that optimum design solutions are different across the four types of assigned jobs.

Table 20.4 Optimized design solutions of an industrial robot for each of the categorized jobs

Type of job	d_1 (mm)	d_2 (mm)	y_d (mm)	t_c (s)	P (Ws)
1	84	54	4.1	0.7*	5400
2	150	63	14.9	0.8*	46600
3	90	45	0.5*	0.93	4950
4	85	44	0.1*	1.09	4800

* the upper bound value

20.5 SUMMARY

Fundamentally, the most necessary factor for designing a machine product is to grasp the environment of the product. In this chapter, practical procedures of product designs considering product environment were discussed using applied examples. Product life cycle issues are classified into three categories which focus on the three aspects of "manufacturing products", "selling products" and "using products". After comprehending the enviroments of the products, concurrent processing of information and knowledge related to the product decision making is conducted so that the product design satisfies the requirements of the product environments as much as possible.

REFERENCES

Cohon, L.L. (1978) *Multiobjective Programming and Planning*, Academic Press.

Eschenauer, H., Koski, J., Osyczka, A. (1990) *Multicriteria Design Optimization*, Springer-Verlag.

Stadler, W. (1988) *Multicriteria optimization in Engineering Sciences*, Plenum Press.

Yoshimura, M., Itani, K., Hitomi, K. (1989) Integrated Optimization of Machine Product Design and Process Design, *International Journal of Production Reseach*, **27** (8), 1241-1256.

Yoshimura, M., Takeuchi, A. (1991) Multiphase Decision-Making Method of Integrated Computer-Aided Design and Manufacturing for Machine Products, *International Journal of Production Reseach*, **31** (11), 2603-2621.

Yoshimura, M. (1993) Concurrent Optimization of Product Design and Manufacture, *Concurrent Engineering* (Editted by H.R. Parsaei, W.G. Sullivan), Chapman & Hall, 159-183.

Yoshimura, M., Takeuchi, A. (1994) Concurrent Optimization of Product Design and Manufacturing Based on Information of Users' Needs, *Concurrent Engineering: Research and Applications* **2**, 33-44.

Yoshimura, M., Kanemaru, T. (1995) Multiobjective Optimization for Integrated Design of Machine Products Based on Working Environment Information, *Proceedings of the IUTAM Symposium on Optimization of Mechanical Systems*.

A META-METHODOLOGY FOR THE APPLICATION OF DFX DESIGN GUIDELINES

Brett Watson; David Radcliffe; Paul Dale

This chapter presents a meta-methodology for addressing the issue of how the information contained within Design for X (DFX) techniques can be organised such that the implications of decisions are proactively evaluated. By applying only those guidelines that are of value to the emerging design at a particular stage during the product development process a more efficient methodology for guideline utilisation is synthesised. This meta-methodology also determines the relative importance of each guideline and the nature of the interactions that occur between competing DFX techniques. To place the methodology into context an illustrative example is presented. The example is used to provide a more detailed understanding of how to apply the methodology and to draw out some of the issues that arise through its application.

Many DFX techniques are based on design guidelines that when followed improve a products performance within the life-cycle phase under examination. When a single DFX technique is employed only local life-cycle cost minimisation is achieved. If the global life-cycle cost is to be minimised a number of DFX techniques need to be applied. Meerkamm (1994) suggests that the application of DFX techniques in fact constrains the design solution and that if numerous DFX tools are utilised that the constraints are often contradictory in nature. Consequently it becomes increasingly difficult to find an optimal solution. As the guidelines tend to be the most flexible aspect of the DFX toolbox they present the best opportunity for determining the nature of interactions between DFX techniques and in ultimately developing an optimal solution.

21.1 BACKGROUND

With anecdotal evidence supporting the claim that the concurrent use of more than one DFX technique results in conflicting and contradictory recommendations a number of independent studies have begun to examine these interactions utilising various frameworks for the analysis. A number of major areas of study can be identified and are briefly outlined below.

Watson and Radcliffe (1995) have developed a theoretical model that evaluates the usefulness of a DFX tool by design phase. We found that DFX tools do have a varying degree of impact depending upon when during the product development process they are applied. Where Norell (1993) and Willcox and Sheldon (1993) found that the application of a Design for Assembly (DFA) methodology was in general most valuable at the conceptual stage, we found that the analysis tool component, which is often marketed as the major part of the package, was more valuable during the embodiment and detailed phases. The analysis tool was in fact a very unreliable tool to use at the conceptual stage as the level of design information required to perform the analysis was as yet unavailable. If the analysis tool is incapable of acting effectively at the conceptual stage the only other part of the DFX techniques that could provide these benefits are the design guidelines. By maximising the potential for improved design through the application of DFX guidelines it follows that the relative improvement achieved through the use of the analysis tool will be reduced, hence the amount of redesign will be minimised.

At present little research has been invested in determining how to deal with the conflicting recommendations that occur when multiple DFX use occurs. Thurston (1991) provides a methodology for modelling the result of design decisions on the overall product worth. The technique was developed to rank order design alternatives and to quantify the trade-offs so that informed decisions can occur. The method is a powerful tool for decision making in engineering design where multiple objectives exist. Unfortunately the method is overly complex and time consuming for most applications, particularly those within small to medium sized organisations which represent the majority of engineering firms. To adapt this method to rank the most valuable guidelines would be unnecessarily tedious as the model used by Thurston is significantly more complex than required in this application.

A simpler method of conducting a trade-off analysis between multiple DFXs is to use a matrix approach. Quality Function Deployment (QFD) uses such an approach to determine relationships between customer requirements and technical requirements. In the case of QFD the matrix is referred to as the "House of Quality". The guidelines being compared in this chapter are somewhat similar in nature to those used in QFD. This is with regard to the information content of the guidelines and the process of doing a qualitative comparison. QFD uses a fixed comparison index and a weighted sum to determine the most promising areas for design improvement. A weighted matrix approach is consistent with the time and simplicity requirements of much of industry and therefore presents a useful starting point for the comparison of DFX guidelines.

21.2 META-METHODOLOGY

21.2.1 Overview of Meta-Methodology

The meta-methodology presented in this section uses a weighted matrix approach to determine interactions between competing DFX techniques. This is done using the design guidelines associated with DFX techniques. The matrix method provides two useful indices as an output. The first of these indices determines if there are any major areas of conflict between the DFX techniques and the second determines how the value of a particular guideline is modified by interactions with the guidelines of the competing DFX techniques. Figure 21.1 describes the format of the matrix.

21.2.2 Procedure of Meta-Methodology

The methodology for evaluating and ranking the competing guidelines requires a number of distinct steps to be taken. These are described in the flowchart presented in figure 21.2. Sections 21.2.3 through to 21.2.6 describe these steps in greater detail.

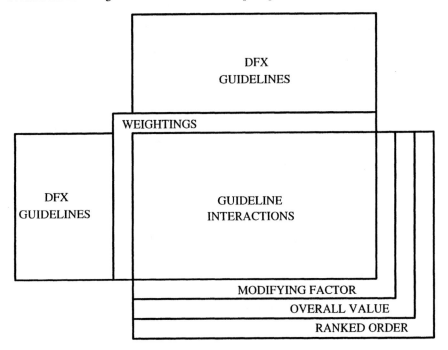

Figure 21.1 Layout of Comparison Matrix.

STEP 1

Select the DFX techniques that will be most valuable for the given circumstance from those available to the designer. Those selected must then be weighted in terms of how effective they are in reducing the overall life-cycle cost. The selection and weighting of the DFX techniques is done using cost estimates for each life-cycle area.

⇩

STEP 2

For each DFX technique the guidelines are categorised as either general design rules or specific design strategies. A further categorisation is made as to what product development phase the rules and strategies can best be utilised in. This results in a tree diagram for each product development phase.

⇩

STEP 3

The rules and strategies are then weighted according to their relevance to the given application. The unmodified strategy weight is calculated as a function of the DFX weight, the rule weight and the strategy proportion.

⇩

STEP 4

Using a matrix approach the design strategies of competing DFX techniques are evaluated for whether they are supporting or conflicting. This occurs for each product development phase. From this comparison it is possible to determine if any areas of severe conflict exist. If so these conflicts must be resolved prior to any further evaluation.

⇩

STEP 5

Using the algorithm presented a modifying factor for each strategy is calculated based upon the nature and number of interactions between strategies. This factor is used to determine the overall strategy value. From these values a ranked list can be generated that identifies those strategies of most importance to the designer.

⇩

STEP 6

While designing constantly refer to the ranked list that contains the strategies for the particular product development phase the design is currently at.

Figure 21.2 Flowchart of Meta-Methodology.

21.2.3 Selecting and Weighting of DFX Techniques

The first step involves selecting and determining the relative importance of the DFX techniques to be utilised. It is readily apparent that the relative worth of a particular DFX technique will vary depending upon the context of its use. The context of use can be defined by specific company requirements, manufacturing capabilities, industry expectations, customer requirements, the nature of the market and numerous other considerations. These contextual factors are used to roughly estimate the life-cycle cost of the product under development. The total life-cycle cost is made up of a number of life-cycle cost areas including design, manufacture, assembly, serviceability, disassembly, recycling and disposal. By examining these areas the proportion of the total life-cycle cost that a particular DFX technique affects can be estimated. These values are an indicator to the level of impact that a particular DFX tool may have in reducing the overall cost. If for example the major cost areas were assembly and manufacture it would be reasonable to assume that tools which minimise these costs will be of most benefit in reducing the total product cost. Therefore these techniques should be incorporated into the product development process.

Other issues that do not always add directly to cost can also be considered, for example environmental impacts and safety. The incorporation of these other DFX techniques should be considered in terms of the qualitative benefits that they provide. The implications of designing for safety or the environment may have very different requirements depending upon the nature of the product and industry. These requirements can only be addressed in terms of risk reduction to the companies overall profitability. That is, what are the possible implications of not explicitly designing to minimise safety hazards or environmental risks.

Weighting of the DFX techniques follows on directly from the method used to select the most promising tools to apply. This weighting factor must reflect the designers evaluation of the potential of each technique to minimise the total life-cycle cost. The weighted value of a DFX technique W_{DFX} has a normalised value in the range from 0 to 1, with 1 indicating the most valuable. W_{DFX} is calculated as follows;

$$W_{DFX} = \frac{Cost\ x}{Cost\ max} \tag{1}$$

where

$Cost\ x$ = The overall cost of life-cycle area x.
$Cost\ max$ = The cost of the life-cycle area with the greatest cost.

The weighting of the qualitative techniques must rely upon the experience and intuition of the designer to select a suitable weight. This should be done after the quantitative weighting has been performed as a point of comparison now exists on which the evaluation can be based. In the case of the qualitative criteria it is possible to select a weight greater than the previously stated maximum of 1. However, if this is the case the weights should be re-scaled to reduce the maximum back to 1. This is necessary in order to bound the final strategy values within suitable limits.

21.2.4 Using a Tree Diagram to Categorise Design Guidelines

The second step in applying the meta-methodology involves two tasks. These are the categorisation of the product development process and establishing the hierarchical level of the design guidelines. Watson and Radcliffe (1995) showed that the DFX analysis tools could provide significantly improved performance if used during the product development phase that contained the type of information that the technique could most effectively manipulate. As in most cases the DFX guidelines evolved out of repeated use of the analysis tool it would be expected that if a particular guideline was applied during all product development phases that it will have a varying degree of impact. Most DFX packages fail to make this distinction and merely provide a list of recommended design rules with little direction on when and how they should be used.

The Pahl and Beitz (1988) model is used as a descriptor of the product development process. Only the broad phases of task clarification, conceptual design, embodiment design and detailed design are required for this method. Though design might not always occur in the sequence outlined by Pahl and Beitz their model provides a detailed description of all the stages that a product must progress through during its development. Table 21.1 contains some examples of guidelines from the Design for Assembly methodology (Boothroyd and Dewhurst (1989)) and where there use is most appropriate within the product development process. In the case of some guidelines it is apparent that they are equally valid over a number of product development phases, e.g. minimise the number of parts. These particular guidelines tend to be more general in nature than those that are only applicable during one phase.

Table 21.1 Sample DFA guidelines by design phase

PHASE	GUIDELINES
Task Clarification	1. Standardise a products style.
Conceptual	1. Minimise the number of components. 2. Reduce the number of components between input and output function. 3. Eliminate features that do not add value to the customer. 4. Standardise a products style.
Embodiment	1. Minimise the number of components. 2. Eliminate the need for conduits and connectors. 3. Design multi-functional parts. 4. Standardise a products style.
Detailed	1. Avoid slight asymmetry of components. 2. Non self securing parts should locate immediately. 3. Use pilot point screws to avoid cross threading. 4. Use standard components where possible.

The second task in the categorisation is to arrange the design guidelines into a hierarchical tree. The DFX techniques consist of a number of high level design guidelines, called design rules. Each rule contains a set of low level design guidelines, called design strategies. As with QFD, for a meaningful comparison to be performed all parameters should contain roughly the same information content. To achieve this a tree diagram can be employed. The guidelines can be placed on three levels of detail. The primary level incorporates the whole DFX tool as weighted by the technique presented in section 21.2.3. The second level represents the broad design rules as outlined above, and the third level contains the specific design strategies for implementing these rules. A tree for each product development phase should be created. This involves examining the guidelines to determine when during the product development process that the design will have progressed to a state of development such that the guideline will have a valuable impact on the upcoming design decisions. This becomes a relatively easy task with a little experience and an understanding of the product development process. Table 21.2 contains an example using the Boothroyd and Dewhurst (1989) DFA guidelines employed during the embodiment phase.

Table 21.2 Tree structure for DFA guidelines at embodiment phase.

DFX Tool	Design Rules	Design Strategies
DFA	Reduce part count and part types.	Test each parts need for existence as a separate component
		Eliminate separate fasteners
		Eliminate parts that act as conduits and connectors
		Design multi-functional parts
		Do not follow piece-part producibility guidelines
	Strive to eliminate adjustments	Reduce the number of parts between input and output function
		Move critically related surfaces close together
		Follow kinematic design principles
	Ensure adequate clearance and unrestricted vision	Ensure adequate clearance for hands, tools and subsequent processes
		Ensure that vision of the process is not restricted
	Minimise reorientations during assembly	Minimise the need for reoientations during assembly

21.2.5 Determining Guideline Weighting Levels

The third step in applying the meta-methodology requires that both the design rules and strategies be weighted. The weighting levels must be determined separately for each phase as the importance of the guideline may change depending on when during the product development process it is applied. The approach for doing this also varies depending upon whether it is a design rule or strategy.

The second level design rules provide general direction for the designer when designing for X. The designer is afforded greater flexibility in avenues for design improvement if there is a large number of design rules, hence the overall design will generally perform better with regard to the life-cycle area under consideration. It is for this reason that the design rules are weighted independently of their number. The weighting of the design rules is done using a scale from 1 to 10. The weighting values must be determined by the design team that will use them and be based upon the particular circumstances of the project. This weight will be influenced by the number and quality of the design strategies it contains. However, this should not be used as the sole indicator as to the value of the design rule.

The design strategies for achieving the objectives outlined within the design rules must be approached slightly differently. As the benefit comes from the successful implementation of the design rules the actual strategies for achieving this add no extra value to the DFX method. It is for this reason that the design strategies must be weighted as a proportion of the design rule it falls under. Therefore each strategy must be rated between 0 and 1 such that the sum of the weightings under each design rule sums to 1. The overall unmodified weight of each strategy is determined by multiplying the overall DFX weight, the design rule weight and the strategy proportion. Hence the initial overall weight of each strategy can range between 0 and 10. However, values in the range from zero to six are most likely. The illustrative example in section 21.3 will demonstrate the application of this approach.

$$W_{strategy} = W_{DFX} \times W_{rule} \times P_{strategy} \qquad (2)$$

where

$W_{strategy}$ = Unmodified weight of the strategy,
W_{rule} = Weight of the design rule,
$P_{strategy}$ = Proportional value of the strategy.

21.2.6 Determining Guideline Interactions

The fourth step in the meta-methodology involves determining the interactions between the strategies and displaying them in the matrix. From these interactions the severity of any conflicts can be gauged. The matrix method presented in this section can be used to compare any number of strategies from any number of DFX techniques. However, as with most matrix methods the process of determining each relationship can become tedious for a large number of guidelines. It would be expected that generally not more than three DFX tools would be used at any one time and that perhaps each tool would provide up to a maximum of twenty strategies per phase.

The process of filling in the matrix is done using a fixed scale. Table 21.3 describes the comparison values and how these values are to be interpreted. By examining the information content of the strategies it is a relatively easy task to determine the nature of the interactions. The example in section 21.3 will explore this issue further.

Table 21.3 Guideline interaction types and values

C Values	Interpretation
+10	Strategies interact very positively. i.e. the strategies make almost identical recommendations as to how the design can be improved.
+5	One strategy tends to support the other in a broad sense, or some positive overlap in the information content of the strategy occurs.
0	No form of interaction exists between the strategies.
-5	Some conflict exists as to the direction the design should take. It is likely that if one strategy is followed that some loss in performance will occur in the other life-cycle area.
-10	The strategies are almost completely contradictory in nature. If one strategy is followed then the designs performance in terms of the other life-cycle criteria will be significantly reduced.

From the matrix it is possible to determine whether any two strategies have conflicted so severely that special consideration must be applied when dealing with them. This is done using the specific conflicts index. The equation used to calculate this index is only applied when strategy sets have a negative interaction. Those strategy sets identified as having a specific conflict need to be examined separately before any further evaluation can proceed. The specific conflict index is calculated as follows;

$$SC = W_{strategy} \times W_{comp} \times C \qquad (3)$$

$$\text{if} \quad SC < -90 \quad \text{then examine conflict}$$

where

SC = Specific Conflicts,
W_{comp} = Weight of other strategy being compared,
C = The interaction index for the two strategies being compared.

If the value of the specific conflicts index exceeds a value of negative ninety it can be assumed that a conflict of considerable consequence has occurred. The critical value of ninety corresponds to two strategies of value three completely contradicting one another. The likelihood of such interactions occurring is rare. The subsequent resolution of this conflict must be guided by the particular circumstances of the products use. Some tactics for the resolution or avoidance of this conflict include;

1. If the specific conflict index is close to ninety the conflict could be ignored on the basis that the conflict will down weight the strategies in the ranked list. This will effectively reduce the relative importance of these guidelines.
2. Examine the specific details of the conflict and develop a design methodology that will minimise the areas for negative interaction. This approach is particularly useful for cases where only partial conflict has occurred.
3. If a significant difference in the weight of the strategy being compared exists the one with the lower value can be eliminated from the matrix. If the difference is large it is likely that the ranking equation will eliminate it anyway.
4. Eliminate both strategies and design according to function and the remaining guidelines. This approach should only be used as a last resort as it results in a significant loss of information.

21.2.7 Determine Ranked List

The fifth step of the meta-methodology involves determining the overall value (V_{TOT}) of a particular design strategy based upon the strategies weight and its interactions with other strategies. Positive interactions increase the strategies overall value and negative ones decrease it. The process is based upon the fact that each strategy has a self worth equal to its total weighted value ($W_{strategy}$) and this is then modified by interactions with other strategies. This modifying factor is a function of the interaction index and the weight of the guideline it is being compared with. By summing the modifying value over all of the interactions a global modifier is determined. The maximum amount any one strategy can modify another by has been set as one third of the guidelines self worth. This value has been determined empirically and reflects the fact that the dominating effect should be the information content of the strategy itself. The interactions should have the capability to change the ranked position of any one strategy but not in such a manner that extreme changes are possible. However, this value can be easily modified for specific situations by adjusting the scaling factor. Equations (4) & (5) describe the complete method.

$$V_{TOT} = W_{strategy}\left(1 + \delta V\right) \tag{4}$$

$$V_{TOT} = W_{strategy}\left(1 + \sum \frac{W_{comp}.C.f}{300}\right) \tag{5}$$

where;

$$f = \frac{15}{Wstrategy} \qquad \text{if } W_c > W_{gd} \text{ and } C < 0$$

$$f = 1 \qquad \text{else.}$$

where;

V_{TOT} = total modified value of strategy,
δV = the total modifying factor over all strategies and DFX techniques,
f = the special case modifier,
300 = scaling factor.

In equation (5) the special case modifier takes account of instances when a design strategy of low weight conflicts with a strategy of high weight. In this case the amount the self worth of the strategy is modified by will be significantly larger in magnitude than due to any other interaction. This requirement is fairly obvious as it would be pointless to follow an unimportant strategy at the extreme detriment to more highly weighted strategies. Hence the self worth can be modified by over an order of magnitude as opposed to the previously mentioned maximum of one third. This situation will be examined further within the illustrated example presented in section 21.3.

If more than two DFX techniques are being applied to the design a matrix must be set up for each possible set of interactions. The total modifying factor must be summed across all matrices and interactions for each strategy. In some cases this may involve summing across a row and then down a column in the next matrix.

Having determined the overall modified weight for each strategy a ranked list can be created from these values. Any strategies that have a negative overall value should be deleted from the list and not utilised at all. If these negative strategies are followed it would be expected that a reduction in life-cycle performance would result due to its conflicting interactions with other strategies. This ranked list is then utilised during the design process.

21.3 AN ILLUSTRATIVE EXAMPLE

This section describes how the meta-methodology has been utilised within a real design project. The example illustrates how meta-methodology works and highlights some of the related issues. The meta-methodology is tailored to the particular context of the case study.

A local ophthalmic surgeon commissioned one of the authors to develop a prototype device to explore the parameters for a new process to increase the success of Corneal transplants. As the device at this stage is primarily a research tool it was intended that there would be a maximum amount of flexibility within the variability of the experimental parameters. It also meant that a complicated and unique product would be designed. As the device was to be operating on a small object with extreme precision at a controlled rate it was apparent that geometry would be a major design factor. It was also apparent that the cutting loads were so small that any component stresses would be negligible.

Within the medical industry cleanliness and sterilisation are major issues. In the case of this device it was required that every component be autoclaved (placed in a steam oven at high temperature and pressure) after every use. This operational requirement results in the device being completely disassembled and re-assembled after each use. In terms of service life it was expected that no major parts should fail throughout the devices life. The addition of certain disposable items would be acceptable as when the product is developed for the general market a substantial amount of profit can be made through the supply of disposable items. The final requirement was that the device was to be designed and manufactured on a very limited budget.

The current state of development and the associated intellectual property and commercial potential of the device prevent the inclusion of detailed technical information. Instead, some of the issues involved are discussed in broad terms.

Table 21.4 DFA Design rules and strategy weights by product development phase

Phase	Design Rules	Wt	Design Strategies	Wt	Tot Wt
Conceptual Design	1. Reduce part count and part types.	10	1. Eliminate parts that act as conduits or connectors.	0.4	4
			2. Eliminate any product features that do not add value to the customer.	0.6	6
Embodiment Design	1. Reduce part count and part types.	10	1. Test each part's need for existence as a separate component.	0.4	4
			2. Eliminate separate fasteners.	0.05	0.5
			3. Eliminate parts that act as conduits and connectors.	0.25	2.5
			4. Design multi-functional parts.	0.25	2.5
			5. Do not follow piece-part producibility guidelines at this stage of the design.	0.05	0.5
	2. Strive to eliminate adjustments.	2	1. Reduce the number of parts between the input and output function.	0.35	0.7
			2. Move critically related surfaces close together to facilitate tolerance control.	0.3	0.6
			3. Follow kinematic design principles.	0.3	0.6
	3. Ensure adequate access and unrestricted vision.	8	1. Ensure adequate clearance for hands, tools and subsequent processes.	0.4	3.2
			2. Ensure that vision of the process is not restricted.	0.6	4.8
	4. Minimise the need for reorientation's during assembly.	1	1. Minimise the need for reorientation's during assembly.	1	1
Detailed Design	1. Design parts to be self-aligning and self-locating.	7	1. Non self securing parts should locate immediately upon assembly.	0.2	1.4
			2. Provide parts with built in alignment.	0.25	1.8
			3. Allow generous clearances but avoid parts jamming up during insertion.	0.25	1.8
			4. Ensure parts locate before release.	0.2	1.4
			5. Use pilot point screws to avoid cross-threading.	0.1	0.6
	2. Ensure the ease of handling of parts from bulk.	3	1. Avoid the use of sharp or fragile parts.	0.1	0.3
			2. Avoid parts that require special grasping tools.	0.9	2.7
	3. Design parts that cannot be installed incorrectly.	9	1. Provide obstructions that will not permit incorrect assembly.	0.35	3.1
			2. Make mating features asymmetrical.	0.15	1.8
			3. Make parts symmetrical so that orientation is unimportant.	0.3	3.7
			4. If incorrect assembly can occur ensure no further assembly is possible.	0.1	0.9
			5. Mark parts with the correct orientation	0.05	0.5
			6. Eliminate flexible parts that can always be assembled incorrectly.	0.05	0.5
	4. Maximise part symmetry or highlight asymmetry.	2	1. Maximise part symmetry to ease handling.	0.5	1.5
			2. Avoid slight asymmetry.	0.5	1.5

Table 21.5 Design rules and strategy weights by PDP for DFD

Phase	Design Rules	Wt	Design Strategies	Wt	Tot Wt
Conceptual Design	1. Improve the products structure for disassembly.	10	1. Subdivide the product into manageable subassemblies.	0.5	5
			2. Minimise the number of components.	0.5	5
			3. Standardise the products style.	0	0
	2. Improve the disassembly planning.	2	1. Avoid long disassembly paths.	1	2
Embodiment Design	1. Improving the product structure for disassembly.	10	1. Subdivide the whole assembly into manageable subassemblies.	0.25	2.5
			2. Minimise the number of connections between subassemblies.	0.35	3.5
			3. Minimise the number of components.	0.4	4
			4. Standardise the products style.	0	0
	2. Improve access and vision for disassembly.	2	1. Make sure that components are accessible.	1	2
	3. Improve disassembly planning.	3	1. Reduce the number of changes in direction required in a removal operation	0.3	0.9
			2. Avoid long disassembly paths.	0.7	2.1
	4. Material compatibility.	0	1. Subassemblies that are difficult to disassemble should be made of the same or compatible material.	1	0
Detailed Design	1. Component design rules.	1	1. Integrate components with the same material and avoid the combination of different materials.	0.3	0.3
			2. Mark materials permanently to assist sorting.	0.3	0.3
			3. Design in predetermined fracture points that allow rapid removal of components.	0.3	0.3
	2. Design and selection of connectors.	10	1. Make connectors of a compatible material to avoid the need for disassembly	0.0	0
			2. Minimise the type and number of connection forms.	0.35	3.5
			3. Select easy to disassemble connectors.	0.25	2.5
			4. Use connectors with fracture points for difficult situations.	0.0	0
			5. Ensure connectors can be removed with standard tools.	0.4	4
	3. Maximise end of life value of the product.	1	1. Standardise components.	0.4	0.4
			2. Design for long life and reuse.	0.6	0.6

21.3.1 The Selection of DFX Techniques (step 1)

In this case design for analysis, assembly, disassembly and serviceability techniques were available for use. The process for selecting which techniques to use was straight forward. As no significant loads existed it was not necessary to develop any detailed models for stress analysis, hence design for analysis was not required. As the product was to be assembled and disassembled after every use, designing to facilitate serviceability would be pointless. If a component failed the whole device must be disassembled, not just those components surrounding the failure. Therefore design for serviceability would not be required. This left only the assembly and disassembly techniques which would account for a considerable proportion of the product's life-cycle cost. Hence both these techniques were selected.

As both the assembly and disassembly process occurred equally often and no real information existed on the likely times for complete assembly and disassembly it was decided that the techniques would be rated evenly. Hence both techniques received a rating of one.

The DFA technique selected for this project was that developed by Boothroyd and Dewhurst (1989). The technique has been refined over the past ten years to provide a reliable analysis tool with sensible guidelines presented in a structured format. The analysis tool uses the same basic structure to analyse for manual, robotic and automatic assembly with different data tables for the various processes. For this project the manual assembly method was adequate. The major difference between this product and the usual type of product that a DFA technique would be applied to was that assembly and regularly re-assembly would occur. This difference has a substantial affect on how the guidelines would normally be interpreted and rated.

The Design for Disassembly technique used for this project is that being developed by a team of researchers at the Manchester Metropolitan University. (See Simon, Fogg and Chambellant (1992), Zhang, Simon and Dowie (1993) or Simon and Dowie (1992)). The technique developed is primarily aimed at disassembling to facilitate recycling and is still in its embryonic stage. Little consideration is given to the disassembly aspects of serviceability or general disassembly. This restriction tends to limit the usefulness of the tool in this particular application. However a number of general DFD guidelines that form a part of this tool were readily applicable to the corneal project. The actual DFD design analysis techniques are also under-developed hence no tool existed that could provide reliable feedback. The tools for evaluating the actual disassembly sequence are quite powerful but are of little value in this application as complete disassembly is required. It is for this reason that no DFD analysis was performed during the product development process.

21.3.2 Guideline Weights by Design Phase (steps 2-3)

Tables 21.3 and 21.4 list the design rules and strategies by product development phase for the design for assembly and disassembly techniques.. They also provide a list of the rule weights, strategy proportions and unmodified strategy weights. The following four points illustrate some of the considerations that influenced the weighting process.

1. *Ensure that vision of the process is not restricted.* Due to the small size of the device and the difficult geometry it was felt that as a general design rule ensuring adequate access and vision would be a serious issue to address, thus its weighting of eight. As only two guidelines were presented for this rule their relative proportion of the total must be large, hence the overall strategy weighting of 4.8.

2. *Eliminate separate fasteners.* Normally the elimination of separate fasteners would be a high priority when designing for assembly. However, in this case, due to the requirement of high accuracy and easy disassemblability deliberate attempts to minimise these fasteners may increase costs and create problems. It is for this reason that the low weighting of 0.5 was calculated for this strategy.

3. *Minimise the number of components.* For DFD the minimisation of components is just a design strategy while for DFA it is a design rule. In the case of DFD the minimisation of components is still a significant component in rationalising the products structure hence the high rating. It can be expected that many positive interactions will exist between this and a number of DFA strategies.

4. *Make connectors of a compatible material to avoid the need for disassembly.* This strategy results from the recycling focus of the DFD methodology. Obviously disassembly in this application must be complete and recycling requirements are inconsequential. Hence this strategy has no value resulting in the weight of zero.

21.3.3 The Guideline Interaction Matrix (steps 4 & 5)

As only two DFX techniques were being used only one matrix for each product development phase was required for the comparison and ranking process. As this is the case the total modified strategy weights and rankings have been included as extra rows and columns in the matrix. Each product development phase will be examined separately in the following sections.

CONCEPTUAL PHASE	DFD STRATEGIES	Subdivide product into sub-assemblies	Minimise no. of parts	Standardise a products style	Avoid long disassembly paths			
DFA STRATEGIES	Wstrat	5	5	0	2			
Eliminate conduits & connectors	4	5	5		5	-0.2	3.2	4
Eliminate unnecessary features	6		5			0.08	6.5	1
		-0.07	0.167	0	0.067	del V		
		4.67	5.83	0	2.13		V tot	
		3	2	6	5			Ranking

Figure 21.3. Comparison Matrix for Conceptual Phase.

EMBODIMENT PHASE

DFA STRATEGIES	Wstrat	Subdivide product into sub-assemblies	Minimise connections between subassemblies	Minimise the number of components	Standardise the products style	Make sure that components are accessible	Reduce direction changes in removal	Avoid long disassembly paths	Make problematic subassemblies of same materia	del V	V tot	
		2.5	3.5	4	0	2	0.9	2.1	0			
Test each parts need for existance	4	-5	5	10				10		0.15	4.60	3
Eliminate separate fasteners	0.5		5	10			-5	10		-0.19	0.41	15
Eliminate conduits & connectors	2.5	-5	10	10			-5	5		0.24	3.10	6
Design multi-functional parts	2.5			10				5		0.15	2.88	7
Ignore piece-part prod. gdlines	0.5									0.00	0.50	14
Reduce no. of parts between input & output	0.7	-5	5	10				5		-0.67	0.23	17
Move critically related surfaces close together	0.6						5			0.00	0.60	13
Follow kinematic design principles	0.6									0.02	0.61	12
Ensure adequate clearance	3.2					10				0.07	3.42	5
Ensure unrestricted vision	4.8					10				0.07	5.14	2
Minimise reorientations	1									0.00	1.00	11
del V		-0.45	0.17	0.34	0.00	0.27	-0.69	0.11	0.00	0.00	V tot	
V tot		1.38	4.10	5.36	0.00	2.54	0.28	2.33	0.00	0.00		
		10	4	1	-	8	16	9	-			

Figure 21.4 Comparison matrix for Embodiment Phase.

DETAILED PHASE

DFA STRATEGIES	Wstrat	Avoid the combination of different materials	Mark materials permanently to assist sorting	Design in predetermined fracture points	Make connectors of compatible materials	Minimise the type and number of connection forms	Select easy to disassemble connection forms	Use connectors with fracture points for difficult situatio	Ensure connectors can removed with standard tools	Use standard components	Design for long life and reuse	del V	V tot	Ranking
		0	0	0.3	0	3.5	2.5	0	4	0.4	0.6			
Non-self securing parts locate immediately	1.4											0.00	1.40	10
Provide parts with built in alignment	1.8											0.00	1.80	8
Allow generous clearances, but avoid jamming	1.8											0.07	1.93	7
Ensure parts locate before release	1.4											0.00	1.40	10
Use pilot point screws to avoid cross-threading	0.8									5		0.01	0.61	12
Avoid sharp or fragile components	0.3			-5								-0.01	0.30	16
Avoid parts that require grasping tools	2.7					5	5		10	5		0.24	3.35	4
Provide obstructions that stop incorrect assembly	3.1								5			0.00	3.10	5
Make mating features asymmetrical	3.7											0.00	3.70	2
Make parts symmetrical	1.8											0.00	1.80	8
If incorrect assembly - no further is possible	0.9											0.00	0.90	11
Mark parts with the correct orientation	0.5											0.00	0.50	14
Eliminate flexible parts	0.5											0.00	0.50	14
Maximise part symmetry to ease handling	1.5											0.00	1.50	9
Avoid slight asymmetry	1.5											0.00	1.50	9
	del V	0.00	0.00	-0.01	0.00	0.05	0.05	0.00	0.12	0.06	0.00			
	V tot	0.00	0.00	0.30	0.00	3.68	2.63	0.00	4.48	0.42	0.60			
	Ranking	-	-	16	-	3	6	-	1	15	13			

Figure 21.5 Comparison matrix for detailed phase.

The Conceptual Phase Matrix

As there are few strategies that have any impact during the conceptual phase the comparison matrix is relatively small. Due to the small number of strategies per rule the overall strategy weights tend to be large. Figure 21.3 contains the matrix.

As highlighted in the matrix two strategies have conflicted to such a degree that special consideration is required before they can be ranked. In this case the conflict occurs when the designer simultaneously attempts to minimise the number of conduits and connectors while attempting to sub-divide the product into sub-assemblies. As the specific conflicts index only slightly exceeds the critical value of ninety it was decided that the conflict could be ignored until specific cases became apparent. However, due to the complexity of the design it was expected that some form of additional undesirable coupling between sub-assemblies would be inevitable. Hence the assembly and disassembly of the product would be adversely affected.

The Embodiment Phase Matrix

Figure 21.4 contains the matrix for the strategies that are useful during the embodiment phase of the product development process. In this case there are a number of conflicting strategies but none are so severe that special consideration is required.

The Detailed Phase Matrix

As seen in figure 21.5 the comparison of design strategies that act during the detailed product development phase resulted in very few interactions and only one negative interaction. The specific conflicts index for this negative interaction was well below requiring further examination. Consequently the majority of strategies ended up with an overall value equal to their initial weight.

21.3.4 Results of Guideline Use (step 6)

The Role of Guidelines During the Conceptual Phase

During the conceptual phase the implications of the top three ranking strategies were considered regularly. However these strategies must only be applied after technically feasible solutions have been developed. In this case the practicality of a number of solution principles reduced the available options drastically. Consequently the value of the strategies in this particular situation was somewhat limited.

The Role of Guidelines During the Embodiment Phase

Having determined the basic concept of the overall solution the initial stage of the embodiment phase involves addressing how the function of the device will be achieved. Though the guidelines generally should not be considered when doing this the designer subconsciously applies them while making initial sketches and models. This application of the guidelines tends to quickly tie down the solution to a few acceptable paths. While the consideration of other preliminary layouts would have been desirable the constraints of time and a limited budget restricted the exploration of other solutions.

In this case the designer possessed a greater understanding of the DFA guidelines, hence these tended to take priority over the application of the DFD guidelines. However this preference merely forced the more detailed consideration of the DFD requirements. This was highlighted when attempting to minimise the number of components within the assembly. As regular assembly and disassembly was required the process of part minimisation was significantly altered. In some cases if only DFA was being applied a number of components could be removed, however constant assembly and disassembly would have a damaging effect on the product if designed this way. Consequently the process of part minimisation was restricted to within certain bounds.

The Role of Guidelines During the Detailed Phase

As time constraints became increasingly tight towards the end of the project the process of considering the application of the DFX strategies became less consistent. This resulted in only the highest ranked strategies being applied rigorously and the others more haphazardly. It was also apparent that the manufacturing cost of the device was becoming excessive. For this type of product and manufacturing method the cost is drastically affected by the details of component features. This implication forced the omission of a number of strategies that tend to increase part complexity. Some examples include providing obstructions to ensure correct assembly, designing components that are symmetrical and making mating features asymmetrical.

The Assembly Efficiency

At three points during the product development process the Boothroyd and Dewhurst Design for Assembly evaluation technique was utilised. These were at the mid point of the embodiment phase, the end of the embodiment and the end of the detailed phases. This was done by hand using the manual assembly charts. Figure 21.6 contains a plot of the efficiency by product development phase utilising the results as calculated at the time they were applied to the product. As more detail becomes available the understanding of how components are manipulated and inserted will change, hence the overall time and efficiency will vary. The interpretation of what the minimum number of components is also varies during the product development process, for example, a design with an identical layout could be evaluated for assembly efficiency by the same designer at different points during the product development process and end up with different results. However the differences are indicative of the level of design refinement that guides the mental model of the product that the designer uses.

Figure 21.6 supports the view that a significant shift or improvement in the understanding of how the components would be handled and installed throughout the product development process occurred. This is characterised by both the actual and minimum number of components reducing from the middle to the end of the embodiment phase. This was followed by a drop in efficiency through the detailed phase. This suggests that as more component detail became available that the selection of handling and insertion times changed while the layout of the design remained stable. Consequently the overall assembly efficiency was only improved by 5 percentage points from 35% to 40% and the actual number of components reduced from 16 to 13.

Another interesting comparison occurs when the assembly analysis is performed with the assumption that constant disassembly and assembly is not required. Table 21.6 presents a comparison of these values at the end of the detailed product development phase.

As the table shows the ratio of the minimum number of components to the actual number and the assembly efficiency remain fairly constant. However a substantial difference exists between the assembly times.

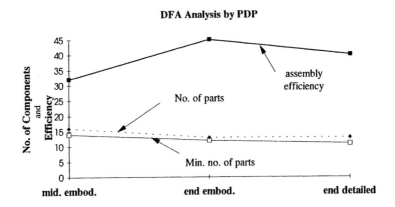

Figure 21.6 DFA efficiency, actual values.

Table 21.6 Impact of DFD requirements on DFA efficiency

	With DFD Requirements	No DFD Requirements
Efficiency	37.2%	38.7%
Minimum No. of Parts	11	4
Actual No. of Parts	13	5
Assembly Time	89 secs	31 secs

21.4 ISSUES OF INTEREST

21.4.1 Weighting Guidelines

It is readily apparent that the weighting of any parameter is a subjective process as two different designers or design teams may weight the same guideline differently. This difference results from the circumstances of use, the designers experience and interpretation of what the guideline means. These differences do not adversely affect the meta-methodology presented as the guidelines will only be used in the manner that the designer interprets them. Hence two designers may achieve vastly different results through the use of the same guideline simply because they have varying interpretations of how to use it. It would be expected that this difference in interpretation would also appear in how the guidelines were weighted. Therefore it is important that when working in a design team that all participants have a common understanding of how each guideline is to be interpreted. In fact this type of discussion should increase the range of applications in which a design rule or strategy can be employed.

Given that the weighting process is somewhat subjective some general trends can still be identified. In the case of some highly weighted strategies the information content was less focused or applicable as the weighting would suggest. This usually results from having a moderately weighted design rule with only a few strategies for implementation. Often these strategies only repeat the overall aim of the rule, hence providing little new information. Therefore it is important to distinguish between the desire to achieve a result and the ability to do so. In cases such as this the weighting of the design rules and strategies must be done concurrently to ensure that sensible values result.

This consideration of desire and ability should also be taken into account when weighting some other more focused strategies as well. Given ideal conditions for product development some strategies would still be difficult to employ effectively. These strategies are completely clear and focused on what and how they are to achieve, yet the actual process of doing so is technically difficult or overly time consuming. An example of such a strategy is incorporating features that stop incorrect assembly. Hence highly desirable but technically difficult strategies should be weighted in accordance with the limitations of the designer and his/her working environment.

Throughout the duration of the design project the designer's interpretation of a number of strategies changed. This change was brought about by the experience of attempting to actually apply the strategies and gaining a detailed understanding of their strengths and limitations. Not only was the understanding of the strategy modified but in some cases completely altered to fit the current circumstance of use. An example from this project occurred when attempting to achieve the tolerance requirements of the device. The interpretation of the strategy "minimise the number of components" was modified from minimising assembly costs to reducing tolerance build up. In this case their was a definite distinction between this strategy and "moving critically related surfaces close together" and "minimising the number of parts between the input and output function". With such changes in perception the designer tends to inherently modify the strategies position within the ranked list. This change does not necessarily have to be made to the matrix or list for the effect to be achieved. However, for the purposes of retaining experience within an organisation such changes should be noted.

21.4.2 DFX Interactions

Through the development of the various comparison matrices a number of issues can be examined. In the case of the specific conflict that occurred during the conceptual phase the solution was ultimately based on functional requirements. As the ranked list suggested the subdivision of the product into manageable subassemblies was of higher priority than minimising the number of conduits and connectors this general approach was taken. By incorporating kinematically sound connectors within the subassemblies the modular structure was maintained with some adjusting required during assembly.

When examining the number of interactions that occurred in the embodiment and detailed phase matrices a number of generalisations about the nature of the guidelines is possible. At the embodiment phase a reasonable number of interactions occur. This suggests that the two DFX techniques provide guidelines with a similar focus on how the product should be improved. In this case it is through the rationalisation of the product structure towards part minimisation and accessibility. However, in the comparison of strategies that act during the detailed phase very few interactions occur. This suggests that there is a fundamental difference between the techniques in terms of how they improve the product. In the case of DFA the focus is placed on specific component features, while DFD focuses upon the detail of the connection forms between components. Hence little overlap exists in the information content of the strategies.

When the assembly efficiency was calculated for the product without the disassembly requirements an important difference can be seen. While the assembly efficiency remained fairly constant there was a large change in the assembly time required. This suggests that the disassembly requirement has significantly reduced the effectiveness of the products assembly process. However, the fact that the efficiency has remained unchanged supports the notion that in the original analysis the other DFX requirements have manipulated what would be considered the minimum number of components for the product. Consequently the overall measure of product assemblability has been modified by the other important life-cycle cost areas.

21.4.3 Structured Application of Design Strategies

The analysis of the product for assembly efficiency provides a number of insights into the mechanism and role of guideline use within the product development process. Unlike Willcox and Sheldon (1993) who through the application of DFA achieved an average part count reduction of 47%, this project achieved a reduction of around 20% having implemented DFX principles from the beginning. Though the circumstances of this project may be different to those examined by Willcox and Sheldon the structured application of DFX design guidelines has reduced the impact of the assembly efficiency calculation by eliminating parts as the design progressed. Consequently the amount of redesign required after the analysis has been significantly reduced. This reduction in rework can be equated to an increase in efficiency of the design process.

What is also apparent is how the DFA efficiency changed from the embodiment to the end of the detailed phase but the number of components did not. This reflects the way in which the design guidelines improved specific product features to facilitate assembly. Hence the guidelines have provided information that improves the design of the product in a more accessible format than through the calculation of an assembly efficiency. By incorporating this information throughout the product development process significant reduction in rework has been achieved.

21.5 SUMMARY

This chapter has presented a meta-methodology that identifies the DFX techniques to use and the relative importance of the design guidelines within each technique for a given application. The methodology identifies and accounts for interactions between DFX techniques and provides a ranked list of design strategies for each product development phase.

The meta-methodology is based on the fact that the information content within the guidelines is of most value at certain stages during the product development process. By identifying the type of information the guideline contains it can be categorised as either a design rule or strategy. This categorisation allows the interactions between strategies to be explicitly examined.

The use of a matrix approach is a suitable method for the comparison of design strategies. Through the creation of a ranked list at each stage of the product development process DFX techniques can be deployed in a strategic and cost effective manner.

The illustrative example demonstrated that the meta-methodology could be easily implemented and used within a relatively small project. Through applying the DFX strategies in a real situation the designer's understanding and interpretation of the guidelines and their role within the product development process both grows and changes. Consequently in future applications the implementation of the methodology will be quicker, easier and more effective.

This proposed meta-methodology is suitable for application to any industry sector and size of enterprise. This restriction has ensured that the tools are relatively quick and simple to use and provide clear direction to the designer with little capitol expenditure.

REFERENCES

Boothroyd, G., Dewhurst, P. (1989) *Product Design for Assembly*, Boothroyd Dewhurst Inc. Wakefield, USA.

Meerkamm, H. (1994) Design for X - A Core Area of Design Methodology, *Journal of Engineering Design*, **5** (2), 145-163.

Norell, M. (1993) The Use of DFA, FMEA and QFD as Tools for Concurrent Engineering in Product Development Processes, *Proceedings of the 9th International Conference on Engineering Design*, The Hague, Netherlands. **2**, 867-874.

Pahl, G., Beitz, W. (1988) *Engineering Design: A Systematic Approach*, (English version Edited by K. Wallace), The Design Council, London.

Simon, M., Dowie, T. (1992) Disassembly Process Planning, *Technical Report DDR/TR2*, The Manchester Metropolitan University.

Simon, M., Fogg, B., Chambellant, F. (1992) Design for Cost Effective Disassembly, *Technical Report DDR/TR1*, The Manchester Metropolitan University.

Thurston, D. (1991) A Formal Method for Subjective Design Evaluation with Multiple Attributes, *Research in Engineering Design*, **3** (2), 105-122.

Watson, B., Radcliffe, D. (1995) A Comparison of DFX Evaluation Tools, *Proceedings of the 10th International Conference on Engineering Design*, Praha, Czech Republic.

Willcox, M., Sheldon, D. (1993) How the Design Team in management terms will handle the DFX tools, *Proceedings of the 9th International Conference on Engineering Design*, The Hague, Netherlands, **2**, 875-881.

Zhang, B., Simon, M., Dowie, T. (1993) Initial Investigation of Design Guidelines and Assessment of Design for Disassembly, *Technical Report DDRTR03*, The Manchester Metropolitan University.

DESIGN FOR TECHNICAL MERIT

Tim N. S. Murdoch; Ken M. Wallace

This chapter introduces a method of design evaluation based on technical merit. With the aim of supporting decision making from the early stages of the design process, technical merit is a measure of the proximity of a product to forecast limits of performance, reliability and economy. The accuracy of the calculations and forecasts are used to define a measure of risk. The chapter is divided into the following sections. Section 22.1 provides a background for Design for Technical Merit and briefly introduces some current evaluation techniques. In section 22.2, ideas drawn from these techniques are combined with a generic model of system development to define technical merit and methods for its calculation. Sections 22.3 and 22.4 outline a case study in design for technical merit, using a manual method and a computer-based tool, respectively.

Design for technical merit emanates from an investigation into configuration optimisation at the Cambridge Engineering Design Centre. In this context the term *configuration* refers to the basic functional behaviour of a system and is reflected in the choice and layout of certain key components. Other terms such as *concept* and *solution principle* also convey a similar meaning. The phrase *configuration optimisation* refers to the manipulation of a configuration in order to change certain characteristics. In this case the optimisation will proceed by changing design *parameters* such as shape, size, material and component type. The *quality* of a design is determined by its functionality, its parameter values and also its derived *attribute* values such as weight and cost evaluated against requirements listed in the design specification. During the optimisation process, a large number of trial solutions will be produced. Those that meet the minimum design requirements are referred to as *feasible* solutions. Those solutions shown also to have the highest quality are referred to as the *state-of-the-art* and define the upper limits of the *design envelope* for a given design task.

22.1 BACKGROUND

The performance, reliability and economy of a technical system depend on decisions made during the design process. A few key decisions taken early in the process determine the overall configuration of a product and establish limits within which the product can be developed. A decision to refine an existing product configuration may place limits on the maximum performance that can be achieved. A decision to develop a new product configuration may introduce high costs, reduced reliability and high risk. High quality decisions therefore require robust measures of evaluation that demonstrate the trade-offs between performance, reliability, economy and risk for the underlying configuration of each proposed design.

Existing evaluation techniques attempt to define a single measure by combining the properties of a product according to the requirements of the design task. An extensive discussion of evaluation techniques in design can found in de Boer (1989). These techniques range from those that use limited abstract data to those that use detailed models of the task and its solutions. Methods in the first category include Pugh charts (Pugh, 1990), value profiles (Pahl and Beitz, 1988) and the ideal point method (Hwang and Masud, 1979) . They provide a coarse comparison of alternative solutions to a specific design task and are generally most applicable at the conceptual and embodiment stages of the design process. Methods in the second category include multi-attribute utility theory (Keeney and Raiffa, 1976) and interactive trade-off methods (Sen and Yang, 1993). They focus on building detailed models of both the task and its proposed solutions and are generally applicable later in the design process. Methods in the first category are usually applied before the configuration has been determined. Methods in the second category are usually applied to specific configurations in order to establish good parameter values.

22.2 TECHNICAL MERIT: CONCEPT

Design for technical merit is based on trends found to be characteristic of many technical systems (Murdoch and Wallace, 1992). The graph shown in Figure 22.1 is a generic representation of results found by studies of product and component development by Byworth (1987), Coplin (1989) and others. Past designs from two product configurations are plotted against a key performance attribute and time. The markers represent example solutions from configurations A and B. The two curves form design envelopes and demonstrate the underlying trend of an increasing performance over time for each configuration. In both cases, after a first proof of concept there is a period of consolidation before a first production version and then a rapid improvement as understanding for the product grows. As the product matures, the law of diminishing returns takes over and performance levels reach limits imposed by the choice of product configuration.

Hindsight shows us that configuration B provides the 'better' product in the longer term. However a designer may be required to chose between configurations A and B without this knowledge. In this case configuration B was introduced when changes to configuration A were still delivering improvements in its performance, and indeed the first production version appears to under-perform with respect to the current version of configuration A. Thus configuration B would be seen as a high risk option with only small potential benefits. Indeed, studies have shown that around 75% (Pahl and Beitz, 1988) of all mechanical design work is the re-development of existing product configurations. This means that decisions made in determining the basic configuration of a product, often early in the design process, have consequences beyond the life of that particular product and often beyond the current understanding of the customer requirements. A strategy of risk reduction tends to commit the designer to refine an existing configuration and an approved set of components and materials. Therefore it is important that the best possible product configuration is identified early in the life of a product.

The curves shown in Figure 22.1 can be regarded as limits placed upon the designer once the configuration of a product has been defined. With hindsight, alternative product configurations may be analysed by comparing the limitations that they impose upon the design. More importantly, accurate forecasts of future limits would enable the designer choose the configuration of a product based on both current and future needs. It is the idea of a limiting design envelope combined with generic evaluation criteria that form the basis for the definition of technical merit (Murdoch, 1993).

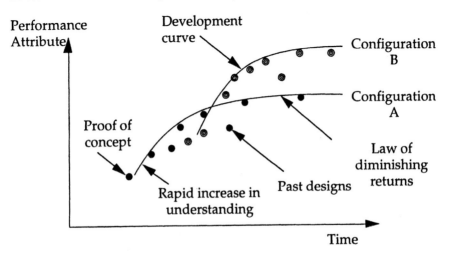

Figure 22.1 Trends in product development.

22.2.1 The Definition of Technical Merit

State-of-the-art systems are associated with products that possess either the highest performance and reliability or the lowest cost and so define the limitations of current technology. Imagine that the design requirements for the device are to transfer the highest possible torque over a predefined distance for the lowest weight and cost. Three solution principles can be identified from existing applications in Figure 22.2. Solution A is a carbon fibre tube from ultra-light aircraft design; solution B is a solid steel bar used in many mechanical devices; and solution C is a magnetic device used in laboratories to stir chemicals. Figure 22.3 shows hypothetical plots for performance (in terms of low weight and high torque) and cost for a number of examples of the three torque transfer solution principles. The two axes define an evaluation space and are normalised with respect to a datum solution, in this case a steel shaft design. The three curves define state-of-the-art design envelopes. Rather than representing trends over time as in Figure 22.1, these curves represent forecasts of the *current* limitations of performance and economy for each solution principle. The shaded regions within each envelope represent areas of feasible design solutions. In the absence of other higher performing or lower costing solution principles, the higher portion of the carbon fibre tube design envelope and lower portion of the steel shaft one combine to define the state-of-the-art and can be regarded as a boundary dividing feasible and unfeasible regions of the evaluation space.

Figure 22.2 Three torque transfer solution principles.

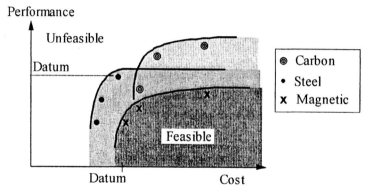

Figure 22.3 Performance and economy trade-offs.

A full definition of the evaluation space is given by the three generic criteria of performance, reliability and economy. Three merit indices are defined to represent these criteria:

Duty	a measure of the degree to which a technical system fulfils key performance criteria.
Reliability	a measure of the degree to which a technical system maintains its performance and economy attributes over its pre-specified life.
Cost	a measure of the degree of difficulty in manufacturing, assembling, maintaining and disposing of a technical system.

Each merit index is calculated with respect to a pre-selected benchmark which provides a datum for comparison. The datum is used both to normalise each merit index and to help determine utility functions for each design objective using methods similar to those described above. Methods of modelling product trade-offs may provide estimates of the limiting duty and cost indices such as those shown in Figure 22.4 for the three torque transfer concepts. Once established, these limits may be used to compare the three alternative configurations in two ways. Firstly, through the calculation of a single merit index to rank the nine individual solutions. Secondly, and perhaps more importantly, through the comparison of the design limitations of each configuration.

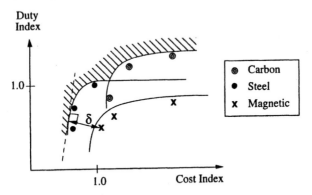

Figure 22.4 Measurement of the proximity to the state-of-the-art.

A single measure of technical merit may be calculated for each design by measuring its proximity to the current state-of-the-art. Measured as the perpendicular distance from the overall design envelope to individual solution, the normalised axes ensure that the scalar value (d) may be used regardless of gradient. Using a linear method of calculation, technical merit is defined as:

$$Technical\ Merit = (1.0 - \delta) \times 100$$

providing a score of 100 for the state-of-the-art and 0 for a poor solution one normalised unit away from the design envelope. This gives the following equation for the technical merit of the cheapest magnetic device shown in Figure 22.4:

$$Technical\ Merit = (1.0 - \delta) \times 100 \approx (1.0 - 0.4) \times 100 = 60$$

The design limitations of each configuration may be compared by analysing the design envelopes. For example, for this particular task the design envelope of the magnetic induction device is entirely dominated by either the steel or the carbon shaft designs. The steel shaft design provides the lowest cost solutions, but with a low limit on performance. On the other hand the carbon fibre tube provides higher levels of performance, but only at the expense of increased cost. Similar to the visual presentation of the value profiles described earlier, perhaps the most important aspect in reducing the amount of data presented to the designer, is the inclusion of trade-offs between each of the key criteria.

An important addition to the criteria of performance, reliability and economy is the issue of risk in design. When a design is already in operation and its attributes can be measured directly, the risk of its failing to achieve those attribute values is minimal. However, when a design is still on paper and is either complex, or incorporates new technology, there is a high risk that it may fail to achieve its design targets. Clearly, this is an important part of robust decision making and therefore a measure of risk must also be presented to the designer. This criterion is represented by the factor of confidence which is defined as:

Confidence   *a measure of the risk of failing to achieve the levels of performance, reliability and economy forecast for a technical system.*

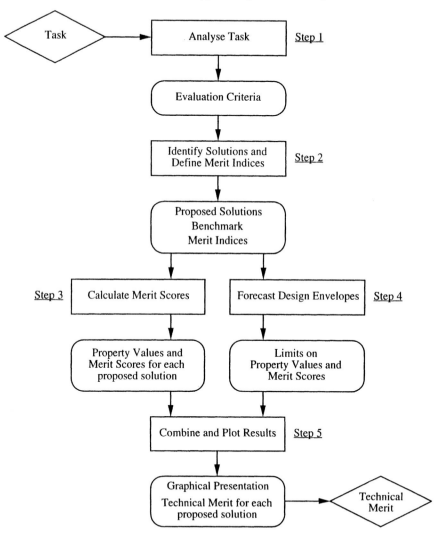

Figure 22.5 Procedure for calculating technical merit.

In practice the confidence factor is determined by estimating the error associated with plotting the merit indices of individual solutions and forecasting the design limitations of each configuration. For example, the weight, maximum torque and cost of a given steel shaft can be measured easily and therefore can be plotted in the evaluation space with confidence. On the other hand the limiting weight, maximum torque and cost of magnetic induction devices can only be estimated. The design envelope of this solution concept may therefore lie in a range of positions, all of which should be presented to the designer. The confidence factor is used in two ways. Firstly, it provides a measure of the error margin in the calculation of a single value of technical merit. Secondly, it provides visual indications of error for each point and curve in the design space, thus demonstrating the robustness of any ranking of solutions or solution principles.

22.2.2 Procedure For Calculating Technical Merit

The procedure for calculating technical merit involves the five following steps (Shown as a flow chart in Figure 22.5):

(1)   analyse the task and identify the performance, reliability and economy criteria;
(2)   establish possible solutions and solution principles, identify a benchmark solution and define the merit indices;
(3)   measure property values, estimate the risk associated with them and calculate merit indices;
(4)   forecast the design envelope for each solution principle; and
(5)   combine results from (3) and (4) and rank individual solutions and general solution principles according to technical merit.

In a similar way to the product trade-off methods (Sen and Yang, 1993), design for technical merit focuses on the presentation and comparison of design envelopes in terms of the three evaluation criteria of performance, reliability and economy; qualified by a measure of risk. As with product trade-off methods, the most difficult task is forecasting the design envelope for each solution principle.

Step 1 - The Task

The design task is analysed to identify the key requirements which are then placed under the headings of performance, reliability or economy criteria as appropriate. The definition of the merit indices and separate check lists may be used to support this process. Where constraints exist in the design task these are noted against the appropriate requirements. Some coarse ranking of the requirements can also be made at this stage, but a more detailed analysis using bench-marking is required to define full utility functions.

For example, the key requirements of the torque transfer device are high torque, low weight and low cost. The low weight and high torque requirements fall into the performance category, and low cost into the economy category. Further information, such as the distance between torque input and output, or environmental requirements such as corrosion resistance define constraints which must be met in order to establish a feasible solution. Similarly, further requirements such as the level of acceptable reliability may also be sought and used to define the merit indices. Where the design task is relatively well defined the low weight and high torque requirements may be ranked according to their relative importance.

Step 2 - The Solutions

The designer then identifies possible solutions to the design task. These may include a specific design requiring evaluation, or a range of alternatives which require ranking. The solutions may come from a wide range of sources including those created during the current design process, and those taken from past products. One 'high quality' solution is selected as a benchmark based on subjective judgement.

For example, three solution principles for the torque transfer device have already been identified. Where specific examples of these exist (on paper or otherwise) these may also be considered for evaluation. The benchmark solution, chosen earlier to be a specific steel shaft, was chosen as that which appeared to the designer to most closely fulfil the requirement criteria.

Equations defining the merit indices are determined using data from step 1, the benchmark solution and designer interaction. The requirements listed under each criterion heading in step 1 are used to collect property data from the benchmark solution. This data is used to elicit trade-offs between the requirements and so define utility functions. The benchmark data and utility functions are then combined to define each merit index according to the general equation:

$$Merit\ Index = \sum_{\forall x_{max}} \left(Ux_{max} \times \frac{X_{max}}{Datum} \right) + \sum_{\forall x_{min}} \left(Ux_{min} \times \left(2 - \frac{X_{min}}{Datum} \right) \right)$$

where X_{max} is the value of a property to be maximised and X_{min} is one to be minimised. Ux is the utility function linking the normalised property value to criterion score. The degree of uncertainty associated with each utility function is recorded separately as a percentage error for later use.

For example, the two performance criteria for the torque transfer device are high torque and low weight. Given a datum torque of say 70 Nm and a datum weight of 1.3 kg the duty index would be defined as:

$$Duty\ Index = U_{torque} \times \frac{torque}{70} + U_{weight} \times (2 - \frac{weight}{1.3})$$

where U_{torque} and U_{weight} are defined according to the specific design task. Given a datum cost of £1.54 and reliability of 5 failures per 1000 hours use the remaining indices are defined as:

$$Cost\ Index = \frac{cost}{1.54}$$

$$Reliability\ Index = 2 - \frac{failures\ per\ 1000\ hrs}{5}$$

Step 3 - Solution Properties

Each of the solutions identified in step 2 is analysed to establish both its functionality and its attribute values. The scope of this analysis varies according to the description of each solution. For example, the properties of an existing product can be measured, whereas those of a design in progress can only be forecast with varying degrees of accuracy.

Feasibility is established by comparing a solution to the design task. If the functions are fulfilled and the constraints met, then a solution may be considered for evaluation. If not, then it is not directly comparable and may not be plotted in the evaluation space.

The merit indices are then calculated for each solution by entering the appropriate attribute data into the equations defined in step 2. The resulting values of each merit index determine the position of a solution in the evaluation space. A factor of confidence accompanies each point which is calculated by combining the predicted margin of error for both the utility functions and the attribute values. This data is used to plot graphs similar to that shown in Figure 22.6 where examples of each of the three solution principles are plotted against the indices of duty and cost, each one with a predicted margin of error.

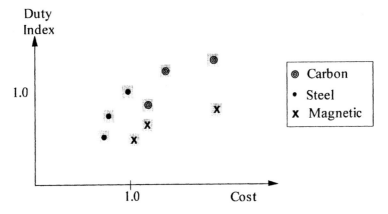

Figure 22.6 Individual solutions plotted in the evaluation space.

Step 4 - The State-of-the-Art

Establishing the limitations of the state-of-the-art is the most difficult part of any evaluation technique which incorporates product trade-off curves. There are two ways in which the design envelopes can be forecast: (1) by experience; and (2) by parametric studies.

Firstly, skilled designers may use their experience to estimate future trends in system development and then, noting the law of diminishing returns, combine these estimates to forecast the ultimate limitations of a given solution concept. For example, magnetic flux density is limited by certain material properties and technical constraints thus defining the maximum torque of a torque transfer device based on magnetic induction. Secondly, the design envelope of a solution concept may be investigated by parametric studies where small changes are made to given configurations in order to explore different performance, reliability and economy attributes. For example, the parameters defining the layup of carbon fibre tubes can be varied and the maximum torque and weight determined either by experimentation or according to standard design models. The parameter values yielding the highest torque and lowest weight solutions will define the highest quality carbon fibre designs.

Manual methods are rarely exhaustive and may fail to investigate certain high quality component combinations and material choices. Therefore, forecasts based on these methods will have various levels of confidence depending on both skill of the designer and the nature of the product. Computer-based optimisation techniques may also be used, but as they rely on well defined models of the design and they are generally only available for known systems within mature product configurations. However, being a computer-based technique, a large number of parametric studies can be performed exploring a wide range of possible solutions. Forecasts supported by such analysis may provide a higher level of confidence.

The results from this step provide further data points which can be plotted in the evaluation space. The points shown to be the lowest cost, highest duty and highest reliability for each solution concept define curves shown in Figure 22.7 which describe the state-of-the-art. The shading around each curve shows the level of confidence of the forecasts. In this case the steel and carbon design envelopes combine to define the overall boundary separating feasible and unfeasible regions of the evaluation space.

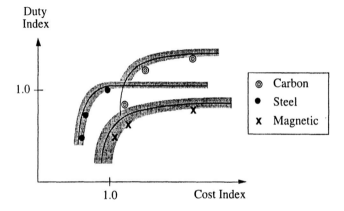

Figure 22.7 Forecast limits on the evaluation space.

Step 5 - Technical Merit

The results from step 4 can be used to support design decision making in two ways. Firstly, through the calculation of technical merit for individual solutions. Secondly, through the comparative analysis of competing solution concepts and parameter values.

The technical merit of the cheapest magnetic device was calculated earlier. Incorporating the factor of confidence in the calculation gives the following:

$$Technical\ Merit = (1.0 - \delta) \times 100 \approx (1.0 - 0.4) \times 100 = 60$$
$$Confidence\ Factor = (0.05 + 0.1) \times 100 = 15$$

where δ is approximately 0.4, the margin of error in plotting the magnetic device is 5% and the margin of error in forecasting the overall design envelope is 10%. Thus this device may be ranked above any other solution with an upper limit of technical merit lower than 52.5. Where the range of technical merit for one solution overlaps with another there will be a reduced level of confidence depending upon the size of overlap. On their own these values of technical merit and confidence factor are difficult to judge. A designer will only be able to consider them as "good" or "bad" in the context of the design problem and by comparison with results from other devices.

The relative technical merit of competing solution principles are determined by analysing the design envelopes. For example, the low cost portion of the overall boundary shown in Figure 22.14 is dominated by the low cost steel design and the high duty portion by the low weight carbon fibre design. Thus, a high performance design requirement will lead to the selection of a thin walled carbon fibre tube, the exact description of which will be determined by the parameters of the nearest solution on the envelope.

The results from analysing either individual solutions or their underlying solution principles can then be used to support decision-making for both immediate use and long term development. For example, the technical merit of individual solutions may aid the selection of a particular solution. The relative dominance of solution principles over various portions of the overall boundary may aid the selection of a particular concept.

22.3 MANUAL APPLICATION OF DESIGN FOR TECHNICAL MERIT

This section outlines manual application of the design for technical merit method. The case study used during this section is the design of a pin-jointed model bridge which forms part of the first year undergraduate engineering course at Cambridge University. Manual methods are used to evaluate and rank 60 bridge designs.

Step 1 - The Task

The task is to design a pin-jointed bridge structure to span a distance of 820 mm and support a mid-span working load of 3.5 kN and a collapse load of 7 kN. The bridge must be as light and as cheap as possible. The materials that can be used for this are angle bars and plate of either all steel or all aluminium construction, fixed by rivets and bolts. Supporting information includes buckling data, bolt failure stresses, costings, etc.

Steps 2 & 3 - The Solutions And Their Properties

Five basic configurations are frequently observed in the undergraduate designs. These are shown in Figure 22.8 and comprise three basic levels of complexity and are of either a box or a triangular cross-section. The figure shows only the principle members of each configuration and the location of the working load. In almost all cases, bracing members are used to reduce buckling effects in the compression members. Out of the 60 undergraduate designs investigated 19 used configuration 1 and 22 used configuration 2. Configuration 1 is a simple and potentially low cost design where the main load passes through the top fixing plate. Configuration 2 is a more complex design with shorter and therefore stronger compression members. Relatively few undergraduates used either a triangular cross-section design or the more complex configurations 3 and 5.

The collapse loads and cost data for both steel and aluminium versions of configurations 1 and 2 are shown in table 22.1 and Figure 22.9. Configuration 1 is shown by the triangular markers (steel) and crosses (aluminium). The four curves represent the design envelopes of the two configurations in steel and aluminium. The steel bridges show a clear advantage over the aluminium versions in both cost and collapse load, with configuration 1 dominating the overall design envelope. It must be noted, however, that the collapse load that each bridge must be designed for is 7 kN and is shown on the graph as the higher of two dotted lines, the first being the working load. While the majority of the bridges achieve the working load of 3.5 kN, less than half successfully meet the collapse load constraint. This is a failure of a key performance requirement, and therefore only those with a collapse load greater than 7 kN can be evaluated for technical merit.

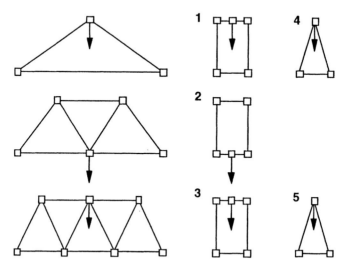

Figure 22.8 Common bridge configurations.

The benchmark solution was selected as the design with lowest cost (450 units) that achieved the collapse load (10 kN) from configuration 2. This was designed by group 33. The merit indices were defined in terms of weight, cost and collapse load. Once the minimum load constraint has been met, the duty index is defined only in terms of low weight, giving:

$$Duty\ Index = U_{weight} \times \left(2 - \frac{weight_x}{weight_{datum}} \right) = 2 - \frac{weight_x}{1.6}$$

where 1.6 kg is the weight of the benchmark solution and U_{weight} is 1.0. The cost of each bridge was determined by analysing its bill of materials and manufacturing process. Using criteria given to the undergraduates, the total cost is calculated according to the components used, their size, and the amount of customisation required. For example, a rivet is costed at 1.1 units, an angled bar at 4 units and a single cut at 1 unit. These costs are combined with a material cost according to the total weight of the bridge at 60 units per kg for steel structures and 500 units per kg for aluminium ones. The total cost is summed to determine the cost index according to the equation:

$$Cost\ Index = \frac{cost_x}{cost_{datum}} = \frac{cost_x}{450}$$

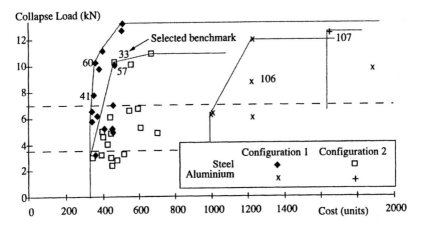

Figure 22.9 Collapse load verses cost for 41 examples of configurations 1 & 2.

During the course of the project it was found that the reliability of each bridge is highly dependent upon the quality of manufacture. This, taken with the fact that collapse loads are taken from measured data, the reliability of each bridge attaining its collapse load is assumed to be the same. Thus:

$$Reliability\ Index = 1.0$$

As the performance, reliability and economy information comes from measured data, the margin of error associated with each calculation and is negligibly low.

This analysis leads to the results shown in Figure 22.10, using the same symbols as before. This shows that only 15 designs achieved the collapse load requirement. The benchmark design, by group 33, appears at datum position. Selected as the lowest cost it also has the highest duty of all the successful steel versions of configuration 2 and so defines the configuration's performance and cost limits. An aluminium version of this configuration by group 107 shows a 5% increase in duty, but also a 250% increase in cost. Both these versions of configuration 2, however, are completely dominated by configuration 1. A steel design of configuration 1 by group 41, shows both a 20% increase in duty, and a 25% reduction in cost over the best design of configuration 2. Similarly, an aluminium version of configuration 1, by group 106, shows a marked improvement over the aluminium version of configuration 2.

Table 22.1 Design data for selected examples of configurations 1 & 2

Group No.	Configuration No.	Collapse Load (kN)	Cost Units	Weight (kg)
33	2	10.2	450	1.6
41	1	7.8	351	1.28
57	1	10.1	450	1.87
60	1	10.2	364	1.68
106	1	8.7	1239	1.12
107	2	12.5	1545	1.52

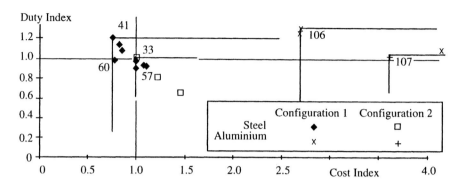

Figure 22.10 Duty and cost for 15 successful examples of configurations 1 and 2.

Step 4 - The State-Of-The-Art

The best designs from each configuration (33 and 107 for configuration 2, and 41 and 106 for configuration 1) define limiting design envelopes. In the absence of other information these examples provide minimum forecasts of the technical limitations imposed on the bridge design. It is important to note, that while 41 example designs have been analysed in some detail, each design envelope is dominated by a single solution. This is due to the strong correlation between weight and cost in the bridge designs, given a particular configuration and material.

The design envelope of Figure 22.10 may be investigated further by analysing each of the good designs for 'spare capacity' in terms of over performance, or redundancy. For example, Figure 22.9 shows that the three steel configuration 2 bridges that achieve the collapse constraint do so with at least 30%, whereas the best of configuration 1 shows only a 7% over performance margin. Analysing individual solutions within a configuration shows that in general, given a minimum collapse load of 7 kN, the greater the actual load capability of a bridge design the lower its duty index. This leads to the possibility of relaxing the limiting duty in all cases except perhaps for configuration 1 in steel. Analysing the cost breakdown of each bridge for over complexity would provide similar results, but unfortunately this data was not collected.

The design envelope in Figure 22.10 is dominated by relatively efficient examples of configuration 1; group 41 in steel (7% over performance) and group 106 in aluminium (10% over performance). Using the results from these two designs to forecast the position of the overall design boundary, the steel one gives a maximum duty index of 1.2 and minimum cost index of 0.75 with a confidence factor of about 7%, and the aluminium one a maximum duty index of 1.3 and minimum cost index of 2.7 with a confidence factor of about 10%.

Step 5 - Technical Merit

Using these forecasts, the technical merit of the benchmark design by group 33 and the best aluminium version of configuration 2 by 107 are calculated as:

$$Technical\ Merit_{33} = (1.0 - 0.3) \times 100 = 70$$

$$Technical\ Merit_{107} = (1.0 - 0.99) \times 100 = 1$$

where δ for group 33 is the distance to group 41 (0.3 units), and δ for group 107 is the distance to 106 (0.99 units). These values are low compared to the technical merit of 100 for both groups 41 and 106 and a confidence factor of about 7% and 10% respectively.

The relative merit of each configuration is established by comparing the four design envelopes. This results in three basic conclusions:

(1) configuration 1 in steel dominates the low-cost portion of the constraint boundary;
(2) configuration 1 in aluminium dominates the high-performance portion; and
(3) configuration 1 entirely dominates configuration 2.

These conclusions lead to the clear decision to develop a steel design of configuration 1 in almost all cases, except those where very high performance is required regardless of the costs incurred. In this case an aluminium version of the same configuration should be considered.

22.3.1 Summary

This section has demonstrated the application of the design for technical merit method in evaluating existing bridge designs. The results show that it is possible both to rank individual solutions using a single measure of merit and to compare alternative configurations according to the limitations they impose on performance, reliability, economy and risk. The spread of the results show that there may be a strong correlation between duty and cost within a configuration. This, coupled with the shape of the design envelopes, often dominated by a single solution, indicates that the overall design limits may be accurately forecast from the results of a few high quality solutions.

22.4 COMPUTER-AIDED APPLICATION OF DESIGN FOR TECHNICAL MERIT

This section describes a computer-aided application of the design for technical merit method. Using the same bridge design task as a case sudy, optimisation techniques are used to forecast the limitations of the state-of-the-art and the results compared to the 60 undergraduate designs. It concludes with an analysis of the technical merit of the underlying bridge configurations.

A computer-based design tool called KATE (Knowledge-based Assistant for Technical Evaluation) has been built to enable the calculation and presentation of technical merit in a limited number of domains (Murdoch, 1993). Currently tested using small scale pin-jointed structures, it is being developed to support other domains including aero-engine design. The tool is specifically aimed at supporting step 4 of the design for technical merit method (forecasting the state-of-the-art) and is based on functional modelling techniques to structure a knowledge-base of components and to model configurations.

Functional modelling techniques are similar to bond graphs in that they build up a configuration in terms of entities and their relationships. In this work the entities are components such as a steel gusset plate, and the relationships are interfaces defining both physical and functional interactions. The components and interfaces within the knowledge-base are defined in terms of functions (such as force transfer), physical parameters (length and material) and attributes (weight and cost). Each component contains mathematical routines and empirical data which maps changes in parameter and function values to changes in attribute values.

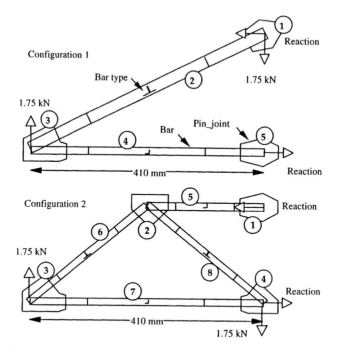

Figure 22.11 Bridge configurations 1 and 2 as represented in KATE.

A specific configuration is modelled by selecting components from the knowledge-base and linking them together with predefined interfaces. The resulting design space is defined in terms of parameters which are manipulated using computer-based optimisation techniques.

Focusing only on steps 4 and 5 from the design for technical merit method, there are four main aspects to this part of the case study:

(1)   representing design knowledge;
(2)   modelling a specific design task and product configuration;
(3)   optimising each configuration to search for the limits; and
(4)   presenting the results to calculate technical merit.

22.4.1 Representing design knowledge

Figure 22.11 shows KATE's schematic representation of bridge configurations 1 and 2. Symmetry has been used to reduce the complexity of the bridge models from two sides in 3D to half of one side in 2D. For configuration 1, three 'pin_joint' and two 'bar' components are selected from the knowledge-base and connected by linking geometric parameters and force transfer functions. The bar is represented by a simple rectangle divided by a number of points of interest that may be used to attach bracing members. The icon at the centre of the bar shows the bar type in cross-section. The pin_joint components represent a gusset plate with associated fixings. The size and shape of the gusset plate is determined by the number of bolts or rivets used to attach each bar.

22.4.2 Modelling The Design Task

Having built up a specific configuration the remaining aspects of the design task are defined by constraining key geometric parameters and applying the collapse load. In this case the half span constraint is defined by fixing the lower bar (component 4 in configuration 1) length to 410 mm and angle to zero degrees. A fixed load of 1.75 kN representing a quarter of the collapse load is applied to the peak pin_joint (component 1) and a similar reaction force is applied to the right-hand pin_joint (component 3). The remaining reaction forces are calculated during the optimisation. The design space is further bounded by limiting the height of pin_joint 1 to less than twice the half span (820 mm). Configuration 2 is defined by four pin_joint (numbers 1 - 4) and four bar components (5 - 8). The load is applied to pin_joint 4 and reaction force to pin_joint 3. The compression is taken by the relatively short bars 5 and 6.

As with the undergraduate designs, the duty index is defined in terms of reducing weight, giving the following indices:

$$\text{Duty Index} = U_{weight} \times (2 - \frac{weight_x}{weight_{datum}}) = 2 - \frac{weight_x}{weight_{datum}}$$

$$\text{Cost Index} = \frac{cost_x}{cost_{datum}}$$

The benchmark solution selected from configuration 1 provides a datum weight of 0.23 kg and cost of 50 units. The benchmark solution from configuration 2 provides a datum weight of 0.55 kg and cost of 175 units. The reliability of each configuration has not been analysed.

22.4.3 Optimising Each Configuration

Each configuration requires a separate optimisation sequence iterating through five steps. The first step is to input a trial set of values into the parameters that define the layout of the configuration and the details of the components. The layout is then analysed to establish the functional interactions between components and calculate the forces within each bar and at each pin_joint. These results are then used to determine the attributes of each component, starting with feasibility (ie buckling for bars or failure for rivets) and then addressing weight and cost. These values are then entered into the equations defining each merit index and determine the technical merit of the trial solution. Finally, the optimisation algorithm uses the merit score to manipulate the trial values and the process iterates until no higher scoring parameter values can be found.

In this case 17 independent parameters define the search space for configuration 1 and 30 for configuration 2. In order to establish both the highest duty and lowest cost solutions, the optimisation algorithm is run several times searching different portions of the design envelope. Starting with low cost regardless of duty, the importance weightings are then changed and the algorithm directed to 10% cost and 90% duty. This continues until the final search direction is for highest duty regardless of cost.

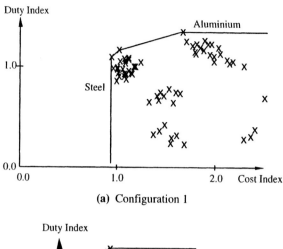

(a) Configuration 1

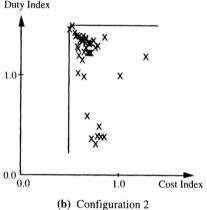

(b) Configuration 2

Figure 22.12 Evaluation spaces.

22.4.4 Presentation Of Results

The resulting evaluation space for each configuration is plotted in Figure 22.12. The crosses (X) denote trial solutions taken during each of the optimisation runs, the dominant ones being defined by those with the lowest cost and highest duty. The design envelope for the first configuration shows that the optimisation process has been able to either reduce the cost index from the configuration 1 benchmark by 9% and increase the duty index by 9%, or increase the duty index by 33% but with the cost penalty of a 66%. While the benchmark solution input by the designer was feasible, it was not a state-of-the-art example of configuration 1. The trial solutions demonstrate the general trend in weight and cost for designs of this configuration. Two distinct clusters are present in graph (a): one on a north-east, and one on a south-west plane. Interrogating the trial solutions shows that the north-east cluster represents designs made predominantly of aluminium whereas the south-west cluster represents those made with (the heavier but cheaper) steel.

A similar investigation of the results for configuration 2 shows that the evaluation space is dominated by steel designs, and that aluminium provides no performance advantage. A comparison of the two design envelopes on the same duty and cost scales in Figure 22.13 shows that configuration 1 completely dominates configuration 2. Indeed, the lowest cost steel versions of configuration 2 are as expensive as the best aluminium versions of configuration 1, with a 50% performance disadvantage. The technical merit of the benchmark solution from configuration 1 which was input by the designer is:

$$Technical\ Merit_{Benchmark1} = (1.0 - 0.15) \times 100 = 85$$

where the proximity of the benchmark to the design envelope of configuration 1 is 0.15 units. While this value is relatively high, it falls short of the overall best and demonstrates the difficulty of a designer successfully selecting the correct configuration and then manipulating 17 separate design parameters to achieve the required collapse load and the two objectives of low weight and low cost. In conclusion, configuration 1 bridge designs provide both the cheapest and lightest solutions to the design task. Where low cost is important steel is the most appropriate material choice. Where high performance is important aluminium versions of configuration 1 will provide the best solution.

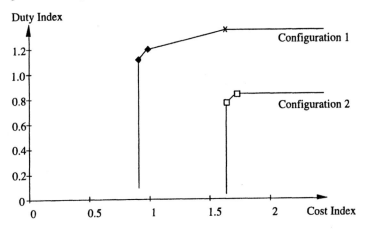

Figure 22.13 Combined results from configurations 1 and 2.

22.4.5 Summary

These forecasts compare closely with the results drawn from the 60 undergraduate designs. Although the computer-based search was based on a model of only one half of one side of the two bridge configurations, the results show a high degree of correlation. The main difference in the results is the apparent low cost of the aluminium designs in the computer model. The differences in results are due to:

1. the redundancy in the collapse load for some of the dominant undergraduate designs.
2. the small number of aluminium bridge designs used to forecast the design envelope.

3.   differences between actual bridges and the two dimensional model of the bridge quarters.

4.   differences between actual components and interfaces and the models and methods of analysis used to represent and "test" them.

The redundancy shown by some of the dominant undergraduate designs may have lead to an over-constrained design envelope at that stage. For example, relaxing the duty constraint from the 1.0 of group 33's benchmark solution by 30% leads to a limiting duty of 1.3 for configuration 2. Similarly, the duty constraint given by the best version of configuration 1 may be relaxed by about 7%. While this does not change the relative positioning of the two curves, the level of dominance of configuration 1 over configuration 2 drops significantly.

The design limits for the aluminium versions of both configurations were established from only 4 undergraduate bridges. Further investigations of either the four existing bridges, or a number of new designs may lead to a relaxing of these constraints in this region.

The configuration models used to forecast the design envelopes in Figure 22.12 were simplified two dimensional versions of one half of one side of the real structures. This discrepancy between the actual and the model is a common one in design, and identifying its effect on the results is an important part of establishing the confidence factor. In this case, the omission of the cross members and more complex 3D pin_joints will lead to an underestimate of the weight and costs involved.

The knowledge-base stores detailed parametric descriptions of the components and associated rules for analysing them. However, these can only be a simplification of the actual dimensions and behaviour and changing these rules to provide more accurate methods may alter the results. For example, the buckling analysis assumed pin jointed bar members, allowing for some rigidity in the model of these joints may provide for larger feasible compression loads perhaps allowing for slender bar members which are lighter and cheaper.

22.5 SUMMARY

The configuration of a product places limits on the maximum performance, reliability and economy that can be achieved. The closer a design is to these limits the higher the risk of failure. High quality decisions therefore require robust measures of evaluation that demonstrate the trade-offs between performance, reliability, economy and risk for the underlying configuration of each proposed design.

This chapter has introduced a method of evaluation based on the property technical merit which is a measure of the proximity of a product to forecast limits of performance, reliability and economy. The method for calculating technical merit involves the five following steps:

(1)   analyse the task and identify the performance, reliability and economy criteria;

(2)   establish possible solutions and solution principles, identify a benchmark solution and define the merit indices;

(3)   measure property values, estimate the risk associated with them and calculate merit indices;

(4)   forecast the design envelope for each solution principle; and

(5)   combine results from (3) and (4) and rank individual solutions and general solution principles according to technical merit.

A case-study has shown how both manual and computer-aided methods can establish forecasts for the current limits on performance and economy and how these forecasts can be used to calculate the technical merit both of individual solutions and underlying product configurations. This work has shown that while technical merit can rank individual solutions, the most important part of the design for technical merit method is the comparison of solution principles according to their design envelope using the four generic criteria of performance, reliability, economy and risk.

REFERENCES

Byworth, S. (1987) Design and Development of High Temperature Turbines, *Turbomachinery International* , May/June, 34-38.

Coplin, J.F. (1989) Engineering Design - A Powerful Influence on the Business Success of Manufacturing Industry, *Proceedings of ICED-89*, Harrogate, IMechE, 1-31.

de Boer, S.J. (1989) *Decision Methods and Techniques in Methodical Engineering Design*, PhD Thesis, University of Twente, The Netherlands.

Hwang, C.L., Masud, A.S. (1979) *Multiple Objective Decision Making - Methods and Applications: a state-of-the-art survey*, Springer-Verlag, Berlin

Keeney, R.L., Raiffa, H. (1976) *Decisions with Multiple Objectives: Preferences and Value Tradeoffs*, Wiley, Chichester.

Murdoch, T.N.S. (1993) *Configuration Evaluation and Optimisation of Technical Systems*, PhD Thesis, Cambridge University.

Murdoch, T.N.S., Wallace, K.M. (1992) An Approach to Configuration Optimisation, *Journal of Engineering Design*, **3** (2), 99-116.

Pahl, G., Beitz, W.(1988) *Engineering Design*, The Design Council, London.

Pugh, S. (1990) Concept Selection - A Method that Works, *Evaluation and Decision in Design*, (Edited by N. Roozenburg, J. Eekels), WDK Heurista

Sen, P., Yang, J.B. (1993) A Multiple Criteria Decision Support Environment for Engineering Design, *Proceedings of ICED-93*, The Hague, Heurista, 465-472.

Action chart 412, 415
Activity hierarchy 404, 407
Activity network 404,408
Activity-based costing (ABC) 5, 321, 398-402
Advising 140
Assembly (DFA), Design for 168, 298
Assembly efficiency 26, 458
Assembly lead time 369
Assembly tolerance analysis 181-3
Attribute values 463
Automated guided vehicle (AGV) 232
Automatic storage and retrieval 232

Benchmarking 137
Bills of materials 113-4
Boothroyd-Dewhurst DFA 19-40, 52-5
Boothroyd Dewhurst DFMA 19-40
    application results 27-35
    procedure 20
    case studies 23-7
Brown & Sharpe 28-9
BS 7000 45
BS 7750 87
BS EN ISO 9000 45
Built-in-test-equipment (BITE) 222
Business Process Reengineering (BPR) 130-2

Cambridge University Engineering Design Centre 245, 463
Car engine coolant container 221
Chrysler 188
Circuit Card Assembly 208
Circuit design 286-90
Civil aircraft in-service inspection 210
Clarity 248, 251-6

Company attitudes 83-85
Competition, Design for 96-106
Competitiveness metrics see performance metrics
Computer aids 291-5
Computer tools 184-8
Consumption intensities 403
Concept and solution principle 464
Concept modelling 90
Concurrent engineering (CE) 1, 20, 43, 98, 133, 217, 233, 399
Confidence factor 472
Configuration optimisation 463
Consumption of activities 403
Convergent product development 98
Coordinate measuring machine 28-9
Cost drivers 9, 403
Critical part 21
Crystal Ball 399
CSC 51

Data collection 121
Defect-free manufacturing 213
Demanufacturing activities 408
Department of Trade and Industry (DTI) 270
Design changes 403
Design decision support system 400
Design decisions 382-3
Design efficiency 27, 65-7
design envelope 463
Design for
    assembly (DFA) 168, 298
    competition 96-106
    dimensional control 173-195
    disassembly 318-334
    EMC 268-97

environment 72-95, 319, 380-97
  inspection 216-29
  manufacture (DFM) 23
  modularity 356-79
  recycling 318-34
  reliability 245-67
  technical merit 463-83
Design for X (DFX) 1-16, 72-95
  applicability 10-2
  barriers 150
  benefits 7-9,
  checklists 122-3,
  development procedure 108
  driven 131
  elements 57
  guidelines 441-62
  history 2
  implementation 130-52
  latest development 13-4
  lookup tables 122-3
  loop 59
  manual 121-3
  mindsets 66
  Off-line 60
  On-line 60
  pattern 4-6
  planning for 148
  shell 107-29
  support team 12-3,
  tool kits 11, 73
  verification 126-8
  workbook 124-6
  worksheets 62, 64, 66, 124-125, 305, 306, 331
Design optimization 424-440
Design parameters 464
Design stages 273-4
Design strategies 271-2
Development capacity 367
Development cost 367
Development lead time 366
DFA 19-40, 41-63, 99
  comparison case study 54-69
  evaluation methods 41-71
  guidelines 69-70
DFMA roadblocks 35-8
DFX/BRP 132-135, *also see* DFX and BPR

Disassembly, Design for 318-334
Diagnosing 139
Digital Equipment Corporation 29
Dimensional Control Systems (DCS) 185-6
Dimensional control, Design for 173-95
Disassembly evaluation chart 321-5
Disassembly task difficulty ratings 303, 326
Disassembly tasks 324
Disassembly tools 325
Divergent product development 98
Divide and conquer 45
Dutch ECOdesign 82

Eco-design 89, *see also* Design for environment
Efficiency audit 148-9
ELDEC Corporation 207
Electromagnetic emissions 269
Electromagnetic immunity 262
Electromagnetic interference (EMI) 268
Electronic Waste Ordinance 318
EMC guidelines 295-6
EMC principles 275-91
EMC, Design for 268-97
EMC, European Directive 268-70
End-of-life analysis 321
Entity-relationship formalism 161
Environmental concerns 382
Environment, Design for 72-95, 319
Environmentally Conscious Design 377-96
Environmental impacts, Design for 380-97
  case study 392-96
  components 386
  computer aids 390-2
  overview 384
  procedure 387
Environment management system (EMS) 68, 85
EPS 99
Evaluation options 274-5

Factor/defect matrix 225
Failure frequency 311
Failure mode 311
Failure source 311
Fault tree analysis (FTA) 246
Feasible solutions 463
Filtering 284-6

Financial audit 148-9
Flow process charts 116-7
FMEA 99, 246, 310-1
Focus requirements 111-2
Ford Motor Company 31, 188
Frequency rank 311
Functionality requirements 109

GAPT 101
GE Automotive 10
General Motors 32, 188, 299-300
GRAI 153-72
    formalisms 158
    grids 160
    integrated methodology (GIM) 115, 153-72
    nets 160
    P_graphs 166
    R_graphs 166
Green design *see* Environment, Design for

Hasbro 34
Hewlett-Packard 33
Highlighting 137-9
Hitachi AEM 48-50
House-service cut-out fuse 54-69

Ideal number of modules 369
IDEF0 115, 159-60
In-service inspection 212
Incineration 387
Influential factors 272-3
Injection molded heater cover 22
Injection-molded part 23
Inspectability 216
Inspection functions 223
Inspection, Design for 216-29
Interface complexity 366

KATE 477-581
Key characteristics 168
Key product characteristics 115

Landfill 388
Layouting 282-3
LiDS wheel 88
Life cycle analysis (LCA) 88
Life cycle assessment 320

Life cycle design 395-421, 424-40, *see also* Design for environment
Life-Cycle Assessment 418
Linearisation method 170
Lucas DFA 50-3

Macro BPR procedure 141
Macro design process 155
Management support 84
MANDECO 99
Manufacture analysis *see* Design for Manufacture (DFM)
Manufacturing (DFM), Design for 23
Material handling equipment 232
Material requirements planning (MRP) 372
Maynard Operation Sequence Technique *see* MOST
McDonnell Douglas Corporation 33-4
Measuring effectiveness 148
Measuring performance 137-8
Measuring results 75
Meta-Methodology 441-62
    case study 452-9
    design rules 446
    design strategies 447
    design tools 446
    guideline interactions 448
    guidelines 446
    interaction matrix 451
    overview 443
    procedure 443-5
    techniques 445
Method of system moments 170
Methods Time Measurement (MTM) 23
Micro DFX procedure 135-40, 143-4
MIM questionnaire 363
Minimum assembly time 22, 26, 54, 65
Modular quality 370
Modularity Evaluation Chart 364-5
Modularity, Design for 356-79
Modularity built 358
Module drivers 362
Module Function Deployment (MFD) 359
Module Indication Matrix (MIM) 361, 374
Monte Carlo simulation 170, 399
Morals and ethics 76
MOST 295, 319-20, 326
Motor-drive assembly 23-7

Motorola 30
Multi-attribute utility theory 465

NCR Corporation 29
Non-destructive inspection 210
North American Uniform Out-of-Service Criteria 214
Northern European industries 98

Obstacles 83-5
Operability requirements 109-11
Operation process charts 117-8
Order picking 236, 240
Over the wall *see* sequential engineering

P_graphs 166
PALLET 237-8
Pareto 141
PARIX 5-6, 107
Part count 21
Pattern matching 124
Patterning 124
PCB assembly process 197, 200
PCB manufacturing process 198
PCB surface mounting 186
Performance benchmark 120-1
Performance characteristics 8
Performance indicators 119
Performance measurements 119-21
Performance metrics *see* performance indicators
Polaroid Corporation 23
Postage stamp 220
Power-on-self-test (POST) 220
Pressure recorder case study 304-10
Printed circuit board (PCB) 196-215
Prioritizing 141
Problem identification matrix 142-3
Problem solving 142-3
Process analysis 136-7
Process audit 148-9
Process capability 168, 170
Process characteristics 112
Process charts 115-8
Process composition 112
Process configuration 112
Process flowchart 200
Process model 201-7

Process modelling 115-8
Product analysis 135
Product audit 148-9
Product characteristics 112
Product composition 112
Product configuration 112
Product costs 367-8
Product design parameters 232
Product environment 426
Product life cycle 79, 80
Product life phases 426
Product model 154
Product modelling 112-5
Product variety 226
Production parameters 203
Project management 142
Pugh's selection matrix 147, 373

Quadrature 170
Quality Function Deployment (QFD) 99, 359, 442

R_graphs 166
Radical reengineering 144-5, 146
Recyclability 371
Recycling 386
Recycling, Design for 318-34
Reference models 156-8
Reliability index 170
Reliability models 250-66
Reliability prediction 246
Reliability testing 246
Reliability, Design for 245-67
Requirement analysis 109-12
Return-to-dig 253
Reusability matrix 367

Scandinavian industries 99
Scania cab 357
Sensitivity chart 415
SEPSON 371-7
Sequential engineering 130, 131
Service, Design for 298-317
    disassembly worksheet 305
    optimisation 310
    procedure 296
    reassembly worksheet 306
    service software 308

service task importance analysis 304
service time breakdown 309
serviceability efficiency 308-9
Shielding 276-80
SIMPICK 238
Simplicity 248
Simultaneous engineering *see* Concurrent
 Engineering (CE)
Snowball effect on cost reduction 28
Socio-Technical Systems (STS) 216
Software considerations 290-1
Solution properties 471
Solution space 471
Space utilization 224, 233
SPC 180
State-of-the-art 463, 471
Storability 231-3
Storability, Design for 230-44
Storage and distribution, Design for 230-44
STORE 237
Sustainable product development *See*
 Design for environment
Swedish industry 96, 99
System costs 368-9
System partition 275

Taguchi method 170
TeamSET 45
Technical merit, Design for 463-83
 case study 473-7, 477-81
 computer aids 477-81
 confidence 468
 cost index 466, 70
 duty index 466, 70
 manual application 473-7
 reliability index 466, 470
Testability, Design for 220
Texas Instruments / Tolerance Analysis
 (TI/TOL) 187
Theoretical minimum number of parts 22,
 25-6, 54, 65
Time standards 301-2
Tolerance accumulation 170
Tolerance analysis and simulation 168
Tradeoff 145, 147
Transplant 450-4

Uncertainty 398

Unit arrangement 225
Unit load 223
Unity 248-9
Universal virtues 56

Variability reduction 167
Variant flexibility 370-1
Variation Simulation Analysis (VSA) 184-5
Visual inspection 210

Warehouse cubic space 228
Winch 371-7
Work measurement 326-8

Xerox 367

Printed in the United States
31220LVS00001BA/74